FOOD CARBOHYDRATES

Chemistry, Physical Properties, and Applications

Edited by
STEVE W. CUI

FOOD CARBOHYDRATES

Chemistry, Physical Properties, and Applications

Taylor & Francis
Taylor & Francis Group

Boca Raton   London   New York   Singapore

A CRC title, part of the Taylor & Francis imprint, a member of the
Taylor & Francis Group, the academic division of T&F Informa plc.

Published in 2005 by
CRC Press
Taylor & Francis Group
6000 Broken Sound Parkway NW, Suite 300
Boca Raton, FL 33487-2742

International Standard Book Number-10: 0-8493-1574-3 (Hardcover)
International Standard Book Number-13: 978-0-8493-1574-9 (Hardcover)
Library of Congress Card Number 2004058621

Library of Congress Cataloging-in-Publication Data

Food carbohydrates : chemistry, physical properties, and applications / Steve W. Cui, editor.
    p. cm.
    Includes bibliographical references and index.
    ISBN 0-8493-1574-3 (alk. paper)
    1. Carbohydrates. 2. Food--Carbohydrate content. I. Cui, Steve W.

TX553.C28F64 2005
664--dc22   2004058621

Taylor & Francis Group
is the Academic Division of T&F Informa plc.

Visit the Taylor & Francis Web site at
http://www.taylorandfrancis.com

and the CRC Press Web site at
http://www.crcpress.com

Preface

Food Carbohydrates: Chemistry, Physical Properties, and Applications is intended as a comprehensive reference book for researchers, engineers, and other professionals who are interested in food carbohydrates. The layout and content of the book may be suitable as a reference or text book for advanced courses on food carbohydrates. The motivation for this book originated from an experience I had six years ago when I was preparing lecture materials for a graduate class on food carbohydrates at the Department of Food Science, University of Guelph. After searching several university libraries and the Internet, I was surprised to find that there was no single book available in the area of food carbohydrates that could serve the purpose, despite finding numerous series and monographs found in the library. When I shared my observation with colleagues who taught food carbohydrates before or who are currently teaching the course, all of them agreed with my thought that a comprehensive book covering carbohydrate chemistry and physical chemistry is in great demand.

As an advanced reference book for researchers and other professionals, the aim of this book is not only to provide basic knowledge about food carbohydrates, but to put emphasis on understanding the basic principles of the subject and how to apply the knowledge and techniques in quality control, product development, and research. There are eight chapters in the book covering basic chemistry of food carbohydrates (Chapter 1), analytical methodologies (Chapter 2), structural analysis of polysaccharides (Chapter 3), physical properties (Chapter 4), molecular conformation and characterizations (Chapter 5), and industrial applications of polysaccharide gums (Chapter 6). Chapter 7 is devoted to starch chemistry and functionality, while Chapter 8 presents the most recent developments in starch modification. Emphasis in the last chapter has been given to the reaction principles, and improved functional properties and practical applications of modified starches.

The uniqueness of this book is its broad coverage. For example, it is rare to find analytical methods and structural analysis of polysaccharides in a regular carbohydrate book; however, these two subjects are discussed in detail in this book. The introduction on conformation and conformation analysis of polysaccharides presented in the book has not been seen in any other food carbohydrate book. Polysaccharides as stabilizers and hydrocolloids have been described in great detail in several books; the most recent and informative one is the *Handbook of Hydrocolloids* edited by G.O. Phillips and P.A. Williams (Woodhead Publishing, 2000). Therefore, the material on polysaccharide gums (hydrocolloids) introduced in Chapter 6 is brief and concise. Information on starch and starch modification is extensive enough to form separate monographs. The two chapters in this book are concise, but with the emphasis on understanding the basic principles and applications of starches.

I would like to acknowledge Dr. Christopher Young for reviewing Chapter 3 and Dr. Robin McKellar for proofreading some sections of the book (both are from the Food Research Program, AAFC, Guelph). My sincere thanks go to Cathy Wang for organizing the references and preparing some figures and tables for Chapter 3 and Chapter 5. I also would like to thank each contributor for the hard work and expertise they have contributed to the book. Lastly, I would like to express my sincere appreciation from the bottom of my heart to my family, Danica, Jennifer (two daughters), and especially my wife Liqian, for their love, patience, and understanding during the course of editing this book.

Steve W. Cui

Editor

Dr. Steve W. Cui is currently a research scientist at the Food Research Program (Guelph, Ontario), Agriculture and Agri-Food Canada, adjunct professor at the Department of Food Science, University of Guelph, and guest professor at the Southern Yangtze University (former Wuxi Institute of Light Industry), Wuxi, China. Dr. Cui is a member of the organizing committee of the International Hydrocolloids Conferences and hosted the 6th International Hydrocolloids Conference at Guelph, Canada. He also sits on the editorial board of Food Hydrocolloids.

Dr. Cui's research interests are on the structure and functional properties of hydrocolloids from agricultural products and their applications in foods. His expertise includes extraction, fractionation, analysis of natural polysaccharides, elucidation of polysaccharide structures using methylation analysis, 2D NMR, and mass spectroscopic techniques. He is also interested in studying the structure-function relationship of polysaccharides by examining their conformation, rheological properties, and functionality (as dietary fiber and stabilizers). He authored a book entitled *Polysaccharide Gums from Agricultural Products: Processing, Structures and Functionality* (CRC Press, 2000) and edited and co-edited two special issues of *Food Hydrocolloids* (2003) and a special issue of *Trends in Food Science and Technology* (Elsevier, 2004) collected from the 6th International Hydrocolloids Conference held in Guelph, Ontario, Canada, in 2002. Dr. Cui holds six patents/patent applications and has published over sixty scientific papers and book chapters in the area of food carbohydrates. He also gives lectures on food carbohydrates in a biennial graduate course in the Department of Food Science, University of Guelph, and has delivered several workshops in Asia on structure and functionality of food hydrocolloids. He is consulted frequently by researchers and food industries on analytical methods and applications of hydrocolloids in foods and nonfood systems.

Dr. Cui graduated from the Peking University (Beijing, China) with a B.Sc. degree in 1983, from the Southern Yangtze University (Wuxi, China) with a M.Sc. degree in 1986, and from the University of Manitoba (Winnipeg, Manitoba) with a Ph.D. degree in food carbohydrates in 1993.

Contributors

Yolanda Brummer, M.Sc.   Research Technician, Agriculture and Agri-Food Canada, Guelph, Ontario, Canada

Steve W. Cui, Ph.D.   Research Scientist and National Study Leader, Agriculture and Agri-Food Canada and Adjunct Professor, Department of Food Science, University of Guelph, Ontario, Canada

Marta Izydorczyk, Ph.D.   Program Manager, Barley Research, Grain Research Laboratory, Winnipeg, Manitoba, Canada and Adjunct Professor, Department of Food Science, University of Manitoba, Winnipeg, Manitoba, Canada

Qiang Liu, Ph.D.   Research Scientist, Agriculture and Agri-Food Canada, Guelph, Ontario, Canada and Adjunct Professor, Department of Food Science, University of Guelph, Ontario, Canada

Qi Wang, Ph.D.   Research Scientist, Agriculture and Agri-Food Canada, Guelph, Ontario, Canada and Special Graduate Faculty, Department of Food Science, University of Guelph, Ontario, Canada

Sherry X. Xie, Ph.D.   NSERC Visiting Fellow, Agriculture and Agri-Food Canada, Guelph, Ontario, Canada

Contents

1

Understanding the Chemistry of Food Carbohydrates

Marta Izydorczyk

CONTENTS

1.1   Introduction

Carbohydrates are the most abundant and diverse class of organic compounds occurring in nature. They are also one of the most versatile materials available and therefore, it is not surprising that carbohydrate-related technologies have played a critical role in the development of new products ranging from foods, nutraceuticals, pharmaceuticals, textiles, paper, and biodegradable packaging materials.[1] Carbohydrates played a key role in the establishment and evolution of life on earth by creating a direct link between the sun and chemical energy. Carbohydrates are produced during the process of photosynthesis:

$$6CO_2 + 6H_2O \xrightarrow{\;h\nu\;} C_6H_{12}O_6 + 6O_2$$

Carbohydrates are widely distributed both in animal and plant tissues, where they function as:

- Energy reserves (e.g., starch, fructans, glycogen).
- Structural materials (e.g., cellulose, chitin, xylans, mannans).
- Protective substances. Some plant cell wall polysaccharides are elicitors of plant antibiotics (phytoalexins). In soybean, fragments of pectic polysaccharides (α-4-linked dodecagalacturonide) induce synthesis of a protein (protein inhibitor inducer factor) that inhibits insect and microbial proteinases. Arabinoxylans have been postulated to inhibit intercellular ice formation, thus ensuring winter survival of cereals.
- Cell recognition moieties. Oligosaccharides conjugated to protein (glycoproteins) or to lipids (glycolipids) are important components of cell membranes and can be active in cell to cell recognition and signalling. It is recognized that oligosaccharide moieties serve as probes through which the cell interacts with its environment. In addition, the environment delivers signals to the interior of the cell through the cell surface oligosaccharides.
- Information transfer agents (nucleic acids).

The first carbohydrates studied contained only carbon (C), hydrogen (H), and oxygen (O), with the ratio of H:O the same as in water, 2:1, hence the name carbohydrates or hydrates of carbon, $C_x(H_2O)_y$, was given. The composition of some carbohydrates is indeed captured by the empirical formula, but most are more complex. According to a more comprehensive definition of Robyt (1998),[2] carbohydrates are polyhydroxy aldehydes or ketones, or compounds that can be derived from them by:

- Reduction of the carbonyl group to produce sugar alcohols
- Oxidation of the carbonyl group and/or hydroxyl groups to sugar acids
- Replacement of one or more of the hydroxyl moieties by various chemical groups, e.g., hydrogen (H) to give deoxysugars, amino groups (NH_2 or acetyl-NH_2) to give amino sugars
- Derivatization of the hydroxyl groups by various moieties, e.g., phosphoric acid to give phosphosugars, sulphuric acid to give sulpho sugars
- Their polymers having polymeric linkages of the acetal type

Food carbohydrates encompass a wide range of molecules and can be classified according to their chemical structure into three main groups:

- Low molecular weight mono- and disaccharides
- Intermediate molecular weight oligosaccharides
- High molecular weight polysaccharides

Nutritionists divide food carbohydrates into two classes:[3]

- Available, or those which are readily utilized and metabolized. They may be either mono-, di-, oligo- or polysaccharides, e.g., glucose, fructose, sucrose, lactose, dextrins, starch.
- Unavailable, or those which are not utilized directly but instead broken down by symbiotic bacteria, yielding fatty acids, and thus not supplying the host with carbohydrate. This includes structural polysaccharides of plant cell walls and many complex polysaccharides, e.g., cellulose, pectins, beta-glucans.

1.2 Monosaccharides

1.2.1 Basic Structure of Monosaccharides

Monosaccharides are chiral polyhydroxy aldehydes and polyhydroxy ketones that often exist in cyclic hemiacetal forms. As their name indicates,

TABLE 1.1

Classification of Monosaccharides

Number of Carbon Atoms	Kind of Carbonyl Group	
	Aldehyde	Ketone
3	Aldotriose	Triulose
4	Aldotetrose	Tetrulose
5	Aldopentose	Pentulose
6	Aldohexose	Hexulose
7	Aldoheptose	Heptulose
8	Aldooctose	Octulose
9	Aldononose	Nonulose

monosacharides are monomeric in nature and cannot be depolymerized by hydrolysis to simpler sugars. Monosaccharides are divided into two major groups according to whether their acyclic forms possess an aldehyde or a keto group, that is, into aldoses and ketoses, respectively. These, in turn, are each classified according to the number of carbons in the monosaccharide chain (usually 3 to 9), into trioses (C_3), tetroses (C_4), pentoses (C_5), hexoses (C_6), heptoses (C_7), octoses (C_8), nonoses (C_9). By adding the prefix aldo- to these names, one can define more closely a group of aldoses, e.g., aldohexose, aldopentose. For ketoses it is customary to add the ending -ulose (Table 1.1).

Various structural diagrams are available for representing the structures of sugars.[2,4] The system commonly used for *linear* (acyclic) monosaccharides is the Fischer projection formula, named after the famous scientist, Emil Herman Fischer (1852 to 1919), which affords an unambiguous way to depict sugar molecules (Figure 1.1), provided the following rules are followed:

- The carbon chain is drawn vertically, with the carbonyl group at the top, and the last carbon atom in the chain, i.e., the one farthest from the carbonyl group, at the bottom.
- All vertical lines represent the (C–C) bonds in the chain lying below an imaginary plane (vertical lines represent bonds below the plane), and all horizontal lines actually represent bonds above the plane.
- The numbering of the carbon atoms in monosaccharides always starts from the carbonyl group or from the chain end nearest to the carbonyl group (Figure 1.1).

Formally, the simplest monosaccharide is the three-carbon glyceraldehyde (aldotriose) (Figure 1.1). It has one asymmetric carbon atom (chiral centre) and consequently, it has two enantiomeric forms. Using traditional carbohydrate nomenclature, the two forms are D- and L-glyceraldehydes.[5] A chiral atom is one that can exist in two different spatial arrangements (configurations). Chiral carbon atoms are those having four different groups attached to them. The two different arrangements of the four groups in space are nonsuperimposable mirror images of each other.

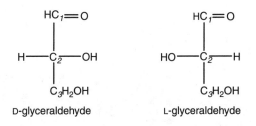

FIGURE 1.1
Structures of D-glyceraldehyde and L-glyceraldehyde.

These two compounds have the same empirical formula, $C_3H_6O_3$, but are distinct, having different chemical and physical properties. For instance, D-glyceraldehyde rotates the plane polarized light to the right (+) and has a specific optical rotation ($[\alpha]_D$) at 25°C of +8.7°; whereas L-glyceraldehyde rotates the plane polarized light to the left (-) and has a different specific optical rotation, $[\alpha]_D$, at 25°C of –8.7°. Carbon C-2 in glyceraldehyde corresponds to the chiral centre. If the OH group attached to the highest numbered chiral carbon is written to the right in the vertical structure as shown above, a sugar belongs to the D-chiral family; if the OH is written to the left, a sugar belongs to the L-chiral family. Since the principal purpose of the D and L symbols is to distinguish between chiral families of sugars, the structural specification should, in fact, be consistent with modern nomenclature of the International Union of Pure and Applied Chemistry (IUPAC):

- *R* (Latin, rectus, right) should be used instead of D.
- *S* (Latin, sinister, left) instead of L.

The higher aldoses belonging to the D- and L-series are derived from the respective D- and L-aldotrioses by inserting one or more hydroxymethylene (–CHOH) groups between the first chiral centre and the carbonyl group of the corresponding isomer. The insertion of the hydroxymethylene group leads to creation of a new chiral centre. The number of chiral carbon atoms (n) in the chain determines the number of possible isomers. Since each chiral carbon atom has a mirror image, there are 2^n arrangements for these atoms (Table 1.2). Therefore, in a six-carbon aldose with 4 chiral carbons, there are 2^4 or 16 different arrangements, allowing formation of 16 different six-carbon sugars with aldehyde end. Eight of these belong to the D-series (Figure 1.2); the other eight are their mirror images and belong to the L-series. The mnemonic ("all altruists make gum in gallon tanks") proposed by Louis and Mary Fieser of Harvard University, is a very convenient way to remember the names of the eight aldohexoses. Since the ketotriose (dihydroxyacetone) has no chiral centre, the first monosaccharide in the ketose series is erythrulose. Again, the higher ketoses are derived by inserting the hydroxymethylene group(s) between the first chiral centre and the carbonyl group of the corresponding isomer (Figure 1.3). It should be noted that the configurational

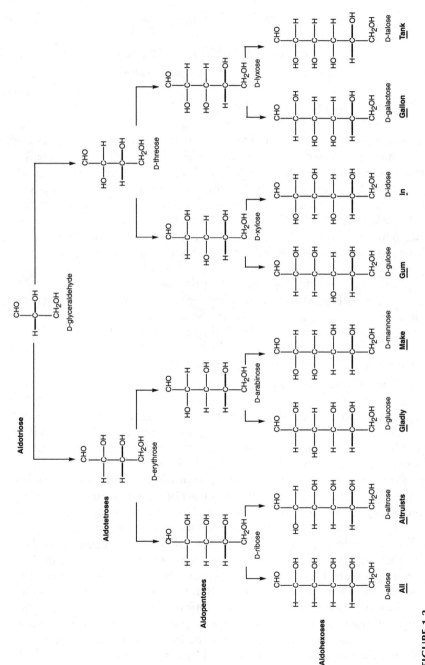

FIGURE 1.2
The D- (R) family of the aldoses.

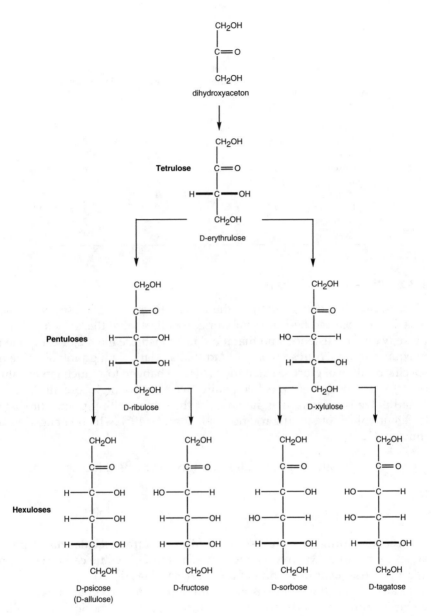

FIGURE 1.3
The D- (R) family of the ketoses.

descriptors D and L do not indicate the direction of rotation of the plane polarized light by monosaccharides. For example, D-glyceraldehyde and L-arabinose are dextrorotatory whereas D-erythrose and D-threose are levorotatory.

TABLE 1.2

Number of Isomers in Monosaccharides

Monosaccharide	Number of Chiral Centers (n)	Number of Isomers (2^n)	Number of Enantiomers (2^{n-1})
Aldose			
Triose	1	2	1
Tetrose	2	4	2
Pentose	3	8	4
Hexose	4	16	8
Ketose			
Triulose	0	1	—
Tetrulose	1	2	1
Pentulose	2	4	2
Hexulose	3	8	4

1.2.2   Ring Forms of Sugars

Even before the configuration of the acyclic form of D-glucose was established, evidence had been accumulating to indicate that this structure is not the only one in existence, and that it did not constitute the major component in equilibrium mixtures.[4] It was found that a relatively high initial value of specific rotation of glucose in solution (+112°) changed to a much lower value (+52°) after a period of time. Eventually, two forms of D-glucose, designated α and β, were isolated; they had almost the same melting point but vastly different values of specific rotation (+112° and +19°), which changed with time, to +52°.

$$\alpha\text{-D-glucose} \leftrightarrows \text{equilibrium mixture} \leftrightarrows \beta\text{-D-glucose}$$

$[\alpha]_D$	+112.2°	$\leftrightarrows$	+52.7°	$\leftrightarrows$	+18.7°
mp	146°C				150°C

These new forms of D-glucose result from an intramolecular nucleophilic attack by the hydroxyl oxygen atom attached to C-5 on the carbonyl group, and the consequent formation of a hemiacetal (Figure 1.4).

Because cyclization converts an achiral aldehyde carbon atom (C-1) into a chiral hemiacetal carbon atom, two new discrete isomeric forms, called anomers are produced; they are designated α and β. In 1926, Walter Norman Haworth (1883 to 1950) suggested that the 6-membered ring may be represented as a hexagon with the front edges emboldened, causing the hexagon to be viewed front edge on to the paper. The two remaining bonds to each carbon are depicted above and below the plane of the hexagon. The six-membered ring is related to tetrahydropyran and is called pyranose. The two new anomeric forms are easily depicted in the Haworth perspective formula (Figure 1.5).

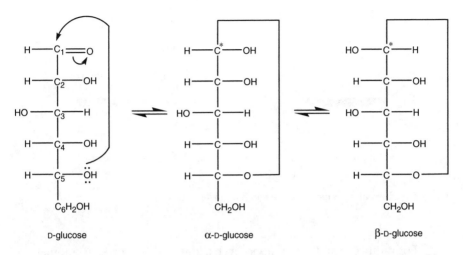

D-glucose   α-D-glucose   β-D-glucose

FIGURE 1.4
Formation of pyranose hemiacetal ring from D-glucose as a result of intramolecular nucleophilic attack by the hydroxyl oxygen atom attached to C-5 on the carbonyl group. The asterisk indicates the new chiral carbon.

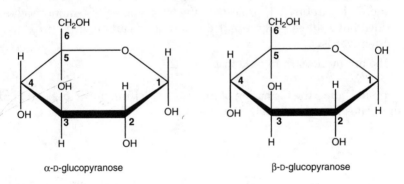

α-D-glucopyranose   β-D-glucopyranose

FIGURE 1.5
Cyclic structures of α-D-glucopyranose and β-D-glucopyranose.

A five-membered ring can also be formed as the outcome of an intramolecular nucleophilic attack by the hydroxyl oxygen atom attached to C-4 on the carbonyl group and hemiacetal formation. The five-membered ring is related to tetrahydrofuran and is, therefore, designated as furanose (Figure 1.6).

Several rules apply when converting the linear form of sugars (Fischer formulae) into their cyclic structures (Haworth formulae):

- All hydroxyl groups on the right in the Fischer projection are placed below the plane of the ring in the Haworth projection; all those on the left are above.
- In D-aldoses, the CH$_2$OH group is written above the plane of the ring in the Haworth formulae; in L-aldoses, it is below.

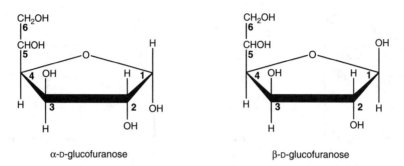

FIGURE 1.6
Cyclic structures of α-D-glucofuranose and β-D-glucofuranose.

- For D-glucose and other monosaccharides in the D-series, α-anomers have the –OH group at the anomeric carbon (C-1) projected downwards in the Haworth formulae; β-anomers have the –OH group at the anomeric carbon (C-1) projected upwards. The opposite applies to the L-series; α-L-monosaccharides have the –OH group at the anomeric carbon (C-1) projected upwards, whereas β-L-monosaccharides have the –OH group at the anomeric carbon (C-1) projected downwards.
- The anomeric carbon of ketoses is C-2.

Figure 1.7 shows the formation of furanose ring from D-fructose and Figure 1.8 illustrates linear and cyclic structures of some common sugars.

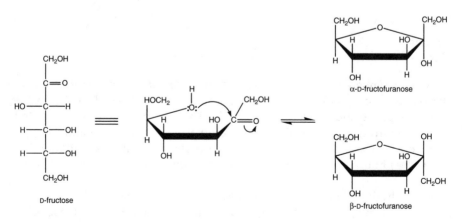

FIGURE 1.7
Formation of furanose ring anomers from D-fructose.

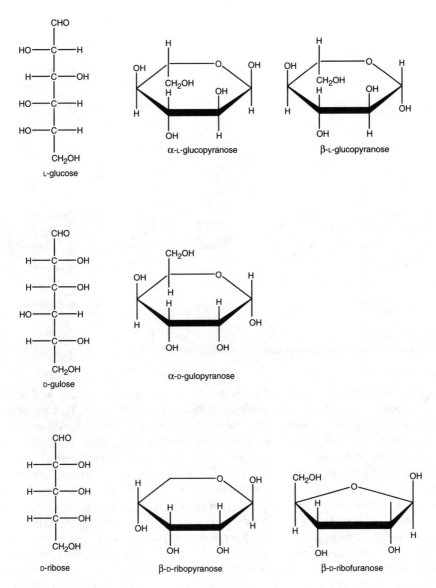

FIGURE 1.8
Linear and cyclic structures of common sugar.

1.2.3 Stereochemical Transformations

1.2.3.1 Mutarotation

When sugar molecules are dissolved in aqueous solutions, a series of reactions, involving molecular rearrangements around the C-1, takes place. These

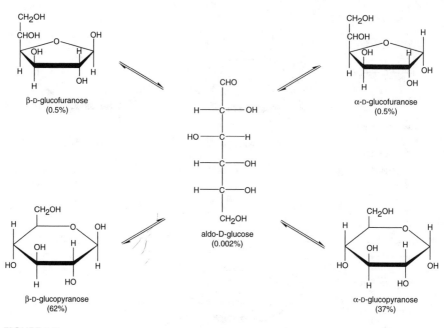

FIGURE 1.9
Tautomeric forms of D-glucose possible in solution.

rearrangements are associated with the change in optical rotation, and lead to formation of a mixture of products that are in equilibrium. This process, first observed for D-glucose, is called mutarotation.[6] If one dissolves α-D-glucopyranose ($[\alpha]_D$ +112°) or β-D-glucopyranose ($[\alpha]_D$ +19°) in water, an equilibrium is formed with the $[\alpha]_D$ of the resultant solution being +52.7°. Theoretically, the mixture contains five different structural forms of glucose: α-D-glucopyranose, β-D-glucopyranose, α-D-glucofuranose, β-D-glucofuranose, and open-chain free aldehyde (Figure 1.9). The four ring structures are transformed into each other via the open chain form. The process will take place if the starting material represents any of the five forms.

The mutarotation process is slow (it may take several hours to reach equilibrium) if conducted in water at 20°C. The rate of mutarotation increases, however, 1.5 to 3 times with each 10°C increase in the temperature. Both acids and bases increase the rate of mutarotation. Certain enzymes, such as mutarotase will also catalyze the mutarotation reactions. The rate and the relative amount of products are also affected by the polarity of the solvent, with less polar solvents decreasing the rate of mutarotation. The reaction begins upon dissolution of sugar molecule and an attack, by either acid or base, on the cyclic sugar. It involves the transfer of a proton from an acid catalyst to the sugar or the transfer of proton from the sugar to a base catalyst as shown in Figure 1.10.

Base (pH 10) catalyzed mutarotation

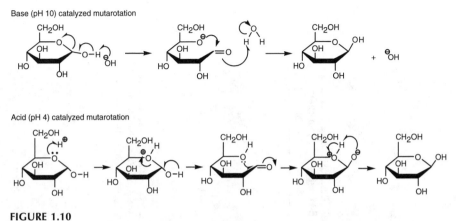

Acid (pH 4) catalyzed mutarotation

FIGURE 1.10
Mechanisms of base and acid catalyzed mutarotation reactions.

The number of various forms present in a measurable amount at equilibrium leads to classification of the sugar mutarotation reactions as either simple or complex. The presence of two major components at equilibrium is the principal characteristic of simple mutarotation, whereas at least three components present in measurable concentration indicate complex mutarotation.

$$\alpha \rightleftarrows \beta \text{ vs. } \alpha \rightleftarrows \gamma \rightleftarrows \beta$$

The distribution of sugar tautomers at equilibrium in water may be calculated (on the basis of optical rotary power and/or conformational free energy) or determined experimentally (gas-liquid chromatography or nuclear magnetic resonance spectroscopy). For complex mutarotation reactions, the percentage distribution of tautomeric forms may be uncertain. While altrose definitely exhibits complex mutarotation, e.g., gulose only probably exists in several forms in solution. Table 1.3 gives the distribution of various tautomeric forms at equilibrium in water solutions. In general, the pyranose ring forms of sugars predominate over furanose rings in solution. In addition to its greater intrinsic stability, pyranose fits better into the tetrahedrally arranged water molecules, and it is stabilized by many sugar-water hydrogen bonds (Figure 1.11). On the other hand, solvents other than water, with a different structure (e.g., dimethyl sulphoxide), may favor the furanose over the pyranose ring.

1.2.3.2 Enolization and Isomerization

In the presence of alkali, sugars are relatively easily interconverted. The transformation involves epimerization of both aldoses and ketoses as well

TABLE 1.3

Distribution of D-Sugars Tautomers at Equilibrium in Water Solution

Carbohydrate	Mutarotation Type	Temp °C	α-Pyranose %	β-Pyranose %	α-Furanose %	β-Furanose %	Open-chain %
D-Glucose	Simple	20	36.4	63.6	—	—	—
		31	38	62	0.5	0.5	0.002
D-Galactose	Complex	20	32	63.9	1	3.1	0.02
		31	30	64	2.5	3.5	
D-Xylose	Simple	20	34.8	65.2	—	—	—
		31	36.5	58.5	6.4	13.5	0.05
D-Fructose	Simple	27	—	75	4	21	
		31	2.5	65	6.5	25	0.8
D-Altrose	Complex	40	27	40	20	13	
D-Gulose	Simple	40	10	88	2	—	
D-Mannose	Simple	20	67.4	32.6	—	—	

Source: From El Khadem, H.S., *Carbohydrate Chemistry: Monosaccharides and their Oligomers,* Academic Press, San Diego, 1988;[5] Shallenberger, R.S., *Advanced Sugar Chemistry, Principles of Sugar Stereochemistry,* AVI Publishing Co., Westport, 1982.[6]

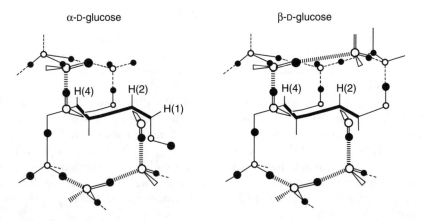

FIGURE 1.11
The pyranose rings of α- and β-D-glucose (indicated by the centrally positioned thick lines) hydrogen-bonded into a tetrahedral arrangement of water (D₂O) molecules above and below the plane of the sugar rings. Oxygen and deuterium atoms are represented by open and filled circles, respectively.

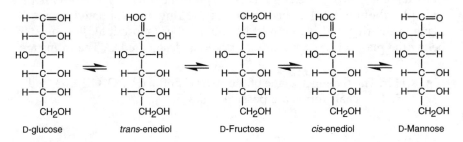

FIGURE 1.12
Enolization and isomerization reactions.

as aldose-ketose isomerization. The mechanism of the reaction is shown in Figure 1.12. The enolization reaction is a general reaction of a carbonyl compound having an α-hydrogen atom. Starting with *aldehydo*-D-glucose, the 1,2 enediol is first formed, which can be converted into another aldose (with opposite configuration at C-2) and the corresponding ketose. Therefore, by enolization and isomerization, D-glucose, D-mannose, and D-fructose can be easily interconverted. Either a base or an enzyme catalyzes isomerization, and it will also occur under acid or neutral conditions, although at a much slower rate.

1.2.4   Conformation of Monosaccharides

Even though the Fischer and Haworth projections for carbohydrates indicate some spacial configuration of the hydroxyl groups, they do not portray the true shapes of these molecules. Three-dimensional model building using the correct bond lengths and angles of the tetrahedral carbons has shown that the pyranose and furanose rings are not flat. Rotation about the sigma bonds between the carbon-to-carbon and carbon-to-oxygen atoms in the ring can result in numerous shapes of the ring in a three-dimensional space. The shape of the ring and the relative position of the hydroxyl groups and the hydrogen atoms in relation to the ring are called conformation. Furanose and pyranose rings can exist in a number of inter-convertible conformers (conformational isomers) that differ in thermodynamic stability from each other.[4–6]

1.2.4.1   *Conformation of the Pyranose Ring*

The recognized forms of the pyranose ring include chair (C), boat (B), half chair (H), skew (S), and sofa forms. The following rules apply to the designation of the different isomeric forms. The letter used to designate the form (for example C or B for chair and boat conformations, respectively) is preceded by the number (superscripted) of the ring atom situated above the plane of the ring and is followed by the number (subscripted) of the atom below the plane of the ring; a ring oxygen is designated O. The forms are:

- Chair: The reference plane of the chair is defined by O, C-2, C-3, and C-5. Two chair forms are possible.

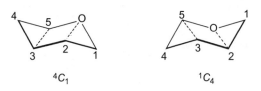

$$^4C_1 \qquad\qquad {}^1C_4$$

- Boat: Six forms are possible, with two shown below.

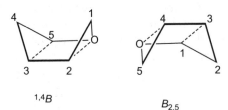

$$^{1,4}B \qquad\qquad B_{2,5}$$

- Half chair: Twelve forms are possible, with two shown below (the reference plane is defined by four contiguous atoms).

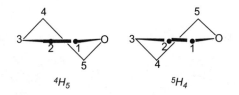

$$^4H_5 \qquad\qquad ^5H_4$$

- Skew: Six forms are possible.

$$^1S_5$$

1.2.4.2 Conformation of the Furanose Ring

The principal conformers of the furanose rings are the envelope (E) and the twist (T). Both have ten possible variations.

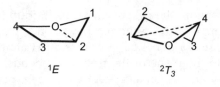

$$^1E \qquad\qquad ^2T_3$$

1.2.4.3 Determination of Favored Pyranoid Conformation

Of the many possible conformations, one conformer is preferred or favored. This conformer is usually the one with the minimum free energy. Conformational free energy is calculated based on the actual attractions and repulsions between atoms in terms of van der Waals forces, polar and hydrophobic interactions, steric interactions, H-bonding effects, solvation effects, and strains associated with the bond's length and angles. The conformation of a sugar molecule in the solid state is not necessarily the same as in solution. For crystalline sugar, a single favored conformation is usually assumed; in solution, an equilibrium of various conformational isomers may exist, with the most favored conformation present in the largest amount. When the molecule is in crystalline state, a single, x-ray structure determination will yield both the molecular structure and the conformation. When the molecule is in a liquid, or solution, ^{1}H-nuclear resonance spectroscopy can usually reveal the conformation.

When D-glucose is dissolved in water, an equilibrium is established, with the β-anomer in the $^4C_1$ conformation as the preferred conformer.

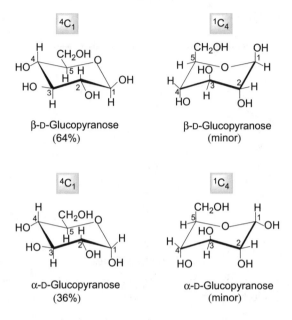

In the $^4C_1$ conformation, the primary hydroxyl group (–CH_2–OH) and all secondary –OH groups of the β-anomer have bonds positioned within the plane of the ring (equatorial bonds), whereas all H atoms have bonds that are perpendicular to the plane of the ring (axial bonds). In the $^1C_4$ conformation, the primary and secondary –OH groups are perpendicular (axial) to the plane of the ring. The most stable, or most favored, conformation is the one that places the majority of the bulky substituents (for carbohydrates, the hydroxyl groups are the bulky groups) in an equatorial position. This spacial arrangement puts the bulky groups far apart from each other and creates a low energy form with a minimum bulky group interactions. Placing the bulky groups in axial positions creates a high energy form with a maximum interaction.

Hassel and Ottar (1947)[7] provided evidence that conformations which placed the –CH_2–OH group and an additional –OH group (at C-1 or C-3) in axial positions on the same side of the ring are very unstable (Figure 1.13). Application of the Hassel and Ottar effect settled the preferred conformation (as $^4C_1$) for β-allose, α-altrose, α- and β-glucose, α- and β-mannose, β-gulose, α-idose, α- and β-galactose, and α- and β-talose.

To extend the predictive powers of the Hassel and Ottar effect, Reeves (1951)[8] proposed that any erected (axial) substituents, other than hydrogen atom, on a pyranose ring introduces an element of instability. In particular, the most important instability factor in the pyranose ring is a situation called delta 2 (Figure 1.14). It arises when the oxygen atom of the OH on C-2 bisects

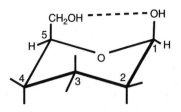 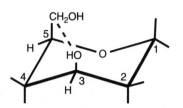

FIGURE 1.13
The Hassel-Ottar effect.

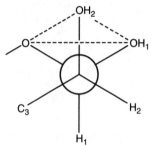

FIGURE 1.14
The delta-2 effect.

the angle formed by the ring oxygen atom and the oxygen atom of the anomeric OH group (on C-1). This occurs in certain β-glycopyranoses, such as mannose and altrose.

Reeves (1951)[8] tabulated the instability factors for all of the aldohexa- and aldopentapyranoses as shown in Table 1.4. It can be seen that for most of the compounds listed in the table, the $^4C_1$ conformation is more favorable (has fewer instability factors) than the $^1C_4$ conformation. However, for some monosaccharides, such as β-D-mannose and β-D-idose, an equilibrium exists between $^4C_1$ and $^1C_4$ conformations.

Another instability factor was recognized by Edward (1955)[9] who noted that derivatization of pyranose sugars at the anomeric center led to a greater stability of the α-D-$^4C_1$ anomers (OH-1 in axial position) than β-D-$^4C_1$ anomers (OH-1 in equatorial position). For example, the methyl β-D-pyranosides of glucose, mannose, and galactose are hydrolyzed more rapidly than α-D-pyranosides. Also, it is known that all D-glycopyranosyl halides have α configurations (halogen in axial) irrespective of whether they have been obtained from α- or β-D-glycopyranoses (Figure 1.15).

This propensity for the formation of the α-anomer over the normally expected β-anomer was termed the anomeric effect by Lemieux.[10,11] The origin of the anomeric effect, which increases with the electronegativity of the substituents and decreases in solvents with high dielectric constant, has been interpreted in several ways; one interpretation posited the existence of

TABLE 1.4

The Instability Factors for Various Aldopyranoses in $^4C_1$ and $^1C_4$ Conformations

Sugar	Instability Factors*		Predicted Conformation
	$^4C_1$	$^1C_4$	
α-D-Allose	1,3	Δ2,4,5	$^4C_1$
β-D-Allose	3	H,1,2,4,5	$^4C_1$
α-D-Altrose	1,2,3	4,5	$^4C_1$, $^1C_4$
β-D-Altrose	Δ2,3	H,1,4,5	$^4C_1$, $^1C_4$
α-D-Galactose	1,4	H,Δ2,3,5	$^4C_1$
β-D-Galactose	4	H,1,2,3,5	$^4C_1$
α-D-Glucose	1	H,Δ2,3,4,5	$^4C_1$
β-D-Glucose	None	H,1,2,3,4,5	$^4C_1$
α-D-Gulose	1,3,4	Δ2,5	$^4C_1$, $^1C_4$
β-D-Gulose	3,4	H,1,2,5	$^4C_1$
α-D-Idose	1,2,3,4	5	$^4C_1$
β-D-Idose	Δ2,3,4	H,1,5	$^4C_1$, $^1C_4$
α-D-Mannose	1,2	H,3,4,5	$^4C_1$
β-D-Mannose	Δ2	H,1,3,4,5	$^4C_1$, $^1C_4$
α-D-Talose	1,2,4	H,3,5	$^4C_1$, $^1C_4$
β-D-Talose	Δ2,4	H,1,3,5	$^4C_1$, $^1C_4$
α-D-Arabinose	1,2,3	4	$^4C_1$
β-D-Arabinose	Δ2,3	1,4	$^4C_1$
α-D-Lyxose	1,2	3,4	$^4C_1$, $^1C_4$
β-D-Lyxose	Δ2	1,3,4	$^4C_1$, $^1C_4$
α-D-Ribose	1,3	Δ2,4	$^4C_1$
β-D-Ribose	3	1,2,4	$^4C_1$
α-D-Xylose	1	Δ2,3,4	$^4C_1$
β-D-Xylose	None	1,2,3,4	$^4C_1$

Note: A number refers to hydroxyl groups in axial position; Δ2 indicates the delta-two effect; H refers to the Hassel-Ottar effect.

Source: From Reeves, P. *Advanced Carbohydrate Chemistry*, 6, 107–134, 1951.[8]

   CHCl$_3$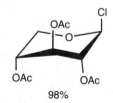

2%   98%

FIGURE 1.15
D-glycopyranosyl halides.

unfavorable lone pair-to-lone pair or dipole-to-dipole interactions between an equatorial hydroxyl substituent at the anomeric centre and the ring oxygen atom (Figure 1.16).

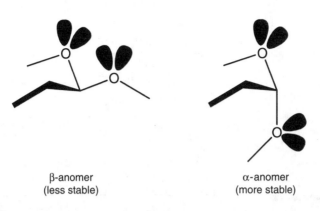

β-anomer   α-anomer
(less stable)   (more stable)

FIGURE 1.16
Anomeric effect.

Manifestations of the anomeric effect are stronger in nonpolar solvents than in polar solvents such as water, with very high dielectric constant. However, they are still felt in aqueous solutions. In the case of D-glucopyranose, for example, even though the β-D-$^4C_1$ anomer (64%) predominates (anomeric OH in equatorial position) over the α-D-$^4C_1$ anomer (36%) (anomeric OH in axial position), the free energy difference between the anomers is only 1.5 kJ mol^{-1}, much smaller than the expected value of 3.8 kJ mol^{-1}. This discrepancy between the actual and theoretical free energy differences is attributed to the anomeric effect in water.

The conformation of sugar molecules must be kept in mind because the three-dimensional shapes often play an important role in the biological functions of carbohydrates as well as affect certain physicochemical properties, such as their chemical reactivity, digestibility, nutritional responses, and sweetness.

1.2.5　Occurrence of Monosaccharides

The majority of the naturally occurring carbohydrates have the D-configuration. L-monosaccharides are less abundant in nature, although L-arabinose and L-galactose are present as carbohydrate units in many carbohydrate polymers (polysaccharides). It is not well understood why and how carbohydrates with only D-configuration were formed on earth. Interestingly, the configuration of the 20 naturally occurring amino acids is opposite (L-configuration) to that of carbohydrates. While the pyranose forms dominate in aqueous solution of most monosaccharides, it is quite common to find furanose form when the sugar is incorporated into a biomolecule.

The popular names of sugars often indicate their principal sources and their optical rotary properties. Synonyms for D-glucose are dextrose, grape sugar, and starch sugar. Synonyms for fructose are levulose, honey sugar, fruit sugar.

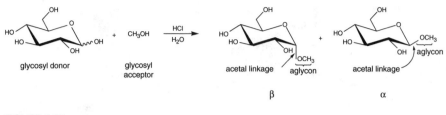

FIGURE 1.17
Formation of acetal linkage.

1.3   Oligosaccharides

1.3.1   Formation of Glycosidic Linkage

The hemiacetal form of sugars can react with alcohol to produce a full acetal[4,12,13] according to the following reaction (Figure 1.17).

The two products are more commonly referred to as glycosides — more specifically, as methyl α- and β-D-glucopyranosides. The carbohydrate (*glycon*) portion of the molecule is distinguished from the noncarbohydrate *aglycon*. The acetal linkage is formed from a glycosyl donor and a glycosyl acceptor.

1.3.2   Disaccharides

Because carbohydrates are polyalcohols with primary and secondary alcohol groups, their alcohol groups can react with a hemiacetal hydroxyl group of another carbohydrate and form a glycoside between two carbohydrate units (Figure 1.18). Disaccharides are therefore glycosides in which the aglycon constitutes another monosaccharide unit.

Taking D-glucose as a starting point, one can form 11 different disaccharides. α-D-Glucose can react with the alcohol group at C-2, C-3, C-4, and C-6 of another glucose unit to give four reducing disaccharides (Figure 1.19). β-D-Glucose can also react with the same alcohol group at C-2, C-3, C-4, and C-6 to give another four possible reducing disaccharides (Figure 1.20). The α- and β-hemiacetal hydroxyl groups (at C-1) of each monosaccharide can also react with each other, giving three more nonreducing disaccharides: α,α-trehalose, β,β-trehalose, and α,β-trehalose (Figure 1.21). The trehaloses are nonreducing sugars because the two hemiacetal hydroxyl groups are engaged in the glycosidic bond and, therefore, no free anomeric groups are available. The acetal linkage between the monosaccharide residues is called a glycosidic linkage. Each of the 11 disaccharides has distinctive chemical and physical properties, although structurally they differ only in the type of glycosidic bond that joins the two moieties. For example, maltose has α-1→4

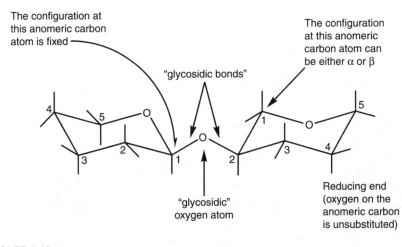

The configuration at this anomeric carbon atom is fixed

The configuration at this anomeric carbon atom can be either α or β

"glycosidic bonds"

"glycosidic" oxygen atom

Reducing end (oxygen on the anomeric carbon is unsubstituted)

FIGURE 1.18
A β-1→2 linked disaccharide.

linkage, cellobiose has β-1→4 linkage, isomaltose has α-1→6 linkage, whereas α,α-trehalose has α-1→1 linkage.

A few of the 11 glucose-glucose disaccharides are quite common. Maltose, although it rarely occurs in plants, can be readily produced by hydrolysis of starch. Maltose is therefore present in malted grains and various food items containing starch hydrolysis products (e.g., corn syrup). α,α-Trehalose occurs in the spores of fungi and it is also produced by yeasts. Isomaltose constitutes the branch point of amylopectin and glycogen. Cellobiose is a product of bacterial hydrolysis of cellulose by enzymes such as endo-cellulases and cellobiohydrolases. Laminaribiose is a repeating unit found in the polysaccharides, laminarin (brown algae), pachyman (fungi), and callose.

Disaccharides can be divided into heterogeneous and homogeneous types, according to their monosaccharide composition, and into reducing or non-reducing disaccharides, depending whether they possess a free anomeric carbon. Homodisaccharides contain two identical monosaccharide units, whereas heterodisaccharides are composed of two different monomers. Reducing disaccharides, in contrast to nonreducing ones, contain a reactive hemiacetal center that can be easily modified chemically (e.g., via oxidation or reduction).

The two most important naturally occurring heterodisaccharides are sucrose and lactose. Sucrose (commonly known as sugar or table sugar) occurs in all plants, but it is commercially obtained from sugar cane and sugar beets. It is composed of an α-D-glucopyranosyl unit and a β-D-fructofuranosyl unit linked reducing end to reducing end, thus it is a nonreducing sugar (Figure 1.22). Its chemical name is α-D-glucopyranosyl-β-D-fructofuranoside.

Sucrose is the world's main sweetening agent and about 10^8 tonnes are produced annually. Sucrose is common in many baked products, breakfast cereals, deserts, and beverages. Sucrose is hydrolyzed into D-glucose and

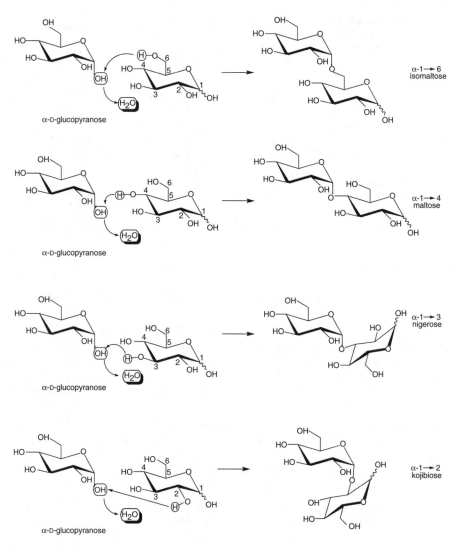

FIGURE 1.19
Formation of α-linked (1→2, 1→3, 1→4, 1→6) D-glucose disaccharides.

D-fructose by the enzyme sucrase, which is present in the human intestinal tract, and therefore can be utilized by humans for energy. Monosaccharides (D-glucose and D-fructose) do not need to undergo digestion before they are absorbed. The plant enzyme invertase is able to hydrolyze sucrose into its two constituent sugars in equimolar mixture of D-glucose and D-fructose; the mixture has a different value of specific rotation, and is termed invert sugar.

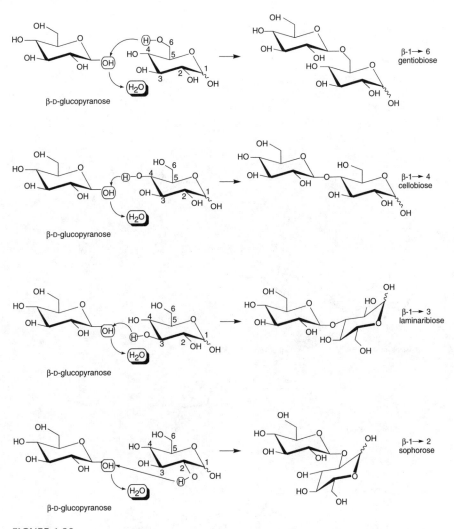

FIGURE 1.20
Formation of β-linked (1→2, 1→3, 1→4, 1→6) D-glucose disaccharides.

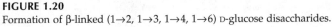

$$\text{Sucrose} \rightarrow \text{D-Glucose} + \text{D-Fructose}$$

Invert sugar

$[\alpha]_D +66°$ $[\alpha]_D -22°$

Lactose occurs in the milk of mammals, where it serves as an energy source for developing mammals. The concentration of lactose in milk may vary from 2 to 8%, depending on the source. Lactose is also a by product in the

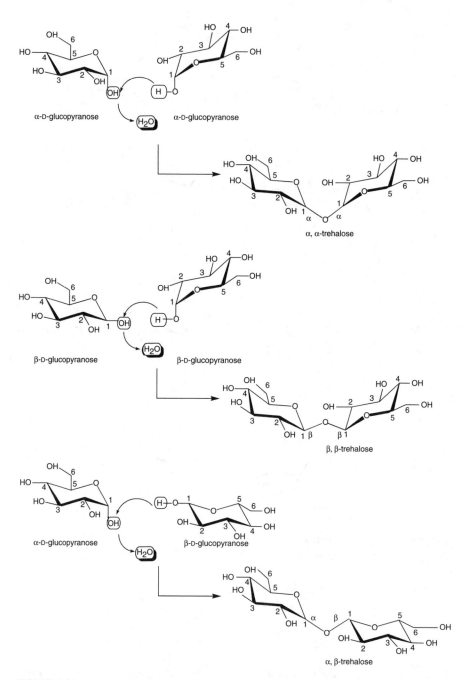

FIGURE 1.21
Formation of nonreducing 1→1 D-glucose disaccharides.

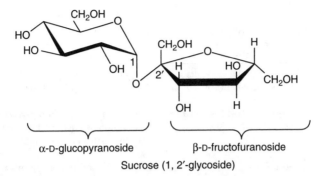

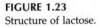

α-D-glucopyranoside   β-D-fructofuranoside

Sucrose (1, 2′-glycoside)

FIGURE 1.22
Structure of sucrose.

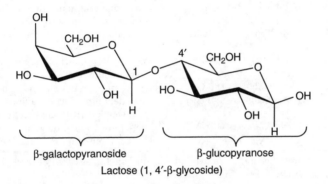

β-galactopyranoside   β-glucopyranose

Lactose (1, 4′-β-glycoside)

FIGURE 1.23
Structure of lactose.

manufacture of cheese. Lactose is composed of two different sugar residues: β-D-galactopyranose and D-glucopyranose linked via β-1→4 linkage (Figure 1.23). Its chemical name is β-D-galactopyranosyl-D-glucopyranoside. Other naturally occurring disaccharides are listed in Table 1.5.

1.3.3 Conformation of Disaccharides

The monosaccharide units in a disaccharide are free to rotate about the glycosidic bonds. The relative orientations between the two participating monosaccharide units are defined by two torsion angles φ and ψ around the glycosidic bonds, as shown in Figure 1.24. When the glycosidic linkage is formed between C-1 of one residue and C-6 of another, there is an extra torsion angle, ω (Figure 1.24). This angle increases freedom to adopt a wide variety of orientations relative to each other.[6]

TABLE 1.5

Structure and Occurrence of Some Natural Disaccharides in Biological Systems

Name	Structure	Occurrence
Cellobiose	O-β-D-Glc$p$-(1→4)-D-Glc$p$	Unit of cellulose
Gentiobiose	O-β-D-Glc$p$-(1→6)-D-Glc$p$	Sugar component in glycosides such as amygdalin
Isomaltose	O-α-D-Glc$p$-(1→6)-D-Glc$p$	Unit of amylopectin and glycogen
Maltose	O-α-D-Glc$p$-(1→4)-D-Glc$p$	Free compound in malt, beer; small amounts in some fruits and vegetables; main unit of starch
Nigerose	O-α-D-Glc$p$-(1→3)-D-Glc$p$	Free compound in honey; unit of polysaccharide nigeran
Laminaribiose	O-β-D-Glc$p$-(1→3)-D-Glc$p$	Free compound in honey; unit of laminaran and of the glucan in yeasts
Kojibiose	O-α-D-Glc$p$-(1→3)-D-Glc$p$	Free compound in honey
Trehalose	O-α-D-Glc$p$-(1→1)-D-Glc$p$	Free compound in mushrooms; in the blood of insects and grasshoppers
Sucrose	O-β-D-Fru$f$-(2→1)-D-Glc$p$	Free compound in sugar cane, sugar beets, in many plants and fruits
Maltulose	O-α-D-Glc$p$-(1→4)-D-Fru$f$	Conversion product of maltose; free compound in malt, beer, and honey
Lactose	O-β-D-Gal$p$-(1→4)-D-Glc$p$	Free compound in milk and milk products
Melibiose	O-α-D-Gal$p$-(1→6)-D-Glc$p$	Degradation product of raffinose by yeast fermentation; free compound in cocoa beans
Mannobiose	O-β-D-Man$p$-(1→4)-D-Man$p$	Unit of polysaccharide guaran
Primverose	O-β-D-Xyl$p$-(1→6)-D-Glc$p$	Free compound in carob tree fruits

Source: Adapted from Scherz, H. and Bonn, G., *Analytical Chemistry of Carbohydrates*, Georg Thieme, Verlag, Stuttgart, 1998.[14]

The linkage conformation of disaccharides is defined by a distinct set of values of torsion angles (sometimes referred to as conformational angles), φ and ψ. The conformational energy associated with a particular pair of φ and ψ can be estimated from the van der Waals, polar, and hydrogen bonding interactions between the two monomers. For example the minimum conformation energy for maltose, with α-1→4 linkage, is calculated for $\varphi = -32°$ and $\psi = -13°$. This conformer is stabilized by internal hydrogen bonds and is the main conformer of the crystalline maltose. In solution, however, due to the solvent-carbohydrate interactions, the minimum conformational energy may be different. Solvents also provide mobility (as opposed to the locked conformation found in crystals); thus the system should be viewed in terms of a dynamic distribution of various conformations, the majority of which populate around conformers with the minimum energy.

1.3.4 Oligosaccharides

According to the International Union of Pure and Applied Chemistry (IUPAC), oligosaccharides are compounds containing 3 to 9 monomeric sugar residues. The number of sugar residues determines the degree of

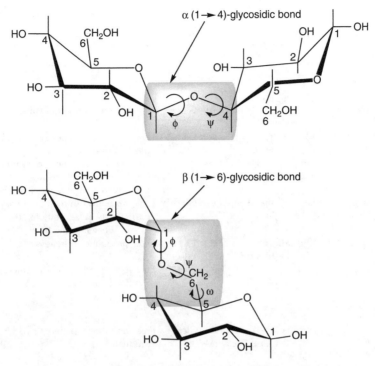

Degree of freedom in (1→ 4) and (1 → 6)-glycosidic bonds

FIGURE 1.24
Rotational angles between adjacent residues. Relative orientations between adjacent residues are defined by the two rotational angles Φ and ψ. 1→6 linkage introduces another possibility of rotation about the C-5 and C-6 bond.

polymerization of oligosaccharides. A number of oligosaccharides occur naturally as free compounds in some plants (Table 1.6). Raffinose is a nonreducing trisaccharide that is formed from sucrose by the addition of α-D-galactopyranose to the C-6 of the glucose moiety of sucrose. It is widely distributed in many plants and commercially prepared from beet molasses and cotton seeds. Stachyose is another derivative of sucrose; it occurs as a free oligomer in many legumes and pulses.

αGalp(1→6)-αGalp(1→6)-αGlcp(1→2)-Fru*f*

Sucrose

Raffinose

Stachyose

TABLE 1.6

Structure and Occurrence of Tri-, Tetra-, and Pentasaccharides in Biological Systems

Name	Structure	Occurrence
Trisaccharides		
Maltotriose	O-α-D-Glcp-(1→4)-O-α-D-Glcp-(1→4)-D-Glcp	Degradation product of starch
Panose	O-α-D-Glcp-(1→6)-O-α-D-Glcp-(1→4)-D-Glcp	Free compound in honey; degradation product of amylopectin
Melezitose	O-α-D-Glcp-(1→3)-O-β-D-Fruf-(2→1)-D-Glcp	Free compound in nectars of many plants; in the sweet exudates of many trees, such as lime, pine, poplars
Raffinose	O-α-D-Galp-(1→6)-O-α-D-Glcp-(1→2)-β-D-Fruf	Free compound in many plants, particularly in leguminous seeds such as soya beans, mung beans, etc.
Erlose	O-α-D-Glcp-(1→4)-O-α-D-Glcp-(1→2)-β-D-Fruf	Free compound in honey
Gentianose	O-β-D-Glcp-(1→6)-O-α-D-Glcp-(1→2)-β-D-Fruf	Free compound in the rhizomes of Gentiana varieties
Tetrasaccharides		
Maltotetraose	O-α-D-Glcp-(1→4)-O-α-D-Glcp-(1→4)-O-α-D-Glcp-(1→4)-D-Glcp	Free compound in starch hydrolyzate products
Stachyose	O-α-D-Galp-(1→6)-O-α-D-Galp-(1→6)-O-α-D-Glcp-(1→2)-β-D-Fruf	Free compounds in seeds of pulses
Pentasacharides		
Verbascose	O-α-D-Galp-(1→6)-O-α-D-Galp-(1→6)-O-α-D-Galp-(1→6)-O-α-D-Glcp-(1→2)-β-D-Fruf	Free compound in seeds of pulses including cow peas, winged beans, lima beans; in the rhizomes of wool plant

Source: Adapted from Scherz, H. and Bonn, G., *Analytical Chemistry of Carbohydrates*, Georg Thieme, Verlag, Stuttgart, 1998.[14]

Fructo-oligosaccharides and inulin are a group of linear glucosyl α-(1→2) (fructosyl)$_n$ β-(2→1) fructose carbohydrates with a degree of polymerization (DP) ranging from 3 to 60 (Figure 1.25).[15] By definition, if these fructans have a DP lower than 9, they are termed fructo-oligosaccharides. The fructans with a higher DP are termed inulin. The main fructo-oligosaccharides are 1-kestose (GF$_2$), nystose (GF$_3$), and fructosylnystose (GF$_4$) (Figure 1.26). Fructo-oligosaccharides are widely found in many types of edible plants, including onion, garlic, bananas, Jerusalem artichoke, asparagus, wheat and rye (Table 1.7). Inulin with GF$_n$ (n>9) is present mainly in chicory and Jerusalem artichoke. Fructo-oligosaccharides can be prepared on a commercial scale from sucrose through the transfructosylating action of fungal enzymes, such as β-fructofuranosidases (E.C. 3.2.1.26) or β-fructosyltransferases (EC

Glucosyl α (1 ⟶ 2) (fructosyl)ₙ β (2⟶ 1) fructose

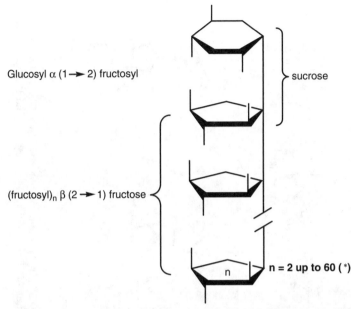

Glucosyl α (1⟶ 2) fructosyl   sucrose

(fructosyl)ₙ β (2 ⟶ 1) fructose

n = 2 up to 60 (*)

(*) in artichoke globe the DP can reach 200

FIGURE 1.25
General chemical structure of fructo-oligosaccharides.

2.4.1.100) (Figure 1.26). They can also be prepared from inulin by partial hydrolysis using endo-inulinase.

The fructo-oligosaccharides have attracted a lot of interest because of their nutritional properties. They are not digested in the human upper intestine, but they are fermented in the colon to lactate and short chain fatty acids (acetate, propionate and butyrate). They stimulate the growth of bifidobacteria while suppressing the growth of some unfavorable bacteria such as *Escherichia coli* or *Clostridium perfringens*. Fructo-oligosaccharides have been shown to be effective in preventing colon cancer, reduction of serum cholesterol and triacylglycerols, and promotion of mineral absorption. They have been used in various functional food products.

Following the success with fructo-oligosaccharides, a number of other indigestible oligosaccharides have been developed, including galacto-oligosaccharides or soy-bean-oligosaccharides (galactosyl-sucrose oligomers).[17] Galacto-oligosaccharides, such as trisaccharides, β-(1→4)-galactosyllactose and β-(1→6)-galactosyllactose, occur naturally in some dairy products as well as in human milk. They can also be produced commercially (Figure 1.27). Many commercially available oligosaccharides are synthetic products resulting from the enzymatic and/or chemical modification of natural disaccharides or polysaccharides (Table 1.8).[18]

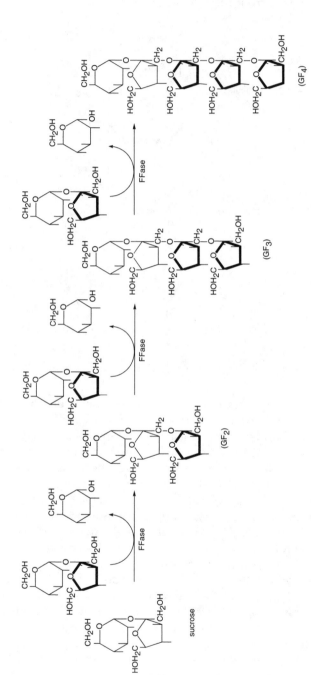

FIGURE 1.26

Enzymatic preparation of fructo-oligosaccharides; 1-kestose (GF$_2$), nystose (GF$_3$), and fructofuranosylnystose (GF$_4$); FFase: β-Fructofuranosidase. (Adapted from Hidaka, H., Adachi, T., and Hirayama, M., *Advanced Fibre Technology*, McCleary, B.V. and Prosky, L., Eds., Blackwell Science, Ltd., Oxford, 471, 2001.)[16]

TABLE 1.7

Occurrence and Distribution of Fructo-Oligosaccharides
in Edible Plants

Plant Material	Amount (%)	Distribution of GF (%)			
		GF$_2$	GF$_4$	GF$_4$	GF$_{5-8}$
Onion	25–40	61	25	10	3
Chicory	15–20	4	5	5	16
Wheat	1–4	30	13	6	50
Jerusalem artichoke	16–20	DP<9 : 50%			
Garlic	25–35	DP<9 : 10-20%			

Source: Adapted from Bornet, F.R.J., *Advanced Dietary Fibre Technology*, McCleary, B.V. and Prosky, L., Eds., Blackwell Science, Ltd., Oxford, 480, 2001.[15]

FIGURE 1.27
The production process of galacto-oligosaccharides; n = 1-7.

Another carbohydrate oligomer which has recently attracted some interest is polydextrose.[19] It was originally developed as a low calorie replacement for sugar and a partial replacement for fat and starch. Polydextrose can be prepared by vacuum thermal polymerization of glucose, using sorbitol and acid as catalyst. Random polymerization results in various glycosidic linkages, with the 1→6 bonds predominating (Figure 1.28). The average degree of polymerization is ~12. According to the IUPAC definition of oligosaccharides (DP ≤ 9), polydextrose is a polysaccharide. However, polydextrose is a low molecular weight compound that is highly soluble in water. It cannot, for example, be quantified like other polysaccharides, by precipitation with aqueous (80%) ethanol. It is, therefore not measured as dietary fiber. Several clinical studies revealed that polydextrose induces physiological effects associated with dietary fiber and fructo-oligosaccharides, i.e., increases fecal bulking, softens stools, is fermented to short fatty acids, increases the amount of beneficial bacteria (e.g., *Lactobacillus* and *Bifidobacterium*) at the cost of detrimental species (*Clostridium*). Polydextrose has a very low glycemic index (15%) compared to glucose (100%).

1.3.5   Cyclic Oligosaccharides

Cyclodextrins are cyclic oligomers derived enzymatically from gelatinized starch.[20,21] The enzymes which catalyze their synthesis are cyclodextrin glycosyltransferases, produced by several bacteria species including *Bacillus*, *Klebsiella*, and *Micrococcus*. The substrate for the enzyme is the linear portion

TABLE 1.8

Commercially Available Nondigestible Oligosaccharides

Name	Chemical Structure	Glycosidic Bond	Origin	Trade Name
Neosugar	Glucosyl (fructosyl)$_n$ fructose (n = 1 – 3)	β (1→2)	Enzymatic synthesis from sucrose	Neosugar® Actilight®
Galactooligo-saccharides (GOS)	Glucosyl (galactosyl)$_n$ galactose (n = 1 – 3)	β (1→6)	Enzymatic synthesis from lactose	Oligomate®
Transgalacto-oligosaccharides	(Galactosyl)$_n$ galactose (n = 2)	α (1→6)	Enzymatic synthesis from lactose	Cup-oligo®
Polydexrose	(Glucose)$_n$ randomly branched (n = 2 – 100?)	(1→2), (1→3), (1→4), (1→6)	Glucose pyrolysis	Poly-dextrose®
Sololigo-saccharides (SOS)	Raffinose and Stachyose		Enzymic synthesis + pyrolysis	Soya-Oligo®
Xylooligo-saccharides (XOS)	(Xylosyl)$_n$ xylose (n = 2 – 4)	β (1→4)	Enzymatic degradation of xylans	Xylooligo®
Lactosucrose	O-β-D-Gal$p$-(1→4)-O-α-D-Glc$p$-(1→2)-β-D-Fru$f$		A raffinose isomer, by action of a fructofurano-sidase on a mixture of lactose and sucrose.	

Source: From Roberfroid, M.B., *Complex Carbohydrates in Foods,* Cho, S.S., Prosky, L., and Dreher, M., Eds., Marcel Dekker, Inc., New York, 25, 1999.[18]

of starch. The synthesis of cyclodextrins involves a formation of a glycosyl-enzyme complex, an intramolecular transglycosylation reaction, and release of a cyclic compound with 6, 7, 8, or 9 glucose unit, referred to as α-, β-, γ-, and δ-cyclodextrin, respectively (Figure 1.29). Cyclomaltodextrins with 10 to 13 D-glucose units have also been obtained.

The cyclodextrins have a doughnut shape and a somewhat conical struc-ture that is relatively rigid and has a hollow cavity of certain volume. All cylodextrins have a depth of 7.8 Å, but the exterior and interior diameter depends on the number of glucose residues in the molecule. The primary and secondary alcohol groups of the glucose residues are projected to the outside of the ring giving the outer surface a hydrophilic character. The cyclodextrins with an even number of D-glucose units are more water soluble than those with an odd number of residues. The internal cavity, on the other hand, contains the glycosidic oxygen atoms and carbon and hydrogen atoms,

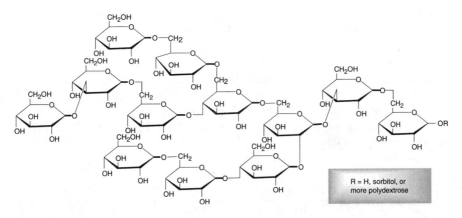

FIGURE 1.28
General structure of polydextrose.

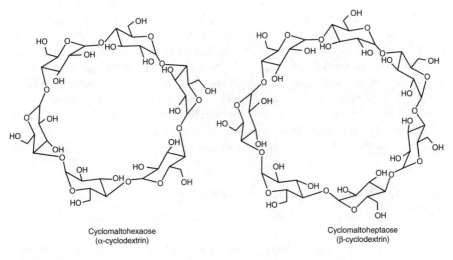

Cyclomaltohexaose
(α-cyclodextrin)

Cyclomaltoheptaose
(β-cyclodextrin)

FIGURE 1.29
Structure of α- and β-cyclodextrins.

which result in regions with a high electron density and hydrophobic character. The internal cavity is responsible for the unique capability of cyclo-dextrins to form inclusion complexes with a variety of organic and inorganic compounds. To bind with cyclodextrins, the inclusion compound must at least partially fit into the cavity. The formation of complexes can stabilize the guest compound against light and/or various chemical reactions, thermal decomposition, and evaporation losses. The solubility of cyclodextrins containing an organic guest may change significantly. The inclusion complexes can be used to control the release, stability, and solubility of various biological compounds including drugs, vitamins, flavors, and odors. Cyclo-dextrins have not yet been approved in foods but have been used for the removal of the bitter compounds from grapefruit juice and other citrus products. The industrial uses of cyclodextrins range from water filters, laundry drier sheets, to perfumes, and various cosmetics.

Since the discovery of cyclodextrins, the structures of several other cyclic oligosaccharides have been elucidated. Cyclosophorans are cyclic β-1→2 glucans, containing 17 to 40 D-glucose residues. Cyclic β-1→6 and β-1→3 glucans, alternating α-1→6/α-1→3 glucans cyclotetraose, and cyclo-L-rhamnosehexaose have been reported to be secreted into the culture supernatants of various bacteria.

1.4   Reaction of Monosaccharides and Derived Carbohydrate Structures

Monosaccharides, being polyhydroxy aldehydes or ketones, possess many reactive groups that can be easily modified, and various derivatives of sugars exist in nature or can be obtained synthetically via chemical reactions. In general, the order of reactivity of the various hydroxyl groups within D-glucopyranose is O1 (hemiacetal OH) > O6 (primary OH) > O2 (secondary OH) > O3 (secondary OH) > O4 (secondary OH). The discussion below highlights only the most important reactions and derived carbohydrate structures. For more exhaustive coverage of the subject the reader is advised to consult other references from the recommended list.

1.4.1   Oxidation and Reduction Reactions

The aldo and keto groups of carbohydrates can be readily reduced in aqueous solutions to give sugar alcohols (alditols). The sugar alcohols retain some properties of carbohydrates, e.g., sweetness, high water solubility, and optical activity. The reduction of D-glucose yields D-glucitol (commonly known as sorbitol), whereas the reduction of D-fructose results in two epimeric alcohols, D-glucitol and D-mannitol (Figure 1.30).

Aldoses and ketoses can be reduced easily in sodium borohydrate at pH between 6 to 10. D-Glucose can be specifically reduced by the enzyme, D-sorbitol dehydrogenase, present in sheep liver, as well as in human eye and sperm. The enzyme is responsible for the high concentration of D-glucitol in the eye that leads to diabetic cataracts. D-Glucitol is present in many fruits and berries, apples, apricots, cherries, and pears. D-Mannitol can be found in plant exudates and in seaweeds. Xylitol also occurs naturally in raspberries, strawberries, and plums. Commercially, glucitol and mannitol are obtained from hydrogenolysis of sucrose. Xylitol is produced by hydrogenation of D-xylose obtained from hydrolysis of hemicelluloses. Sugar alcohols are present in various food products including confectionery products, bakery

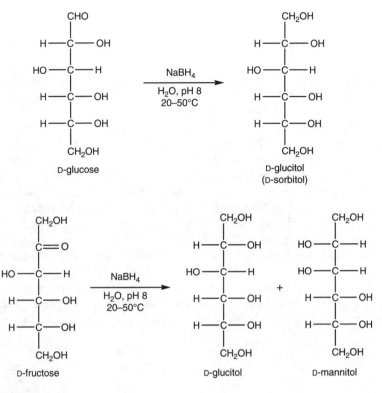

FIGURE 1.30
Reduction of glucose and fructose.

products, desserts, meat products, and pet food. They are slightly less sweet. Addition of polyols to chewing gum provides better texture, a cool sensation in the mouth, and reduces dental caries. Solubilization of xylitol is an endothermic reaction and gives a cooling sensation when xylitol is placed in the mouth. Xylitol is not metabolized by the mouth microflora and, therefore, does not promote formation of plaques. Because the sugar alcohols are absorbed at a slower rate than glucose, they find some uses in the production of low calorie soft drinks. In puddings and jellies, the addition of sorbitol improves texture. Sugar alcohol will also lower water activity in various products and, therefore, helps to prevent mold formation.

The aldehyde group of aldoses can be readily oxidized to carboxylic group, and the reaction leads to formation of aldonic acids (e.g., gluconic acid). This reaction is commonly used to detect and measure sugars. Since during the process of oxidizing the aldehyde group of an aldose, the oxidizing agent is reduced, aldoses have been called reducing sugars. Fehling reagent, an alkaline solution of copper (II) which is reduced to copper (I) during oxidation of sugars, has been used for determining the amount of reducing sugars in foods and other biological materials.

$$H \qquad\qquad O$$
$$| \qquad\qquad ||$$
$$2Cu(OH)_2 + R-C = O \rightarrow R-C-OH + Cu_2O + H_2O$$

All aldoses are reducing sugars since they contain an aldehyde carbonyl group. It should be noted that although ordinary ketones cannot be oxidized, some ketoses have reducing power. For example, fructose reduces Tollen's reagent (Ag^+ in aqueous ammonia) even though it does not contain the aldehyde group. This occurs because fructose is readily isomerized to an aldose in a basic solution by a series of keto $\rightleftarrows$ enol tautomeric shifts (Figure 1.31).

The enzyme glucose oxidase specifically oxidizes the β-D-glucopyranose to give 1,5 lactone (an intramolecular ester) of the gluconic acid. The reaction (Figure 1.32) is used in an analytical procedure for quantitative determination of D-glucose in the presence of other sugars in foods and other biological specimens, including blood (Chapter 2).

The nonanomeric hydroxyl groups can also be oxidized. Oxidation of the primary hydroxyl group at the terminal carbon atom (the farthest from the aldehyde group) to carboxylic acid leads to formation of uronic acids (e.g., glucuronic acid). The oxidation of both the aldehyde and the primary alcohol groups creates aldaric acids (e.g., glucaric acid). The structures of these sugar

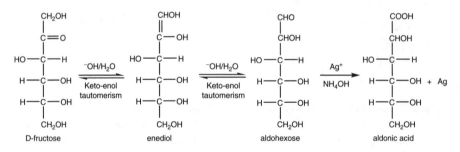

FIGURE 1.31
Keto-enol tautomerism and oxidation of fructose.

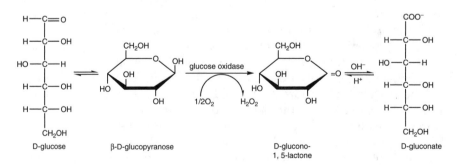

FIGURE 1.32
Glucose oxidase-catalyzed oxidation of D-glucose

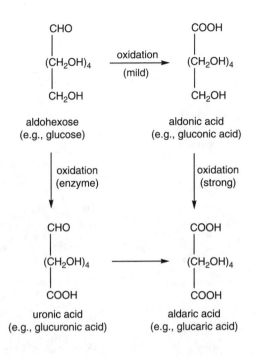

FIGURE 1.33
Oxidation reactions of aldoses.

acids are illustrated in Figure 1.33. D-Gluconic acid is a natural constituent of fruit juices and honey. D-Glucono-delta (1,5)-lactone (GLD) (Figure 1.32) is used in meat, dairy and baked products due to its mild taste and slow acidification properties (when it hydrolyzes to D-gluconate). The uronic acids occur as structural units of many polysaccharides. For example, D-galacturonic acid is the principal component of pectins, whereas D-glucuronic acid is present in glucuronoarabinoxylans.

1.4.2 Deoxy and Amino Sugars

When a hydroxyl group of a carbohydrate is substituted by a hydrogen atom, the outcome is a deoxy sugar. Theoretically, any hydroxyl group can be substituted; however, there are only a few naturally occurring deoxy sugars. D-Ribose substituted in the 2-position results in 2-deoxy-D-ribose (Figure 1.34), the carbohydrate constituent of DNA (deoxyribonucleic acid). The replacement of hydroxyl group at C-6 of D-mannose results in 6-deoxy-D-mannose, commonly called D-rhamnose. The replacement C-6 of D-galactose results in 6-deoxy-D-galactose, commonly called D-fucose (Figure 1.34). Rhamnose and fucose occur in nature more frequently in the L- than D-configuration. They are constituents of many polysaccharides (e.g., pectins, seaweed polymers, gum tragacanth) and glycoproteins. L-rhamnose is found in many plant glycosides (e.g., quabain) and occurs free in poison ivy and sumac.

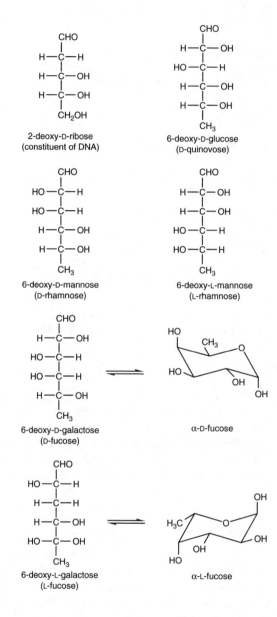

FIGURE 1.34
Structures of common deoxy sugars.

Replacement of a hydroxyl group by an amino group produces an amino sugar. Sometimes the hydroxyl group is replaced by an *N*-acetyl-amino group. The most common amino sugars are 2-amino-2-deoxy-D-glucose, known as D-glucosamine, and 2-amino-2-deoxy-galactose, known as D-galactosamine. *N*-acetylamino derivatives of D-glucose and D-galactose are also known (Figure 1.35). Amino and *N*-acetylamino sugars are found as carbohydrate constituents of glycoproteins. *N*-acetyl–D-glucosamine is a building

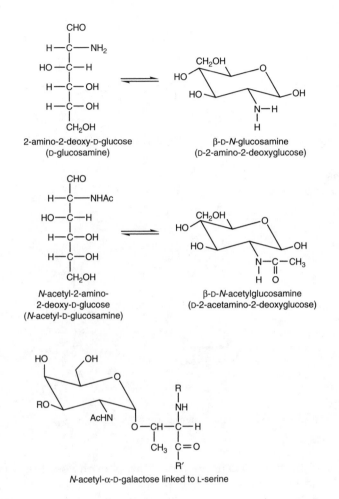

FIGURE 1.35
Structure of common amino and *N*-acetylamino sugars.

block of chitin, the structural polysaccharide that replaces cellulose in the cell walls of lower plants, fungi, yeast, and green algae. Chitin is also the major component of the exoskeletons of insects and shells of crustaceans.

1.4.3 Sugar Esters and Ethers

Hydroxyl groups of carbohydrates can be also converted into esters and ethers. Esterification is normally carried out by treating of carbohydrate with acid chloride or acid anhydride in the presence of base. All the hydroxyl groups react, including the anomeric one. For example, β-D-glucopyranose is converted into its pentaacetate by treatment with acetic acid in pyridine solution (Figure 1.36).

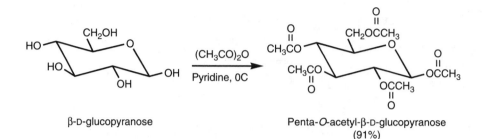

β-D-glucopyranose

Penta-*O*-acetyl-β-D-glucopyranose
(91%)

FIGURE 1.36
Esterification of glucose.

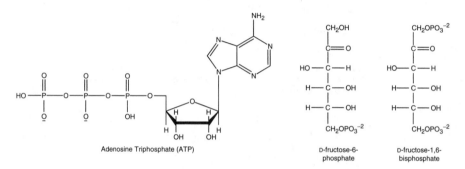

FIGURE 1.37
Structures and names of some common sugar phosphates.

Many carbohydrates occur in nature as phosphate-ester derivatives. D-ribose-5-phosphate is a precursor of RNA (ribonucleic acid) and DNA (deoxyribonucleic acid). The phospho-esters are important intermediates during metabolism and synthesis of carbohydrates by living organisms. Some common sugar phosphates are shown in Figure 1.37. Carrageenans, the red seaweed polysaccharides, are composed of galactopyranosyl residues having sulphate ($-SO_3^-$) groups esterified to hydroxyl groups at carbon atoms C-2 and C-3. Some food polysaccharides are esterified by chemical modification to improve their functionalities. The most common are acetate and succinate esters of starch. Sucrose fatty acid polyesters are molecules of sucrose to which six or eight fatty acids have been esterified by chemical transesterification (Figure 1.38). The product called olestra is a fat substitute. It has been approved by the U.S. Food and Drug Administration (FDA) for use in snacks, primarily as a replacement of conventional fats for frying. It has no calories because the molecule cannot be hydrolyzed by digestive lipases, but olestra can potentially cause some gastrointestinal effects (diarrhea, abdominal cramping, loose stool) and interfere with adsorption of fat soluble vitamins.

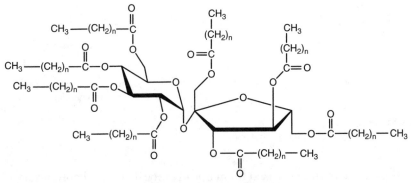

Octa fatty acyl sucrose (Olestra)
where n = 12–24 (in increments of 2)

FIGURE 1.38
Structure of olestra.

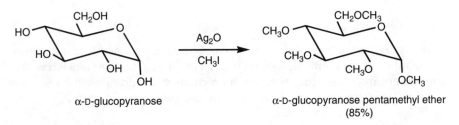

α-D-glucopyranose   α-D-glucopyranose pentamethyl ether
                                                   (85%)

FIGURE 1.39
Etherification of glucose.

Carbohydrates can be converted into ethers by treatment with an alkyl halide in the presence of base. For example, α-D-glucopyranose is converted into its pentamethyl ether in reaction with iodomethane and silver oxide (Figure 1.39).

Ethers of carbohydrates are not very common in nature; however, an internal ether formed between the hydroxyl groups at carbon atom C-3 and C-6 of D-galactose is a building unit of red seaweed polysaccharides, such as agar, fulcellaran, kappa-, and iota-carrageenan (Figure 1.40). The etherification

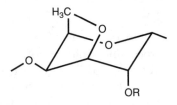

FIGURE 1.40
An internal ether, 3,6-anhydro-α-D-galactopyranosyl residue of red seaweed polysaccharides.

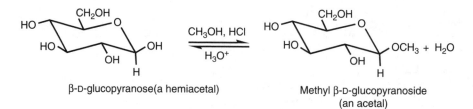

β-D-glucopyranose(a hemiacetal)   Methyl β-D-glucopyranoside
                                              (an acetal)

FIGURE 1.41
Formation of methyl β-D-glucopyranoside.

reaction is also frequently used to chemically modify natural polysaccharides to improve their properties and utilization. For example, methyl ($-O-CH_3$), sodium carboxyl methyl ($-O-CH_2-CO_2^-Na^+$), and hydroxypropyl ($-O-CH_2-CHOH-CH_3$) ethers of cellulose and hydroxypropyl ethers of starch are approved for food uses.

1.4.4   Glycosides

Carbohydrate acetals are called glycosides. They are formed in the reaction of sugar hemiacetal with alcohol (Figure 1.41). Glycosides are stable in water and are not in equilibrium with an open chain form (do not show mutarotation). They can be converted back to the original monosaccharide by hydrolysis with aqueous acid.

Glycosides are distributed widely in nature, and many biologically important molecules contain glycosidic linkages. Digitoxin, for example, is an active component of the digitalis preparations used for treatment of heart disease (Figure 1.42). Other cardiac glycosides are also known. Their sugar moiety usually contains one to four monosaccharide units (L-rhamnose, D-glucose, D-fructose), and the aglycone moiety contains steroid ring systems. Naturally occurring glycosides are found primarily in plants. Some, like digitoxin, ouabain, and arbutin have medicinal effects; others, like sinigrin, may be responsible for characteristic flavor of plant materials or contribute to their color (anthocyainins) (Figure 1.43 and Figure 1.44).

1.4.5   Browning Reactions

1.4.5.1   *Maillard Reaction*

The phenomenon of browning of certain foods is frequently encountered during preparation and processing which involve heat. This phenomenon is called nonenzymatic browning to distinguish it from the enzyme-catalyzed browning reactions occurring usually in freshly cut or bruised fruits and vegetables at the ambient temperature.[22] The formation of brown pigments was first observed by the French chemist Louis Maillard (1912) after

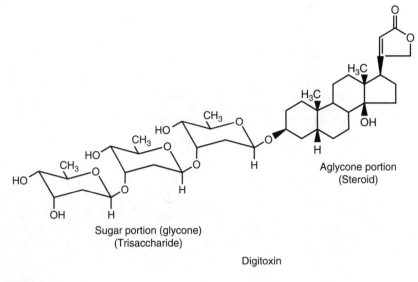

Sugar portion (glycone)
(Trisaccharide)

Aglycone portion
(Steroid)

Digitoxin

FIGURE 1.42
Structure of digitoxin.

heating a solution containing glucose and lysine. When amines, amino acids, or proteins are heated with sugars, aldehydes, and ketones, a series of complex reactions, referred as the Maillard reactions, occurs, leading to formation of flavors, aromas, and dark colored pigments. These reactions may be desirable or undesirable. They are responsible, for instance, for pleasant aroma and color of roasted coffee and baked bread. However, of particular concern are the toxicity and mutagenecity of certain intermediates formed during the reactions.

The first step in the Maillard reaction involves a reversible condensation between the α-amino-group of amino acids or proteins and the carbonyl group of reducing sugars. The condensation compound rapidly loses water to form a Schiff's base, which undergoes cyclization to form the corresponding N-substituted glycosamine. The N-substituted glycosamine is very unstable and undergoes a transition, the Amadori rearrangement, from an aldose to a ketose sugar derivative (Figure 1.45). The reducing sugars are essential in this initial step of the Maillard reaction. The rate of the first stage of the Maillard reaction depends on the kinetics of the sugar ring opening. The order of reactivity is greater for aldopentoses than for aldohexoses, and relatively low for reducing disaccharides. Among the hexoses, the rate is in the order D-galactose > D-mannose > D-glucose.

The reactions involved in the conversion of 1-amino-1-deoxy-1-ketose to brown pigments are complex and not entirely understood. It is proposed, however, that the pigment formation proceeds via three distinct routes (Figure 1.46). One involves enolization of 1-amino-1-deoxy-1-ketose at 1- and

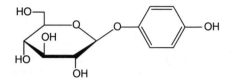

Arbutin present in the leaves of blueberries, cranberries, and pear tree. It is known as a diuretic and anti-infective urinary agent.

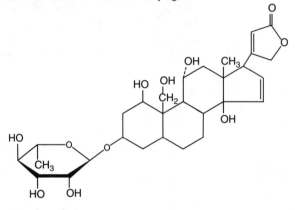

Ouabain is a cardiac glycoside.

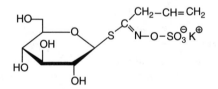

Sinigrin can be isolated from black mustard seeds and from horseradish root. It may contribute to the hot flavor of these materials.

FIGURE 1.43
Structure of some naturally occurring glycosides.

2- position, and formation of 1,2-eneaminol which undergoes several reactions leading to formation of furfural and hydroxymethylfurfurals. The furfurals undergo condensation with amino compounds, followed by polymerization reactions and formation of nitrogenous dark brown polymers, called melanoidins. Alternatively, a high pH favors enolization of 1-amino-1-deoxy-1-ketose at 2- and 3- position, and formation of 2,3-enediols which

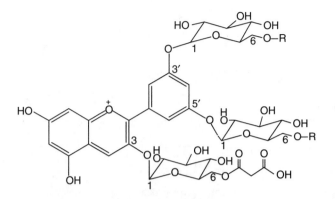

Anthocyanins are glycosides of polyhydroxy or polymethoxy
derivatives of flavylium salts. The sugars most commonly bonded
to anthocyanins are glucose, galactose, rhamanose, arabinose or
their oligosaccharides.

R—functional groups containing sugar molecules and coumaric acid

FIGURE 1.44
General structure of anthocyanins.

are converted to reductones. The intermediate dicarbonyl compounds
formed during this process may undergo decomposition, dehydration, and
fragmentation. Various flavor compounds are formed, such as aldehydes,
furans, maltol, isomaltol, diacetyls. The third route in the Maillard reaction
involves oxidation of amino acids in the presence of dicarbonyl intermediates formed from Amadori compounds. The reaction, called Strecker degradation, leads to formation of various aldehydes, which upon condensation
reactions form many heterocyclic compounds, pyrazines, pyrrolines,
oxazole, oxazolines, and thiazole derivatives responsible for the flavor of
heated foods.

1.4.5.2 Caramelization

Caramelization is another example of nonenzymatic browning reactions
involving the degradation of sugars during heating. Heating of both reducing (e.g., glucose) and nonreducing sugars (e.g., sucrose) without any nitrogenous compounds results in a complex group of reactions leading to a dark
brown caramel. The chemical composition of caramel is complex and
unclear.[23,24] Caramel, produced by heating sucrose solution in ammonium
bisulfite, is used in colas and other soft drinks, baked goods, candies, and
other food products as a colorant and flavor compound. However, if the
caramelization reaction is not controlled, it will create bitter, burned, and
unpleasant tasting products. Intense heat and low pH trigger the reactions.

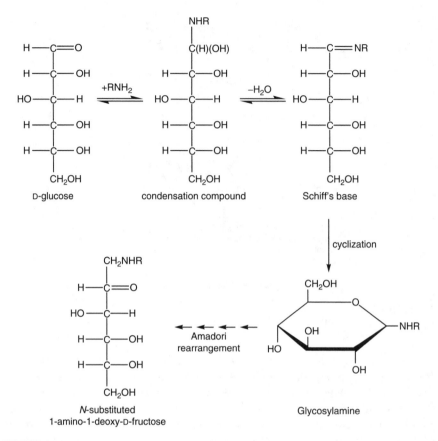

FIGURE 1.45
The carbonylamine reaction and Amadori rearrangement leading to formation of 1-amino-1-deoxy-2-ketose in the Maillard reaction.

The first step involves the conversion of glucose, fructose, and mannose to 1,2-enediols, which upon further heating undergo dehydration reactions, leading to the formation of 5-hydroxymethylfurfurals (Figure 1.47). If the initial sugar was pentose, then the final dehydration product is 2-furaldehyde (furfural). It is believed that polymerization of furfural derivatives leads to formation of the colored pigment. If the heating of sugars occurs under alkaline conditions, then 1,2- and 2,3-enediols are formed, which may undergo cleavage and fragmentation reactions leading to formation of various aroma compounds such as saccharinic acid, lactic acid, 2,4-dihydroxybutyric acid, ethyl alcohol, and aromatic compounds, benzene, maltol, catechol, benzalaldehydes.

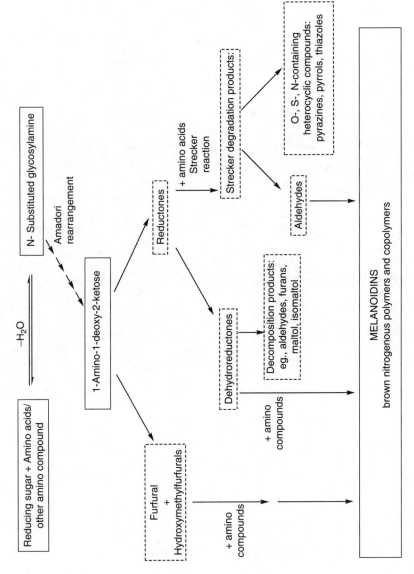

FIGURE 1.46
Nonenzymatic browning reactions.

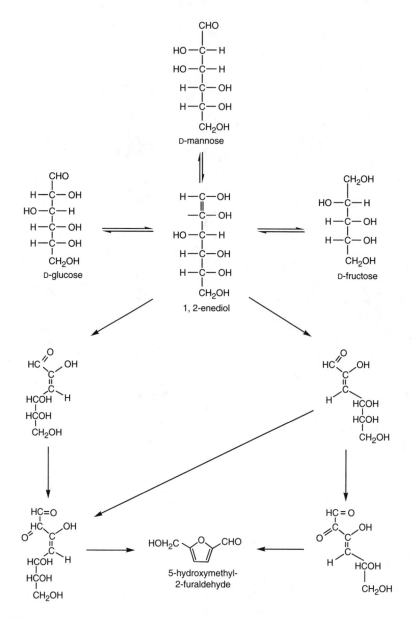

FIGURE 1.47
Mechanism of sugar dehydration and formation of 5-hydroxymethyl-2-furaldehyde, a precursor of caramel.

1.5 Polysaccharides

1.5.1 General Structures and Classifications

"Glycans" is a general term given to polysaccharides in which large numbers of glycoses (monosaccharides) are mutually joined by O-glycosidic linkages.[1,20] Polysaccharides are condensation polymers in which glycosidic linkage is formed from the glycosyl moiety of hemiacetal (or hemiketal) and a hydroxyl group of another sugar unit, acting as an acceptor molecule or aglycone.

A consequence of the monovalent character of glycosyl units but the polyvalent nature of aglycone is that:

- Branching is possible within the polysaccharide chains.
- Intramolecular cross linking by covalent bonds between adjacent chains is impossible through glycosidic linkages.

Polysaccharides may be linear or branched, and with the exception of cyclic polysaccharides known as cycloamyloses, there is a defined chain character from the nonreducing terminus (or termini) to one reducing terminus (Figure 1.48).

Based on the number of different monomers present, polysaccharides can be divided into two classes:[25]

- Homopolysaccharides, consisting of only one kind of monosaccharides
- Heteropolysaccharides, consisting of two or more kinds of monosaccharide units

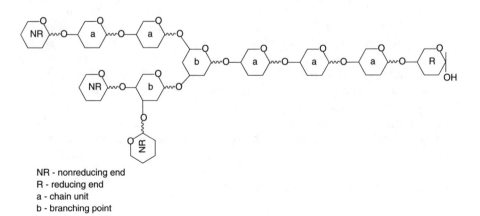

NR - nonreducing end
R - reducing end
a - chain unit
b - branching point

FIGURE 1.48
General structure of polysaccharides.

TABLE 1.9

Homopolysaccharides

Polysaccharides	Repeating Unit: Glycosidic Linkage Type/Glycose Unit
Linear	
Amylose	α-(1→4)-Glc
Cellulose	β-(1→4)-Glc
Xylan	β-(1→4)-Xyl
Inulin	β-(2→1)-Fru
Levan	β-(2→6)-Fru
Laminaran	β-(1→3)-Glc
Chitin	β-(1→4)-GlcNAc
β-Glucan	β-(1→4, 1→3)-Glc
Branched	
Amylopectin	α-(1→4, 1→6)-Glc
Dextran	α-(1→2, 1→3, 1→4, 1→6)-Glc
Levan	β-(2→1, 2→6)-Fru
Pullulan	α-(1→6)-Maltotriose
Scleroglucan	β-(1→3, 1→6)-Glc
Glycogen	α-(1→4, 1→6)-Glc

Homopolysaccharides can be further divided by the type(s) of glycosidic linkages which join the monosaccharide units. The glycosidic linkage may have either α- or β-configuration, and various positions, i.e., α-1→2, α-1→3, α-1→4, and so on, or β-1→2, β-1→3, β-1→4 and so forth. Homopolysaccharides can have homolinkages or heterolinkages with respect to configuration and/or linkage position (Table 1.9). Heteropolysaccharides, in addition to the different kinds of monosaccharide units, can have the same kind of linkage diversity as homopolysaccharides. Thus, heteropolysaccharides can have different types and sequences of monosaccharide units as well as different types and sequences of glycosidic linkages. This allows for an almost limitless diversity in their structure. Some polysaccharides are composed of sugar units only: they are known as neutral polysaccharides (e.g., amylose, amylopectin, cellulose). Polysaccharides containing sugar acids in their structure will carry negative charges, and, therefore, they are anionic polysaccharides. For example, pectins are built up of galacturonic acid residues, whereas alginates contain both guluronic and mannuronic acid residues. There is only one known cationic polysaccharide, chitosan, obtained by modification of the naturally occurring animal polysaccharide chitin. Chitosan is a polymer of β(1→4)-2-amino-2-deoxy-D-glucopyranose, which carries positive charges at pH 6 to 7. Another frequently used classification of polysaccharides is based on their origin (Table 1.10).

Based on the type of sequence of sugar units, polysaccharides can be generally divided into three groups (Table 1.11.):

TABLE 1.10

Classification of Polysaccharides According to Their Origin

Origin	Name
Seaweeds	Agarose, carrageenans, alginates
Higher plants	
Grains, fruits, vegetable (edible and non- edible parts), seeds, tubers, roots, trees	Pectins, arabinoxylans, arabinogalactans, glucuronoxylans, β-glucans, inulin, cellulose, xylans, gum arabic, gum ghatti, gum tragacanth, gum karaya, guar gum, locust gum, tara gum, starches, konjac mannan
Microbial	Xanthan gum, gellan, welan, zooglan, pullulan, dextran, curdlan, levan, scleroglucan
Animal	Glycogen, glycosaminoglycans, chitin, chitosan

TABLE 1.11

Classification of Polysaccharides According to the Sequence Types of Sugar Units in Carbohydrate Chains

Polymer	Occurrence/Function	Sugar Unit	Sequence
Periodic Type			
Amylose	Energy storage material in almost all higher plants	4-linked α-D-glucopyranose (○)	----○-○-○-○-○-○---
Cellulose	Structural material in higher plants	4-linked β-D-glucopyranose (●)	----●-●-●-●-●-●-●----
Interrupted Type			
Alginates	Extracellular and intracellular gel-forming polysaccharides in certain algae, contributing to ionic interactions and physical protection	4-linked α-L-GulA*p* (○); 4-linked β-D-ManA*p* (●)	--○-○-○-○-○-○--●-●-● ●-●-●-●-○-●-○-●---
Pectins	Components of cell walls of many growing plants, providing hydrated, cementing matrix between walls of mature plant tissues	4-linked α-D-GalA*p* (○); 2-linked L-Rha*p* (▲)	--○-○-○-○-○-○-▲-○--▲- ○-▲-○-○-○-○-○-○-○---
Aperiodic Type			
Oligosaccharide chain of glycoproteins or glycolipids; e.g., oligosaccharides in the human blood groups (ABO)	O-linked oligosaccharides that specify the O-type human blood group		

- Periodic types, in which sugar units arranged in a repeating pattern
- Interrupted types, whose chains have repeating sequences separated by irregular sequences (kinks)
- Aperiodic types, characterized by irregular sequences of monosaccharide units, linkage positions, and configurations

Polysaccharides are high molecular weight polymers. The degree of polymerization (DP), which is determined by the number of monosaccharide units in a chain, varies from a hundred to a few hundred thousands. Only a few polysaccharides have DPs below 100. Unlike proteins, polysaccharides are secondary gene products. Those occurring in nature are products of various biosynthetic enzymes (glycosyl transferases) and, therefore, their synthesis is not under strict and direct genetic control. The general biosynthetic pathways of many polysaccharides are known (Figure 1.49), but the mechanisms controlling certain biosynthetic events, such as the density and distribution of branches along the polysaccharide chain or the chain's length, are not fully elucidated. In general, it is agreed that different transferases are needed for addition of each monosaccharide unit to the growing chain. The backbone growth most likely occurs via addition of new sugar residues to the nonreducing end (tailward growth). However, when lipid intermediates are involved, addition of new residues occurs at the reducing end (headward growth). Most polysaccharides are thought to undergo a precise synthesis, in which the backbone growth is concurrent with the addition of side chains. The membrane systems of the endoplasmic reticulum, Golgi bodies, and plasma lemma, are responsible for synthesis and transport of the plant cell wall polysaccharides. It is thought that the rate of transfer and deposition of the newly synthesized polysaccharide in the target tissue may have some role in determining its chain length. Post-polymerization modifications may include esterification and/or etherification of chains, although in some cases these modifications may occur simultaneously with polymerization of the backbone chains. The lack of strict genetic control during synthesis of each polysaccharide chain is responsible for a great degree of heterogeneity of these polymers with respect to their molecular mass (DP) and certain aspects of the molecular structure (e.g., ratio of different monosaccharides, linkage distribution, degree and distribution of branches). Polysaccharides are, therefore, known as polydispersed polymers. However, not all structural characteristics of polysaccharides are equally heterogeneous; for example, the configuration of glycosidic bonds in polysaccharides is the most conservative, whereas the molecular weight is probably the most variable characteristic.

1.5.2   Factors Affecting Extractability and Solubility of Polysaccharides

Proper identification and characterization of carbohydrate polymers require, first of all, isolation and purification of the polymer of interest from its

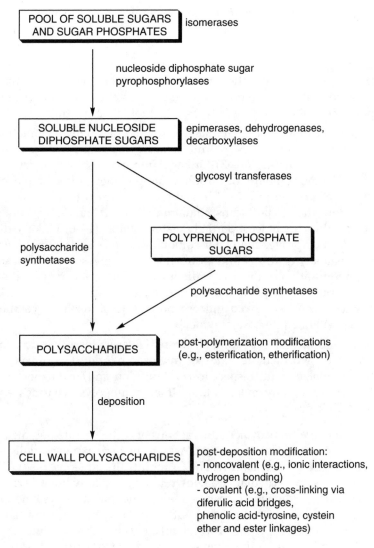

FIGURE 1.49
General overview of polysaccharides synthesis.

original source. The nature of the polysaccharides, more specifically, their chemical composition, molecular structure, and weight, will affect their solubility. Some general structural features affecting the extractability and solubility of polysaccharides in water include:

- Branching. The presence of branches in the polymer chain will disrupt intermolecular association and promote solvation and solubilization. For example, guar gum galactomannans are more soluble than locust bean galactomannans because the former have a higher degree of branching. Also, the highly branched amylopectin or glycogen polymers are more soluble and stable in solutions than linear amylose chains.

- Ionizing groups. The presence of carboxylate (COO^-) or sulphate (SO_4^{2-}) groups in the polymer chains improves solvation and inhibits intermolecular association through electrostatic repulsion. Anionic polysaccharides are generally more soluble than neutral ones.

- Glycosidic bond. (1→4)-Linkage provides a highly symmetrical structure and facilitates intermolecular associations between units of different chains. (1→3)-Linkage imparts less symmetry and increases the solubility of carbohydrate polymers; for example, cellulose, composed of β-1→4 linked Glc residues is totally insoluble in water, whereas β-glucans, containing β-(1→4) and β-(1→3) linked Glc residues are at least partially water soluble. (1→6)-Linkage dramatically improves the water solubility of carbohydrate polymers, for instance, dextrans and pullulans are easily soluble in water. α-Configuration of the glycosidic linkage does not confer total solubility but does improve it compared to β-configuration (e.g., soluble amylose [α-(1→4)-linked glucose polymer] vs. insoluble cellulose [β-(1→4)-linked glucose polymer]).

- Nonuniformity in repeating structure. Heteropolysaccharides, consisting of two or more kinds of monosaccharide units, linked via heterolinkages with respect to configuration and/or linkage position, are generally more soluble than homopolysaccharides with homolinkages.

The covalent and noncovalent interactions of carbohydrate polymers with other components of the tissues from which they are extracted will significantly affect their extractability and solubility in aqueous media. Plant polysaccharides, in particular, are known to be covalently bound (via diferulic acid bridges, phenolic acid-tyrosine or phenolic acid-cysteine bonds, glycosidic, ether, and ester linkages) with such plant constituents as lignin, proteins, and cellulose. These types of covalent linkages will impart water insolubility, and solvents other than water will have to be used for extraction. Noncovalent interactions through ionic bonds (e.g., COO^- groups of rhamnogalacturonans with minerals or proteins) or van der Waals forces and H-bonds of carbohydrate polymers with each other, or with other tissue constituents, will also reduce their extractability and solubility. More details on the effects of structural features on solubility and physical properties of polysaccharides are given in Chapter 4.

1.5.3 Extraction of Polysaccharides

In general, the storage polysaccharides, exudate gums, and the bacterial capsular polysaccharides are easier to extract than the structural or matrix

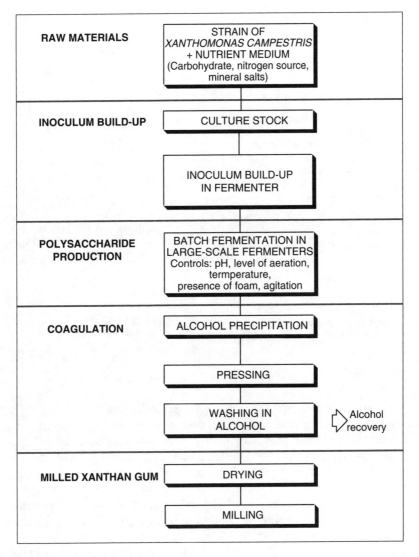

FIGURE 1.50
Commercial process of extraction and purification of xanthan gum.

polysaccharides of the cell walls of plants and microorganisms.[26] For example, the xanthan gum produced by *Xanthomonas campestris* is released directly into the culture medium and can be easily isolated from the culture broth by precipitation with ethanol (Figure 1.50).The cell wall polysaccharides, on the other hand, have to be extracted from the insoluble cell wall material. Although cell wall components have been obtained from extracts of cereal flours or brans, it may be preferable to use purified cell wall preparations from single tissues as the starting material for detailed structural analysis of the polysaccharides of interest. A number of strategies have been used to

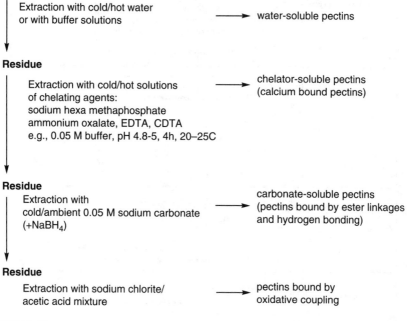

Cell wall material

Extraction with cold/hot water
or with buffer solutions   ⟶ water-soluble pectins

Residue

Extraction with cold/hot solutions   ⟶ chelator-soluble pectins
of chelating agents: (calcium bound pectins)
sodium hexa methaphosphate
ammonium oxalate, EDTA, CDTA
e.g., 0.05 M buffer, pH 4.8-5, 4h, 20–25C

Residue

Extraction with   ⟶ carbonate-soluble pectins
cold/ambient 0.05 M sodium carbonate   (pectins bound by ester linkages
(+NaBH$_4$)   and hydrogen bonding)

Residue

Extraction with sodium chlorite/   ⟶ pectins bound by
acetic acid mixture   oxidative coupling

FIGURE 1.51
Extraction of various fractions of pectins.

separate specific tissues from whole grain, including microdissection of grains, or processing such as milling and pearling if larger quantities are required. Also, prior removal of fat from the starting material is usually necessary, because fat can limit water penetration and negatively affect the efficiency of extraction. Lipid substances are usually removed with polar solvents, such as chloroform-methanol solutions (95:5 v/v), ethanol (90% v/v) or dioxane, and hexane. Refluxing in ethanol is often performed to inactivate the endogenous hydrolytic enzymes present in the biological materials.

Cell wall constituents can be extracted with many solvents, but water at various temperatures is usually the first choice for extraction of neutral polysaccharides. Normally, the extractability of polysaccharides increases with increasing temperature of the aqueous solvent. Acidic polysaccharides, such as pectins, are solubilized by complexing tightly coordinated bivalent metal ions with chelating agents such as ammonium oxalate, sodium hexametaphosphate, EDTA, or CDTA (Figure 1.51). Polar, nonaqueous solvents, such as dimethyl sulphoxide (DMSO) (normally with ~10% of water), are known to solubilize starch granules, and extract *O*-acetyl-4-*O*-methyl glucuronoxylans and *O*-acetyl-galactoglucomannans without stripping these polymers of their labile substituents. *N*-methylmorpholine-*N*-oxide has been reported a good solvent for cell wall polysaccharides including cellulose.

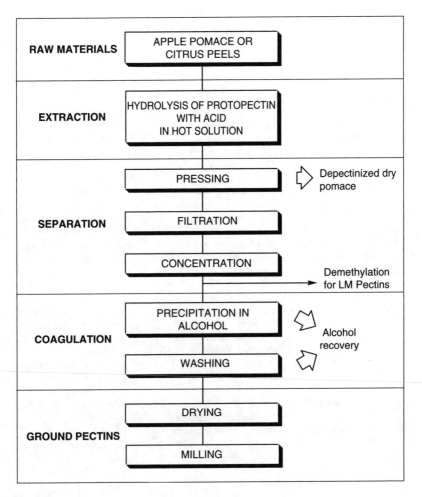

FIGURE 1.52
Commercial process of extraction of pectins from apple pomace or citrus peel.

Acidic solutions are usually avoided for extraction of polysaccharides because of the risk of hydrolysis of the glycosidic linkages. However, commercial preparations of pectins from citrus peel or apple pulp are prepared by extraction with acidic aqueous solutions of pH 1 to 3 at 50 to 90°C (Figure 1.52). Alkali solutions are used extensively to extract the cell wall polysaccharides. Sodium borohydrate is normally added to the extracting solution to reduce the risk of β-elimination or peeling reactions (sequential elimination of the reducing end residues from the polysaccharide chain) which may occur at a high pH. It is thought that under the alkali conditions, the ester and other covalent and noncovalent linkages are broken and the initially unextractable polysaccharides are released from the complex network of the cell walls (Table 1.12). Saturated barium hydroxide is known to specifically extract initially water insoluble arabinoxylans from wheat and barley,

TABLE 1.12

General Scheme for Sequential Extraction of Polysaccharides from the Plant Cell Wall (Fruits and Vegetables)[a]

Treatment	Effect	Polysaccharides Solubilized
H_2O at various temperatures with physical action (stirring, shaking)	Extraction of water-soluble populations of polysaccharides. Disruption of some H-bonding, van der Waals interactions.	Neutral, water-soluble β-glucans, arabinoxylans, arabinogalactans.
CDTA, 50 mM, 20°C	Abstraction of calcium from pectins and disruption of ionic bonds.	Acidic polysaccharides, slightly branched polysaccharides from middle lamella.
Na_2CO_3, 50 mM; + 20 mM $NaBH_4$, extraction at 1°C and 20°C	Abstraction of calcium within the wall matrix and disruption of ionic bonds. Cleavage of some ester bonds.	Highly branched pectic polysaccharides.
KOH, 1 M; + 20 mM $NaBH_4$	Cleavage of phenol-carbohydrate ester and protein-carbohydrate bonds. Disruption of H-bonds.	Highly branched pectins; glycoproteins, initially water insoluble β-glucans, arabinoxylans, glucuronoarabinoxylans, highly branched xyloglucans.
KOH, 4 M; + 20 mM $NaBH_4$	Swelling of cellulose and disruption of H-bonds.	Mainly slightly branched xyloglucans.
Chlorite/acetic acid (0.3% w/v; 0.12% v/v), 70°C	Disruption of phenolic cross-links and mild delignification.	Hydroxyproline-rich glycoproteins and some highly branched pectic polysaccharides.

[a] The exact composition of the extracted polysaccharides may vary depending on the origin of the starting material.

whereas dilute solutions of sodium hydroxide have been used to extract xyloglucans, xylans, β-glucans, and pectins. It should be pointed out that once the polysaccharides are released from the cell walls by extraction with alkali, they become soluble in water.

A different approach has to be taken to isolate highly insoluble polysaccharides, such as cellulose. This polymer can be isolated from raw cotton or wood by successive removal of other components present in the raw materials. For example, waxes and lipids can be extracted with chloroform, benzene, and ethanol. The chlorous acid treatment removes the lignin, and the alkaline treatments solubilize hemicelluloses. Consequently, the remaining residue consists primarily of insoluble cellulose. Cellulose is insoluble in most solvents. However, it can be dissolved in solutions of zinc chloride or Cadoxen, a solution of 5% (w/v) cadmium oxide in 28% (v/v) of ethylenediamine in water.

1.5.4 Purification and Fractionation of Polysaccharides

The extraction methods described above usually result in solutions containing a mixture of components (i.e., other polysaccharides as well as noncarbohydrate material), which have to be further purified to isolate the specific polysaccharide of interest. Noncarbohydrate materials, such as proteins, are often removed by digestion with proteolytic enzymes, papain, or bacterial proteases. Care must be taken to ensure that the enzymic preparations are devoid of glycohydrolases capable of degrading polysaccharide chains. Alternatively, proteins may be removed by precipitation with trichloroacetic acid (TCA) or sulfosalicylic acid, but again there is a risk of co-precipitating some gums, such as alginates or low methoxyl pectins. Nonspecific adsorption of proteins on celite, diatomaceous earth, and various clays has also been employed for de-proteinization of plant extracts. Finally, removal of proteins during the purification of water soluble polysaccharides isolated from plant material may be accomplished by passage of a solution through a strong cation exchange resin in the H^+ form. Since only proteins and not neutral or anionic polysaccharides have a positive charge, they will bind to the column, whereas polysaccharides will pass through and can be recovered from the effluent.

Removal of other contaminating carbohydrate polymers can also be achieved by specific digestions of contaminants with appropriate hydrolytic enzymes, but only highly purified enzymes with no side activities should be used for this purpose. For example, salivary α-amylase or porcine pancreatic α-amylase are often used to hydrolyze contaminating starch polymers because of purity and lack of other enzyme activity in these preparations. α-Amylase preparations from *Bacillus subtilis*, on the other hand, are often contaminated with β-glucanases. Low molecular weight soluble fragments generated during hydrolysis can be removed by dialysis, or the remaining carbohydrate polymers may be precipitated with ethanol.

Several methods based on solubility differences or selective precipitations have been used to separate different polysaccharides. Thus β-glucans are separated from arabinoxylans in aqueous extracts from barley by fractional precipitation with ammonium sulphate; β-glucans can be precipitated at a much lower saturation level of ammonium sulphate (20 to 50%) than arabinoxylans (55 to 95%). Similarly, arabinoxylans and arabinogalactans can be separated by ammonium sulphate precipitation of the former. Arabinogalactans are soluble even in 100% saturated ammonium sulphate solutions. Amylose forms insoluble complexes with butanol, and thus can be separated from solutions containing mixture of both starch polymers, amylose and amylopectin. β-Glucans can also form insoluble complexes with Congo Red or Calcofluor. Most anionic polysaccharides can be made insoluble (i.e., precipitated) by converting them into cetylpyridinium salts. Kappa-carrageenans are made insoluble in the presence of potassium salts, whereas alginates and low methoxyl pectins can be precipitated with calcium ions. Neutral polysaccharides can also form complexes between the suitably dispositioned

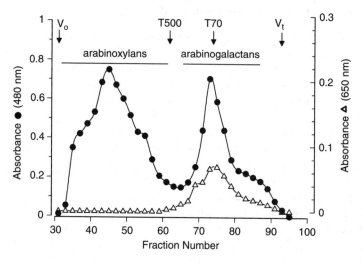

FIGURE 1.53

Separation of wheat water-soluble pentosan components, arabinoxylans and arabinogalactans-peptide, by gel filtration chromatography on Sepharose CL-4B. Absorbance at 480 and 650 nm corresponds to eluting carbohydrates (●) and proteins (△), respectively.

hydroxyl groups (OH) and certain metal ions (e.g., mannans and xylans form complexes with alkaline copper reagent, whereas glucomannans form complexes with barium salts).

Several types of chromatographic techniques can be used for separating polysaccharides from each other and/or from noncarbohydrate contaminants. Gel filtration chromatography is suitable for both neutral and anionic polysaccharides, and separates these biopolymers based on their molecular weight and hydrodynamic volume. For example, arabinoxylans and arabinogalactans can be effectively separated by gel filtration chromatography using Sepharose columns (Figure 1.53). Ion exchange chromatography using the diethylaminoethyl (DEAE)-cellulose is especially suitable for separating neutral and acidic polysaccharides. The former pass through the column without binding, whereas the latter, because of their negative charge, are retained on the column and can be eventually eluted with buffers of increasing ionic strength or pH. Affinity chromatography, based on specific noncovalent interaction between the binding ligand attached to the column and polysaccharides, is one of the most powerful techniques for purification of certain carbohydrate polymers. Lectins are natural plant proteins known for their carbohydrate-binding activity and, therefore, are used as ligands in affinity chromatography columns. Although many lectins recognize and bind to simple sugars, such as glucose, mannose, galactose, N-acetylgalactosamine, N-acetyglucosamine, or fucose, they have higher affinity for oligosaccharides. Concanavalin A (Con A), a lectin obtained from *Canavalla ensiformis*, exhibits specific affinity towards D-mannopyrans.

Finally, the polysaccharides of interest remaining in solutions after various purification steps can be precipitated with alcohol or acetone. Precipitation of polysaccharides with alcohol may, however, be affected by their concentration (polysaccharides do not precipitate from highly diluted solutions) and molecular weight (very low molecular weight polysaccharides may not precipitate in alcohol). Normally, polysaccharides are precipitated by addition of three to four volumes of ethanol (71 to 76% v/v) to polysaccharides solutions. Alternatively, polysaccharide solutions may be freeze dried after extensive dialysis of solutions to ensure removal of small molecular weight salts and/or enzymic degradation products.

1.5.5 Criteria of Purity

The purity of polysaccharide preparations may be confirmed by various means. In many cases, the amount of nitrogen content indicates the level of contamination with proteins. This method works well with polysaccharides such as β-glucans, arabinoxylans, cellulose, xanthan gum, gellan, carrageenans, and alginates, which should contain only minimal levels of contaminating proteins. It does not work with polysaccharides that are known to contain covalently bound peptides, such as gum arabic or arabinogalactans. Constancy in monosaccharide composition and in the ratio of sugars after repeated purification steps indicates the purity of polysaccharide preparations. Finally, the elution profile (number of peaks, size and symmetry of peaks) from either size exclusion or ion exchange chromatography can also indicate the purity of polysaccharide preparations.

Suggested Reading

Bols, M. *Carbohydrate Building Blocks,* John Wiley & Sons, Ltd. Chichester. 1996.

Binkley, R. W. *Modern Carbohydrate Chemistry.* Marcel Dekker, Inc. New York. 1988.

Collins, P. M. and Ferrier, R. J. *Monosaccharides: Their Chemistry and Their Roles in Natural Products.* John Wiley & Sons, Ltd. Chichester. 1996.

David, S. *The Molecular and Supramolecular Chemistry of Carbohydrates.* Oxford University Press, Oxford. 1998.

Lehmann, J. *Carbohydrates: Structure and Biology.* Georg Thieme Verlag, Stuttgart. 1998.

Lindhorst, T.K. *Essentials of Carbohydrate Chemistry and Biochemistry.* John Wiley & Sons, Ltd. VCH, Weinheim. 2003.

McNaught, A. Nomenclature of carbohydrates. *Carbohydrate Research.* 297, 1–92. 1997.

Preparative Carbohydrate Chemistry, edited by S. Hanessian. Marcel Dekker, Inc. New York. 1997.

The Carbohydrates, edited by W. Pigman and D. Horton. Vol. IA, IB, IIA & IIB. Academic Press, New York. 1970–1980.

References

1. Yalpani, M. *Polysaccharides; Synthesis, Modifications and Structure/Property Relationships.* Elsevier, Amsterdam. 1988.
2. Robyt, J. F. *Essentials of Carbohydrate Chemistry.* Springer Advanced Texts in Chemistry. Springer-Verlag, New York. 1998.
3. Southgate, D. A. T. *Determination of Food Carbohydrates.* Elsevier Applied Science, London, 2nd Edition. 1991.
4. Stick, R. V. *Carbohydrates: The Sweet Molecules of Life.* Academic Press, San Diego. 2001.
5. El Khadem, H. S. *Carbohydrate Chemistry; Monosaccharides and Their Oligomers.* Academic Press, San Diego. 1988.
6. Shallenberger, R. S. *Advanced Sugar Chemistry. Principles of Sugar Stereochemistry.* AVI Publishing Co. Westport. 1982.
7. Hassel, O. and Ottar, B. *Acta Chemistry Scandinavia,* 1, 929, 1947.
8. Reeves, P. Cuprammonium-glycoside complexes. *Advanced Carbohydrate Chemistry.* 6, 107–134. 1951.
9. Edward, J. T. Stability of glycosides to acid hydrolysis. *Chemistry Industry.* 1102–1104. 1955.
10. Lemieux, R. U. Rearrangements and isomerizations in carbohydrate chemistry. In *Molecular Rearrangements — Part 2.* de Mayo, P. Ed Interscience Publishers: John Wiley & Sons, New York, 709. 1964.
11. Lemieux, R. U. Newer developments in the conformational analysis of carbohydrates. *Pure and Applied Chemistry,* 27, 527–547. 1971.
12. Overend, W. G. Glycosides, In *The Carbohydrates.* Pigman, W. and Horton, D. eds. Academic Press, London, IA, 279. 1972.
13. Bochkov, A. F. and Zaikov, G. E. *Chemistry of the O-Glycosidic Bond: Formation and Cleavage.* Pergamon Press, Oxford. 1979.
14. Scherz, H. and Bonn, G. *Analytical Chemistry of Carbohydrates.* Georg Thieme Verlag, Stuttgart, 1998.
15. Bornet, F. R. J. Fructo-oligosaccharides and other fructans: chemistry, structure and nutritional effects. In *Advanced Dietary Fibre Technology,* McCleary, B.V. and Prosky, L. Eds. Blackwell Science Ltd., Oxford, 480. 2001.
16. Hidaka, H., Adachi, T., and Hirayama, M. Development and beneficial effects of fructo-oligosaccharides. In *Advanced Dietary Fibre Technology.* McCleary, B. V. and Prosky, L. eds. Blackwell Science Ltd., Oxford, 471. 2001.
17. Schoterman, H. C. Galacto-oligosaccharides: Properties and health aspects, In *Advanced Dietary Fiber Technology.* McCleary, B.V. and Prosky, L. eds. Blackwell Science Ltd. Oxford, 494. 2001.
18. Roberfroid, M. B. Dietary fiber properties and health benefits of non-digestible oligosaccharides, In *Complex Carbohydrates in Foods.* Cho, S.S., Prosky, L., and Dreher, M. eds. Marcel Dekker, Inc. New York. 25. 1999.
19. Craig, S. A. S., Holden, J. F., Troup, J. P., Auerbach, M. H., and Frier, H. Polydextrose as soluble fiber and comples carbohydrate In *Complex Carbohydrates in Foods.* Cho, S.S, Prosky, L., and Dreher, M. eds. Marcel Dekker, Inc. 229. 1999.
20. Bender, H. Production, characterization and applications of cyclodextrins. *Advanced Biotechnology Processes,* 6, 31–71. 1986.

21. Clarke, R. J., Coates, J. H., and Lincoln, S. F. Inclusion complexes of the cyclo-malto-oligosaccharides (cyclodextrins). *Advanced Carbohydrate Chemistry and Biochemistry* 46, 205–249. 1988.
22. Eskin, N. A. M. *Biochemistry of Foods*. Academic Press, San Diego, 239. 1990.
23. Feather, M. S. and Harris, J. F. Dehydration reactions of carbohydrates. *Advanced Carbohydrate Chemistry,* 28, 161. 1973.
24. Theander, O. Novel developments in caramelization. *Progress in Food Nutrition Science*, 5, 471. 1981.
25. Aspinall, G. O. *The Polysaccharides*. I, Academic Press, New York. 1972.
26. BeMiller, J. N. Gums/hydrocolloids: Analytical aspects, In *Carbohydrates in Food*. Eliasson, A-C. ed. Marcel Dekker, New York. 1996.

2

Understanding Carbohydrate Analysis

Yolanda Brummer and Steve W. Cui

CONTENTS

2.1 Introduction

Carbohydrates are one of the most important ingredients in foods and raw materials. They may occur naturally or be added to food products to provide nutrients and, in most cases, to improve the texture and overall quality of a food product.

Naturally occurring polysaccharides in foods are an intrinsic or innate part of the raw material. For example, starch is the most abundant naturally occurring carbohydrate in food products, followed by pectin, hemicellulose and cell wall materials.

Many polysaccharides are also added to food systems as stabilizers and dietary fiber. For example, locust bean and guar gum are used to stabilize emulsions and prohibit ice crystal growth in ice cream. Many other polysaccharide gums have been widely used in bakery and dairy products to improve the texture and organoleptic properties of baked goods and as gelling agents for making desserts (see Chapter 6 for details). In addition, polysaccharides may be produced as a by-product of bacteria, such as in yogurt. Recent applications of nonstarch polysaccharides in breakfast cereals and snack food products have increased due to their perceived physiological effects and health benefits. This group of components are called dietary fiber. For example, galactomannans and cereal β-glucans can reduce serum cholesterol and attenuate blood glucose levels whereas psyllium and flaxseed gum have been used as laxatives.

Regardless if they were added or present in food indigenously, it is important to know what kind and how much carbohydrates are present in a food system. This chapter describes current methods used for the determination of total carbohydrates and uronic acids, as well as for analyzing oligosaccharides and dietary fiber in food products.

2.2 Total Sugar Analysis

As a group, carbohydrates are quite heterogeneous, differing in primary structure (ring size and shape), degree of polymerization (mono vs. oligo vs. polysaccharides), macromolecular characteristics (linear structure vs. branched compact structure), linkage (i.e., α or β glycosidic linkage, linkage position), and charge (see Chapter 1 for more detailed coverage of carbohydrate structure). The physical and chemical differences give rise to disparate properties, including solubilities, reactivities, and susceptibility to digestive enzymes.

From an analytical perspective, the simplest situation is where there is only one type of carbohydrate present in a sample with minimal interfering compounds — measuring glucose oligomers in corn syrup, for example. In this case the disparate reactivities of sugars based on structure (e.g., ketose or hexose forms), charge, and type (e.g., glucose vs. arabinose) do not present a problem. It is the case most often though, especially with raw materials (e.g., seeds, cereal grains) and food products (ice cream, baked goods), that various types of carbohydrates are present in a sample with other compounds including lipid solubles, proteins, and minerals. The heterogeneity of this group of compounds can make analyzing the total carbohydrate content of a sample quite complex.

2.2.1 Sample Preparation

Before analyzing for any class of carbohydrate, whether it is monosaccharides or insoluble cellulosic material, the sample must be prepared so as to remove substances that can interfere with analysis. For samples that are already essentially sugar solutions (juice, honey) very little sample preparation is required. For other samples, such as oil seeds, cereals or whole foods, fats, proteins, pigments, vitamins, minerals, and various other compounds should be removed prior to analysis. There are several detailed procedures outlined in the literature for the removal of these substances prior to the analysis of simple sugars or polysaccharides.[1,2] It should be noted that the extent of sample preparation required is also dependent on the analytical technique and/or equipment being used. Generally, samples are dried and ground first, followed by a defatting step. Drying can be done under a vacuum, at atmospheric pressure, or for samples that are sensitive to heat, in a freeze dryer. Samples, once ground to a specified mesh size, are defatted using a nonpolar solvent such as hexane or chloroform. Low molecular weight carbohydrates can then be extracted using hot 80% ethanol. The ethanol extract will contain mineral salts, pigments, and organic acids as well as low molecular weight sugars and proteins, while the residue will mainly contain proteins and high molecular weight carbohydrates including cellulose, pectin, starch, and any food gums (hydrocolloids) that may be present. Protein is generally removed from samples using a protease such as papain. Water soluble polysaccharides can be extracted using water and separated from insoluble material by centrifugation or filtration. Depending on the compound of interest in the sample, an enzymatic treatment with α-amylase and/or amyloglucosidase can be used to rid the sample of starch. In this case starch is hydrolyzed to glucose, which can be separated from high molecular weight polysaccharides by dialysis or by collecting the high molecular weight material as a precipitate after making the solution to 80% ethanol. Glucose is soluble in 80% ethanol while polysaccharide material is not.

The chemical methods applicable to neutral sugars, the phenol–sulfuric acid assay and the anthrone assay, are classic methods with a long history of use, although the phenol–sulfuric acid assay is probably the more popular of the two. A chemical method for the analysis of uronic acids is also outlined. Enzymatic methods, while widely used and readily available, are generally specific for only one or more sugars in a sample. Chemical methods such as the phenol–sulfuric acid assay can be utilized to provide an approximate value, but because different sugars in solution react differently with the assay reagents, a measurement of the total sugar would be an estimate based on the reactivity of the sugar used to construct the calibration curve (usually glucose). If an approximate or estimated value is all that is required then this method is sufficient. When the exact amount of each sugar present is required, highly specific enzymatic methods are more appropriate. This is frequently done by instrumental analysis as described in Section 2.3.

2.2.2 Phenol–Sulfuric Acid Assay

2.2.2.1 Reaction Theory

In the presence of strong acids and heat, carbohydrates undergo a series of reactions that leads to the formation of furan derivatives such as furanaldehyde and hydroxymethyl furaldehyde.[1,3] The initial reaction, a dehydration reaction (Figure 2.1) is followed by the formation of furan derivatives, which then condense with themselves or phenolic compounds to produce dark colored complexes. Figure 2.2 displays some furan derivatives and the carbohydrates from which they originate. The developed complex absorbs UV-VI light, and the absorbance is proportional to the sugar concentration in a linear fashion. An absorbance maximum is observed at 490 nm for hexoses and 480 nm for pentoses and uronic acids as measured by a UV-VI spectrophotometer (Figure 2.3).[4]

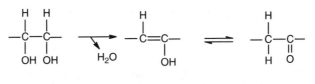

FIGURE 2.1
Dehydration reaction.

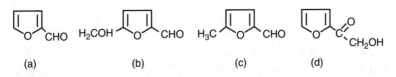

(a)   (b)   (c)   (d)

FIGURE 2.2
Furan derivatives from (a) pentoses and hexuronic acids, (b) hexoses, (c) 6-deoxyhexoses, and (d) keto-hexoses, respectively.

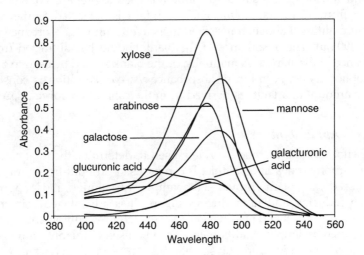

FIGURE 2.3
Phenol–sulfuric acid assay absorbance maxima for hexoses and pentoses.

2.2.2.2 Operating Procedure

Phenol, in a 5 or 80% solution is added to a glass test tube containing a clear sample solution. Concentrated sulfuric acid is added in a rapid stream directly to the surface of the liquid in the test tube. The mixture is thoroughly combined using a vortex mixter and then permitted to stand a sufficient time to allow for color development. The solution absorbance is read at 490 or 480 nm using a spectrophotometer, depending on the type of sugar present. Mixing and standing time should be kept the same for all samples to assure reproducible results.[4]

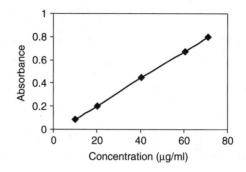

FIGURE 2.4
Phenol–sulfuric acid assay calibration curve.

2.2.2.3 *Quantification*

A calibration curve is constructed using the sugar being assayed. A stock 1 mg/ml aqueous sugar standard solution is used to prepare 5 or 6 standards ranging from 10 to 100 µg/ml. Each standard is subjected to the reaction procedure outlined above, transferred to a cuvette, and its absorbance read at 480 or 490 nm. The absorbance of the blank should be subtracted from the absorbance of the standards manually, or the blank should be used to zero the spectrophotometer. A graph of absorbance vs. concentration is constructed and the amount of analyte is derived from the calibration curve (Figure 2.4).

2.2.2.4 *Applicability*

The phenol–sulfuric method is widely used to determine the total concentration of carbohydrates in a sample. It can be used on lipid-free extracts from cereals, seeds, and plants, provided the sample is in solution, and is appropriate for both reducing and nonreducing sugars. It is advantageous in that the reagents are low cost and readily available, the required equipment is minimal, and the assay is simple. Additionally, it can be used to quantify monosaccharides, oligosaccharides, and polysaccharides. Absorption curves are characteristic for different sugars; therefore, this method provides the most accurate results when applied to samples containing only one type of carbohydrate.

This assay has been used to quantify total sugars in a sample containing more than one type of carbohydrate. In this case, glucose is often used to construct the calibration curve and the results are then approximate only and should be stated as glucose equivalents.

2.2.3 Anthrone–Sulfuric Acid Assay

2.2.3.1 *Reaction Theory*

Similar to the phenol–sulfuric acid assay, the anthrone method is based on the condensation of furaldehyde derivatives, generated by carbohydrates in the presence of a strong acid, with a reagent, in this case anthrone (9,10-dihydro-9-ozoanthracene), to produce colored compounds.[5] The reaction of carbohydrates in a strongly acidic environment with anthrone results in a blue-green color and the absorbance is read at 625 nm.

2.2.3.2 Operating Procedure

A cooled mixture of 2% anthrone in concentrated sulfuric acid is mixed with an aliquot of a clear sample solution containing the sugar being assayed. After incubation in a temperature-controlled environment for sufficient time to allow color development, the solution is poured into an appropriate spectrophotometric cuvette and the absorbance measured at 625nm.[5,6]

2.2.3.3 Quantification

Similar to the phenol–sulfuric acid assay, the anthrone reaction is nonstoichemetric and therefore requires the construction of a standard curve for quantitative purposes.

2.2.3.4 Applicability

The anthrone–sulfuric method is most applicable to solutions containing one type of hexose because even sugars with similar structures result in different rates and quantities of color development. Other sugars, such as pentoses and hexuronic acids, will also react to produce colored compounds that absorb at the same wavelength, but this only becomes a problem if they are present in a solution above a certain level.[5] This assay can also be used for quantitative analysis of oligo- and polysaccharides provided only one type is present in solution.

The anthrone method has been modified for use with a micro-plate, thus permitting the analysis of many samples within a short period of time and reducing the quantity of reagent needed.[7]

2.2.4 Analysis of Uronic Acids

Colorimetric methods for determining uronic acids are similar to the phenol–sulfuric acid and the anthrone assay in that they are based on the reaction of a reagent with carbohydrate derivatives formed in concentrated acid. The carbazole assay reported by Dishe[8] essentially involves mixing a sample containing uronic acid with concentrated sulfuric acid, heating it at 100°C, cooling it and then reacting it with 0.1% carbazole in ethanol. After sufficient time for color development, absorbance is read at 535 nm. Modifications to this assay include alterations to the timing of steps and reagent concentrations.[9] The carbazole assay, while simple, rapid, and sensitive suffers interferences from hexoses and pentoses. The replacement of carbazole with *m*-hydroxydiphenyl[10] increased the specificity and sensitivity of the assay.

2.2.4.1   Reaction Theory — m-Hydroxydiphenyl Method

While all carbohydrates react in concentrated acid to form colored compounds, uronic acids react with *m*-hydroxydiphenyl in a strongly acidic environment to form pink colored complexes. Absorbance measurements read at 520 nm increase linearly with uronic acid concentration from 0 to 100 µg/ml.

2.2.4.2   Operating Procedure

A sample solution is thoroughly mixed with sulfuric acid containing tetraborate in a test tube and placed in a boiling water bath for 5 minutes. After rapid cooling in an ice water bath, *m*-hydroxydiphenyl is added to each sample test tube and vortexed to ensure adequate mixing. The absorbance for each sample is read after allowing the color to develop for 20 minutes. A sample blank (containing a sample solvent) should be prepared at the same time as the samples.

2.2.4.3   Quantification

A solution of an appropriate uronic acid standard (i.e., galacturonic acid) is used to prepare several dilutions and these are used to prepare a standard curve. A graph of concentration vs. absorbance is constructed and the sample absorbance reading plotted along the curve to obtain concentration values. Standard curves typically range from 10 to 100 µg/ml.

2.2.4.4   Applicability

The *m*-hydroxydiphenyl assay can tolerate the presence of nonuronic acid sugars, up to ~200 µg/ml has been reported,[11] but higher concentrations of neutral sugars may artificially increase absorbance readings. Additionally, the presence of protein in a sample may interfere with the absorbances.[12] This method is appropriate for the quantification of pectic material in fruits and vegetables[13] and has been adapted for use with a micro-plate.[12]

2.3   Monosaccharide Analysis

The most frequently used methods for determining the concentration of monosaccharides in a sample are probably gas chromatography (GC) and high performance liquid chromatography (HPLC). These methods can also be suitable for qualitative determinations provided appropriate detection systems and/or standards are used. Unlike enzymatic methods, which tend to be specific for one type of monosaccharide only, chromatographic techniques provide qualitative and quantitative information about one or several monosaccharides in a sample.

Sample preparation is simplest when starting with a relatively pure, dry sample. For an HPLC analysis, the sample needs only to be finely ground, solubilized, diluted if required, and filtered prior to injection onto the column. In some cases, especially when analyzing complex food products, sample preparation becomes somewhat more complex and one or more clean up steps may be required. Lipid soluble compounds are generally removed via extraction with a suitable solvent, such as ether or hexane. Protein can be removed from samples enzymatically using a suitable protease (i.e., papain). The presence of inorganic salts can negatively affect column life and these can be removed using pre-packed preparatory columns that employ a mixed anion/cation exchange resin.

When the sample starting material is a polysaccharide and a quantitative and/or qualitative analysis of its constituent monosacharides is required, the sample must be depolymerized. This is most commonly accomplished using acid hydrolysis.

2.3.1 Acid Hydrolysis

In the presence of a strong acid and heat, the glycosidic bond between monosaccharide residues in a polysaccharide is cleaved. During this reaction, one molecule of water is consumed for every glycosidic linkage cleaved. During an acid hydrolysis released monosaccharides are susceptible to degradation in the presence of hot concentrated acid. However, not all glycosidic linkages are cleaved at the same rate and the hydrolysis time must be sufficient to hydrolyze all linkages in the sample. These two needs must be balanced; the need for hydrolysis of sufficient strength and length to permit complete hydrolysis, but not so long so as to lead to sample degradation.

Sulfuric acid and trifluoracetic acid (TFA) are commonly used for hydrolysis. It has been reported that sulfuric acid is superior to TFA for the hydrolysis of fibrous substrates such as wheat bran, straw, apples, and microcrystalline cellulose.[14] However, sulfuric acid can be difficult to remove post-hydrolysis and its presence can interfere with some analyses. TFA is volatile and can be easily removed prior to an HPLC analysis.

A hydrolysis procedure appropriate for neutral polysaccharide gums using TFA requires heating ~10 mg of polysaccharide material in 1 ml of 1M TFA at 121°C for 1 hour.[2] After cooling to room temperature, the TFA can be removed under a stream of nitrogen. A hydrolysis procedure using sulfuric acid appropriate for water soluble dietary fiber material in foods has been outlined. It requires mixing the sample material with 1M sulfuric acid and heating at 100°C for 2.5 hours.[6] Another recommended procedure for neutral polysaccharides involves mixing 2 to 5 mg of accurately weighed dry sample with 0.1 to 0.25 ml of 2M HCL and heating at 100°C for 2 to 5 hours.[15] While many hydrolysis procedures exist in the literature, when working with a new or unknown polysaccharide, it is best to check that the hydrolysis protocol is not resulting in excessive decomposition and is also cleaving bonds quantitatively. This is done by subjecting the sample to a chosen

TABLE 2.1

Acid Hydrolysis Methods

Substrate (s)	Acid Used for Hydrolysis	Hydrolysis Conditions	Reference
Xylan, wheat bran	H_2SO_4	72% H_2SO_4 1 hr at 30°C	Garleb et al., 1989
		2.5% H_2SO_4 1 hr at 125°C	
Wheat flour	HCL	2M HCL, 100°C for 90 min	Houben et al., 1997
Celluose	H_2SO_4	72% H_2SO_4 1 hr at 30°C,	Adams, 1965
		3% H_2SO_4, 100°C 4.5 hr	Saeman et al., 1963
Neutral polysaccharides	HCL	2M HCL, 2-5 hr	Pazur, 1986

hydrolysis protocol and monitoring the quantity of each sugar present in the solution at timed intervals. As hydrolysis proceeds, the quantity of each sugar should increase until all of that sugar present has been released from the polysaccharide. Degradation of released sugars will be evident as a decrease in monosaccharide concentration as hydrolysis time increases. Table 2.1 summarizes various hydrolysis procedures found in the literature. The following two sections outline the chromatographic techniques most frequently used for the analysis of monosaccharides.

Samples containing acidic sugar residues such as pectins and certain fungal polysaccharides can be difficult to hydrolyze quantitatively using traditional methods that employ TFA or sulfuric acid. The difficulty arises from the disparate susceptibilities to hydrolysis of neutral and acidic sugar residues as well as different linkage types. In cases where quantitative hydrolysis cannot be achieved, qualitative information can be obtained using an appropriate chromatographic technique post–acid hydrolysis (see next sections), and the uronic acid content determined using the spectrophotometric method outlined in Section 2.2.4.

2.3.2 Gas-Liquid Chromatography (GC)

Gas-liquid chromatography is a technique whereby components in a mixture are separated based on their degree of affinity for or interaction with a liquid stationary phase. In the case of GC, the sample components are dissolved in a gas phase and moved through a very small bore column, the interior of which is coated with the stationary phase. These separations occur at high pressure and high temperatures. Components in the sample mixture with a high affinity for the stationary phase will stay in the column longer and elute later than those with less affinity for the stationary phase. The degree of affinity for or interaction with the stationary phase that a molecule has is governed by its structure, properties, and the chemistry of the stationary phase being used.

The prerequisite of a GC separation in the sample must be volatile. Given that monosaccharides are not volatile, they must be derivatized prior to analysis.

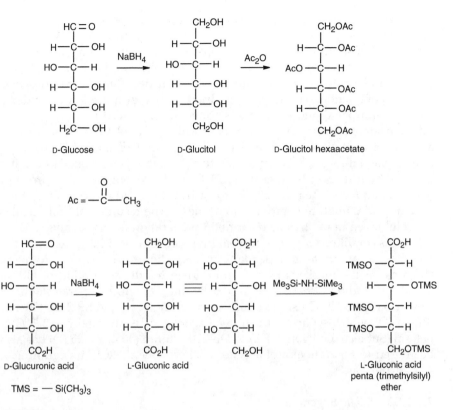

FIGURE 2.5
Alditol acetates and TMS derivatives from neutral and acidic sugars, respectively.

2.3.2.1 Derivatization

Neutral monosacharides are most often derivatized into alditol acetates prior to GC analysis. The age of this derivatization technique and the frequency with which it is used is evident from the number of methods for this procedure available in the literature.[6,16,17] The essential elements of this derivitization procedure are the reduction of neutral sugars to alditols and their subsequent acetylation. The resulting alditol acetates are then dissolved in a suitable solvent and injected onto a GC column. Acidic sugars are treated differently to yield trimethylsilyl (TMS) derivatives. The derivitization process is shown in Figure 2.5 for both neutral (alditol acetates) and acidic (TMS derivatives) sugars.

2.3.2.1.1 Neutral Sugars

The starting material must be a dry sample of one or more monosaccharides. Polysaccharide material must first be hydrolyzed and the acid removed prior to analysis. Trifluoracetic acid works well for this purpose because it is volatile and can be easily removed by rotary evaporation.

The dry sample containing a small amount (~10mg) of accurately weighed monosacharide material and inositol hexa-acetate (internal standard) is mixed with a solution of sodium borohydride in ammonium hydroxide to convert monosaccharides into alditols. Acetic acid is added to acidify the sample and destroy excess sodium borohydride after the reaction has reached completion. The mixture is dried (rotary evaporation or under a stream of nitrogen) and methanol is added and removed by drying with a nitrogen stream several times. Treatment with methanol removes borate ions as volatile methyl borate. When the final portion of methanol has been removed and the sample is dry, the mixture of alditols is acetylated by adding acetic anhydride and heating it at 121°C for a few hours. A few drops of water are added to the reaction vial to destroy any residual acetic anhydride and the entire mixture is brought to dryness.[2] The resulting alditol acetates are solubilized in methylene chloride in preparation for GC analysis.

Various conditions for separating alditol acetates exist in the literature; the one most appropriate is dependant on the column used. There are many columns on the market that are appropriate for the separation of alditol acetates including SPB-1701 (30 m × 0.25 mm ID), which separates alditols isothermally at 220°C using helium as the carrier gas, and SP-2380 (30 m × 0.25 m ID), which separates the derivatized sugars isothermally at 275°C in helium (Supelco, Bellefonte, PA). A DB-210 column (30 m × 0.25 ID) has also been used to separate alditol acetates isothermally at 220°C using nitrogen as a carrier gas.[18]

2.3.2.1.2   Acidic Sugars

As with neutral polysaccharides, polysaccharides containing uronic acids must be hydrolyzed and dried prior to analysis. It is very difficult to obtain a quantitative yield of acidic monosaccharides using acid hydrolysis, and when this is required, it is advisable to use a spectrophotometric method such as is outlined in Section 2.2.4.

A sample containing a small amount (~10 mg) of accurately weighed carbohydrate material is dissolved in sodium carbonate and then treated with sodium borohydride. Acetic acid is added to destroy excess borohydride and borate ions are removed with methanol as described for neutral monosaccharides. The resultant mixture of aldonic acids and aldoses (from neutral sugars if present) is made into TMS derivatives by treating the dry residue with a mixture containing pyridine, hexamethyldisilazane, and trifluoroacetic acid.[2] An internal standard, such as docosane, should be used for quantification.

2.3.2.2   Quantification

A flame ionization detector (FID) is the most commonly used detector. With an FID, quantification requires using an internal standard and the formulation of response factors (RF). Response factors are used to correct for disparate GC response to monosaccharides and losses arising from hydrolysis and derivatization. They are obtained for each monosacharide by subjecting a mixture of standard monosaccharides corresponding to the monosaccharides present in the sample to the same hydrolytic and derivatization conditions and using the following equation:

$$RF = (A_S \times W_M)/(A_M \times W_S)$$

where A_S = peak area of internal standard, W_M = weight in mg of individual monosaccharide standard, A_M = peak area of monosaccharide standard, and W_S = weight in mg of internal standard.

Once response factors have been determined for each monosacharide present, they are used to determine the percent content of each monosaccharide residue, %M, in the sample according to the following equation:

$$\%M = (RF \times A_M \times W_S \times F \times 100)/(A_S \times S)$$

where RF = the response factor for each monosaccharide, A_M = peak area for monosacharide in sample, W_S = weight in mg of internal standard in sample solution, F is a factor for converting monosaccharides to polysaccharide residues (use 0.88 for pentoses and 0.90 for hexoses), A_S = peak area of internal standard in sample solution, and S is the dry weight in mg of starting sample material.

2.3.2.3 Advantages/Disadvantages

GC analysis of carbohydrates is advantageous because it is an established technique and much of the method optimization has been done. It also requires small sample sizes and is very sensitive, a vital advantage when only a small amount of sample is available. The disadvantages of this technique originate chiefly from the preparatory steps. If either the reduction or the acetylation steps do not proceed to completion, the quantity of derivatised sugars will be underestimated. In short, there is ample opportunity for sample loss. Additionally, preparation may appear prohibitively laborious, especially when compared to HPLC techniques (see next section).

2.3.3 High Performance Liquid Chromatography (HPLC)

Similar to GC, HPLC is a separation technique whereby compounds in a mixture are separated on a stationary phase. In this case, the mobile phase (eluant) containing the sample and the stationary phase are both liquids. Unlike GC, HPLC separation is a function of the compatibility (or differing compatibility) of sample components for the eluant and stationary phase. A HPLC separation can be manipulated by changing the concentration and/or makeup of the eluant. For example, a compound that is very compatible with the stationary phase (and therefore one that stays on the column longer and elutes later) can be forced off the column faster by changing the solvent makeup such that sample components favor it vs. the stationary phase.

Under the broad category of HPLC there are several sub-types characterized by the type of stationary and mobile phases employed and therefore the chemistry of the separation. In normal phase chromatography the stationary phase is hydrophilic and the mobile phase varies in its hydrophilic/hydrophobic nature depending on the sample components being separated. In reverse phase chromatography, the stationary phase is hydrophobic. High performance anion exchange chromatography (HPAEC), an increasingly popular choice for the separation of carbohydrates, is characterized by an anionic stationary phase and a mobile phase with a high pH. At high pH (10 to 14), carbohydrate hydroxyl groups ionize and their separation is based on their differing affinity for the oppositely charged stationary phase and the mobile phase.

2.3.3.1 *High Performance Anion Exchange Chromatography (HPAEC)*

A sample containing monosaccharides, either as its natural state or after a hydrolysis step, must be separated chromatographically so that each monosaccharide can be identified and quantified. Over the past few years, high performance anion exchange chromatography has proven invaluable in the analysis of carbohydrate material. This type of chromatography is based on the fact that carbohydrates in a strongly alkaline environment will ionize, thereby rendering them amenable to separation on an ion exchange column. HPAEC columns used for carbohydrates are coated with an anion exchange resin. For example, the Dionex (Sunnyvale, CA) PA1 column, optimized for the separation of mono-, di-, oligo-, and low molecular weight polysaccharides, is composed of 10 μm nonporous beads covered in a quaternary amine anion exchange material. HPAEC systems typically use sodium hydroxide as the eluant to separate mono- and disaccharides, while eluants for larger molecules often include sodium acetate to increase ionic strength. The detector of choice for HPAEC is the pulsed amperometric detector (PAD). In general, amperometry measures the change in current resulting from the oxidation or reduction of a compound at an electrode. In PAD, it is the change in current resulting from carbohydrate oxidation at a gold or platinum electrode that is measured. The advantage of PAD is not only its low detection limits, reportedly in the picomole range,[19] but also its suitability for gradient elution, which provides an analyst with more flexibility when optimizing separation conditions.

Figure 2.6 presents the HPAEC separation profile of arabinose, rhamnose, glucose, galactose, mannose, and xylose. These six standards were separated using a sodium hydroxide gradient. Excellent baseline separation is also achieved for mixtures of mono- and oligosacchrides (Figure 2.7), sugar alcohols,[20,21] and uronic acids.[22] HPAEC-PAD permits the separation and quantification of both neutral and acidic monosaccharides in one analytical run. A sodium hydroxide/sodium acetate gradient and PA1 column (Dionex, Sunnyvale Ca.) has been used to separate and identify neutral sugars (rhamnose, arabinose, galactose, glucose, mannose) and uronic acids (galacturonic and glucuronic acid) from citrus pectin and fungal polysaccharides after

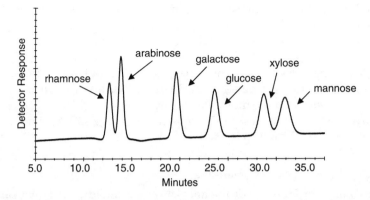

FIGURE 2.6
HPAEC-PAD chromatogram showing separation of rhamnose, arabinose, glucose, galactose, xylose, and mannose.

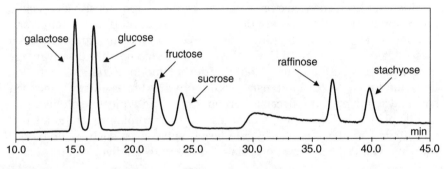
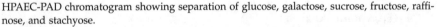

FIGURE 2.7
HPAEC-PAD chromatogram showing separation of glucose, galactose, sucrose, fructose, raffinose, and stachyose.

hydrolysis with TFA, sulfuric acid, and methanolysis combined with TFA.[22] Furthermore, a sodium hydroxide/sodium acetate gradient can also be used to separate neutral and acidic monosaccharides as well as oligogalacturonic acids from strawberries after treatment with pectolytic enzymes.[23] For many samples, preparation entails extraction of the desired components with water, centrifugation to remove precipitated matter, and filtering.[20,21] The final filtering step prior to injection onto the HPLC column is usually through at 0.45 μm filter.

2.3.3.2　Quantification

Quantifying the amount of each individual mono-, di-, oligo-, or polysaccharide from an HPAEC chromatogram is quite simple provided 4 to 5 dilutions of appropriate standards are run with a set of samples. The software associated with the HPLC system will integrate peaks, provide peak areas/ heights, and give concentration values provided standards with known concentrations have been run. It is important to run standards corresponding to sample constituents because PAD response varies with the analyte. Detector response decreases with the increasing degree of polymerization (DP),[19] therefore, the relative percentage of each component in a chromatogram (and therefore the sample) cannot be determined by peak areas alone, especially when the sample contains mono-, di-, oligo-, and polysaccharides. Sugar concentration values obtained for acid hydrolysates need to be further adjusted to account for the molecule of water that was added to each residue upon hydrolysis. For hexoses this requires a conversion factor of 0.90 and for pentoses a factor of 0.88 is used.

2.3.3.3   Advantages/Disadvantages

The advantage of HPAEC for the analysis of carbohydrates is that samples do not require derivatization and the analysis itself is usually quite fast. In the past, disadvantages of HPLC originated with detection systems, which for the most part were not very sensitive. Refractive index detectors have traditionally been the detector of choice, because more sensitive detectors, for example UV or fluorescence detectors, are not appropriate for analyzing carbohydrates since carbohydrates do not contain moieties that respond to these detection systems. HPAEC-PAD has overcome this disadvantage, enabling the separation and quantification of monosaccharides with low detection limits. In addition, using sodium hydroxide as eluant is inexpensive and relatively safe. Because carbohydrates are typically soluble in the mobile phase, derivatization is unnecessary and sample preparation is usually limited to the removal of interfering substances such as lipids and proteins. The PAD is also appropriate for use with gradient elution.

2.3.4   Enzymatic Analysis

Enzymatic methods for determining sugar content rely on the ability of an enzyme to catalyze a specific reaction and employ a suitable method for monitoring the progression of the reaction or the concentration of reaction product. Enzymatic methods are highly specific, usually rapid and sensitive to low sugar concentrations.

There are enzymatic methods for most common sugars, and for some such as glucose several different enzymes can be used, and these assays are available in kit form. Enzymatic methods for the quantitative analysis of glucose, fructose, sucrose, lactose, and galactose are presented below. Enzymatic methods have been developed for the quantification of various polysaccharides, including β-glucan[24] (American Association of Cereal Chemists [AACC] method 32-23; Association of Official Analytical Chemists [AOAC] method 995.16), starch[25] (AACC method 76-13; AACC method 76-11), and the galactomannans locust bean gum and guar gum.[26,27] These generally require enzymatic hydrolysis of the polymer followed by enzymatic determination of the released monosaccharides using one of the methods below.

2.3.4.1 Analysis of Glucose

2.3.4.1.1 Reaction Theory

One of the earliest and most widely used enzymes for the quantitative determination of glucose is glucose oxidase. This highly specific enzyme, which can be obtained from *Penicillium notatum* and *Aspergillis niger*, catalyzes the oxidation (2 hydrogen atoms removed) of β-D-glucopyranose to form D-glucono-1,5-lactone, which is a short lived species that hydrolyzes to yield D-gluconic acid.[28] The reaction of glucose with glucose oxidase also yields H_2O_2 in a ratio of 1:1.

Early methods for detecting the quantity of glucose in a sample after treatment with glucose oxidase relied on the volumetric titration of gluconic acid formed.[28] Today, colorimetric methods based on the use of glucose oxidase are more common. In these reactions peroxidase is used in combination with a chromogen to yield a colored complex in the presence of H_2O_2. Glucose is oxidized to yield gluconic acid and peroxide in the presence of glucose oxidase (EC.1.1.34) and peroxidase catalyzes the oxidation of a chromogen (i.e., *o*-dianisidine, Figure 2.8) in the presence of H_2O_2, thereby enabling quantitative measurement spectrophotometrically when an appropriate standard curve has been established. Detailed methods for this assay

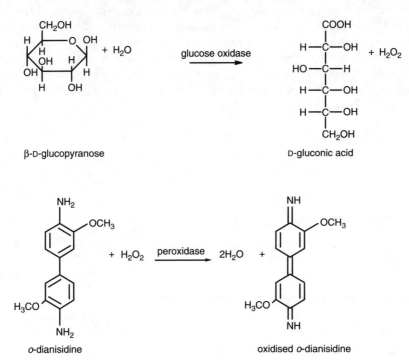

β-D-glucopyranose   D-gluconic acid

o-dianisidine   oxidised *o*-dianisidine

FIGURE 2.8
Glucose oxidase assay reaction.

are available elsewhere[6] (AACC method 80-10) and complete kits containing the method and reagents are available commercially.

Hexokinase is another enzyme frequently used in the quantitative determination of glucose. Glucose reacts with hexokinase in the presence of adenosine triphosphate (ATP) to form glucose-6-phosphate and adenosine diphosphate (ADP). Glucose-6-phosphate reacts with glucose-6-phosphate dehydrogenase in the presence of nicotinamide-adenine dinucleotide (NAD) to produce 6-phosphogluconate and NADH.[6] NADH concentration is measured spectrophotometrically at 340 nm, or the reaction is modified so it is amenable to colorimetric determination.

1. D-Glucose + ATP $\xrightarrow{\text{Hexokinase}}$ Glucose-6-phosphate + ADP

2. Glucose-6-phosphate + NAD $\xrightarrow[\text{dehydrogenase}]{\text{Glucose-6-phosphate}}$ Gluconate-6-phosphate + NADH

2.3.4.1.2 *Operating Procedure — Glucose Oxidase*

Aliquots of sample solution are mixed with a buffered solution containing glucose oxidase, peroxidase, and the chosen chromogen and incubated under temperature controlled conditions for a specified period of time (time and temperature dependant on a chosen analytical method). After sufficient time for color development, the absorbance is read at an appropriate wavelength. For example, when using AACC method 80-10, which uses *o*-dianisodine hydrochloride as chromogen in 0.1 M acetate buffer at pH 5.5, sample solutions are incubated at 30°C for 5 min and absorbance read at 525 mn against a reagent blank.

2.3.4.1.3 *Quantification*

A calibration curve is constructed by plotting absorbance vs. concentration for 5 separate glucose dilutions. Quantification is achieved using the calibration curve.

2.3.4.1.4 *Applicability*

Glucose oxidase is a highly specific enzyme. Therefore this method is applicable for use with samples that contain other sugars.

2.3.4.2 **Analysis of Galactose and Lactose**

D-Galactose has been assayed enzymatically using galactose dehydrogenase[6] and galactose oxidase.[29] In the presence of oxygen, galactose is oxidized to D-galacto-hexodialdo-1,5-pyranose by galactose oxidase (EC 1.1.3.9). This reaction generates hydrogen peroxide, which can be measured colorimetrically in the presence of hydrogen donors. Alternately, D-galactose is oxidized to galactonic acid by NAD when β-galactose dehydrogenase (EC 3.2.1.23) is present, and the resultant formation of NADH can be monitored spectro-photometrically. Lactose can also be detected as galactose after treatment with β-galactosidase, an enzyme that catalyzes the hydrolysis of lactose to D-glucose and D-galactose. Both galactose and lactose may be determined in a sample by correcting the lactose content of a sample obtained post–β-galactosidase digestion by subtracting the free galactose determined in the absence of the enzyme. Enzymatic methods for determining lactose and galactose are available in the literature[6,30] and in kit form (Megazyme, Ireland). The procedure outlined below is adapted from Mustranta and Ostman.[30]

2.3.4.2.1 Reaction Theory

Galactose is oxidized to galactonic acid by NAD in the presence of galactose dehydrogenase. NADH generated by this reaction is present in solution at a concentration proportional to the galactose content.

$$1.\ \text{D-Lactose} + H_2O \xrightarrow{\text{β-Glactosidase}} \text{D-Glucose} + \text{D-Galactose}$$

$$2.\ \text{D-Galactose} + NAD^+ \xrightarrow[\text{Dehydrogenase}]{\text{Galactose}} \text{Galactonic acid} + NADH + H^+$$

2.3.4.2.2 Operating Procedure

Samples containing galactose exclusively are mixed with a buffer containing NAD, the initial absorbance is read at 340 nm and then measured again after incubation with galactose dehydrogenase. Lactose containing samples must first be treated with β-galactosidase to convert lactose to glucose and galactose and then incubated with galactose dehydrogenase. Both galactose and lactose can be determined in one assay using these two enzymes. Blanks (each containing all required enzymes and buffer but no sample) need to be prepared at the same time as the samples.

2.3.4.2.3 Quantification

The quantity of NADH formed is stoichiometric with the amount of galactose present in the sample. The absorption differences for the samples and blanks must be determined first by subtracting initial absorbance values (pre–galactose dehydrogenase addition) from the final absorbance values (post–galactose dehydrogenase addition).

2.3.4.2.4 Applicability

This assay is applicable for determining galactose concentration in solutions that are free of fat and protein. Samples to be analyzed should be clear and less than 0.5 g/L total galactose and lactose. Samples containing protein can be treated with perchloric acid[6] or Carrez reagent,[1,3] which precipitates proteins and absorbs some colored compounds.

The enzymatic assay for lactose and galactose can be used for many different food products, including meat, dairy, and bakery products, provided sample treatment includes steps to extract sugars and remove interfering substances. Galactose dehydrogenase also oxidises L-arabinose, therefore, the presence of arabinose will interfere with the analysis.

2.3.4.3 Analysis of Fructose, Glucose, and Sucrose

2.3.4.3.1 Reaction Theory

Fructose can also be quantitatively assayed using hexokinase. Glucose and fructose can be assayed together using hexokinase, glucose-6-phosphate dehydrogenase, and phosphoglucose isomerase (PGI) to catalyze specific reactions. In the presence of ATP and hexokinase, glucose and fructose are phosphorylated to glucose-6-phosphate and fructose-6-phosphate (F-6-P) respectively. Adding NAD and glucose-6-phosphate dehydrogenase oxidises G-6-P to gluconate-6-phosphate and results in the formation of NADH which can be measured at 340 nm. Adding phospho-glucose isomerase changes F-6-P into G-6-P, which is then oxidized to gluconate-6-phosphate in the presence of NAD and leads to the formation of NADH. Sucrose may also be assayed using hexokinase or glucose oxidase (Section 2.3.4.1.) by first treating it with invertase (EC.3.2.1.26) to release glucose and fructose.

1. D-Fructose + ATP $\xrightarrow{\text{Hexokinase}}$ Fructose-6-phosphate + ADP

2. Fructose-6-phosphate $\xrightarrow{\substack{\text{Phospho-Glucose}\\\text{Isomerase}}}$ Glucose-6-phosphate

3. Glucose-6-phosphate + NAD $\xrightarrow{\substack{\text{Glucose-6-Phosphate}\\\text{Dehydrogenase}}}$ Gluconate-6-phosphate + NADH

2.3.4.3.2 Operating Procedure

Buffer, NAD, ATP, and sample solution containing glucose and fructose are combined in a test tube. A mixture of hexokinase and glucose-6-phosphate dehydrogenase is added and after sufficient time for the reaction to proceed, the solution absorbance is read at 340 nm. The absorbance is read again after adding phospho-glucose isomerase. A sample blank should be subjected to the same procedure as the samples and used to zero the spectrophotometer. A detailed procedure to determine both glucose and fructose is available in the literature[6] and in kit form (Megazyme, Ireland).

2.3.4.3.3 Quantification

Glucose is quantified from absorbance values obtained after adding hexokinase and glucose-6-phosphate dehydrogenase. To quantify fructose in the same sample, absorbance values read after adding phospho-glucose isomerase are corrected for initial glucose absorbance values. The quantity of NADH produced is stoichiometric with the quantity of glucose and fructose.

2.3.4.3.4 *Applicability*

This method is appropriate for the determination of glucose and fructose in many different varieties of food stuffs including jam, honey, and ice cream, as long as they have been treated to remove interfering substances such as lipids and proteins. Samples should be clear, relatively colorless, and free from precipitated material. Solutions that are turbid or contain interfering matter can be filtered or treated with Carrez reagents for clarification.

2.3.4.4 **Analysis of *(1→3) (1→4)-β-D-Glucans***

2.3.4.4.1 *Reaction Theory*

(1→3) (1→4)-β-D-glucans are cell wall polysaccharides found in greatest abundance in cereal grains such as oats and barley, while smaller quantities are also found in wheat and rye grains. β-Glucans from cereal grains have been extensively researched over the past several decades based on their postulated health benefits, including attenuating blood glucose levels[31] and lowering serum cholesterol levels.[32,33] The analysis of β-glucan requires enzymatic hydrolysis using (1→3)(1→4)-β-D-glucan-4-glucanohydrolase (E.C. 3.2.1.73) (lichenase) to yield oligosacchrides and subsequent hydrolysis with β-glucosidase (E.C.3.2.21.). Released glucose is assayed using glucose oxidase as outlined in Section 2.3.4.1. This assay is available in kit form from Megazyme (Ireland).

1. $(1 \rightarrow 3)\ (1 \rightarrow 4)$-β-D-Glucan $\xrightarrow{\text{Lichenase}} 1 \rightarrow 3$ linked cellodextrins

2. $(1 \rightarrow 3)$ linked cellodextrins $\xrightarrow{\text{β-glucosidase}}$ Glucose

2.3.4.4.2 *Operating Procedure*

An accurately weighed amount of dry sample (usually flour or milling fraction) is mixed with phosphate buffer (pH 6.5), boiled briefly, and mixed thoroughly. The mixture is incubated with high purity lichenase at 50°C, combined with sodium acetate buffer (pH 4.0) and centrifuged. A portion of the supernatant is treated with glucosidase and another portion with acetate buffer to serve as a blank. After incubation at 40°C with glucosidase the samples are diluted and glucose determined enzymatically using the glucose oxidase-peroxidase method (Section 2.3.4.1). Standards containing 50 and 100 µg/ml glucose, reagent blanks (containing acetate buffer and glucose oxidase-peroxidase reagent) and flour controls with known β-glucan values are prepared and run with each set of samples.

2.3.4.4.3 Quantification

Absorbance values from glucose standards are used to evaluate percent glucose in the sample. When calculating the percent glucose it is important to include a conversion factor of 0.9 to account for the difference in molecular weight of free glucose vs. glucose in polysaccharides.

2.3.4.4.4 Applicability

This assay is appropriate for use with dry flours and milling fractions of cereal grains such as rye, oats, barley, wheat, and unsweetened cereals provided all of the above have been milled to pass a 0.5 mm mesh. This assay may also be used for samples containing simple sugars by extracting the sample with 50% ethanol to remove these sugars (which can artificially increase measured glucose values) prior to lichenase treatment. Assay steps have been modified and optimized for different applications including flours and milling fractions from cereal grains[24] (AACC method 32-22 and 32-23) and β-glucan in malt, wort, and beer.[34]

2.3.4.5 Analysis of Galactomannans

Galactomannans are polysaccharides consisting of a β-1→4 linked mannosyl backbone substituted to varying degrees at the C-6 position by an α-galactose unit. The level of substitution varies depending on galactomannan source; for example, galactomannan from guar has a galactose:mannose (G:M) ratio of approximately 1:2 while the ratio for locust bean gum (carob gum) is 1:4.[35] An enzymatic method for quantitatively determining galactomannans has been developed that is based upon the known G:M ratios of common galactomannans.[26,27]

2.3.4.5.1 Reaction Theory

The concentration of galactomannan in a sample is determined from the quantity of galactose released by α-galactosidase after β-mannanase digestion. Galactose is quantified spectrophotometrically after treatment with NAD and galactose dehydrogenase.

$$1. \text{Galactomannan} \xrightarrow{\text{β-mannanase}} \text{Oligosaccharide mixture}$$

$$2. \text{Oligosaccharide mixture} \xrightarrow{\text{α-galactosidase}} \text{Galactose} + \text{manno-oligosaccharides}$$

2.3.4.5.2 Operating Procedure

Dry flour samples from guar or locust bean gum are extracted with 80% ethanol to remove simple sugars and oligosaccharides which may contain galactose. The galactomannan present is solubilized by boiling the flour in a buffer followed by incubation at 40°C with β-mannanase. Sample mixtures are centrifuged to separate insoluble material and portions of the supernatant are incubated with α-galactosidase to release galactose and another portion is treated with an acetate buffer (blank). Galactose is quantitatively determined in the sample after treatment with NAD and galactose dehydrogenase.

2.3.4.5.3 Quantification

Blank absorbance values are subtracted from sample absorbances and the resulting values used to quantitatively determine galactose by accounting for sample dilutions, the conversion of free galactose to anhydro-galactose, and a factor to relate free galactose to galactomannan content (×0.90). The average percentage of galactose in guar and locust bean gum is 38 and 22%, respectively.[36]

2.3.4.5.4 Applicability

This method can be used for analysing seed flours provided the galactose:mannose ratio of the galactomannan is known. It can also be used when low molecular weight sugars are present in the sample provided they are removed via ethanol extraction prior to analysis. The presence of 1→6 linked α-D-galactose units in some oligosaccharides, such as raffinose and stachyose, are also susceptible to α-galactosidase; these oligosaccharides will artificially increase the gum content if they are not removed before enzyme treatment.

2.4 Oligosaccharide Analysis

Strictly speaking, oligosaccharides are carbohydrates composed of between two and ten monosaccharide residues glycosidically linked.[37] Practically speaking, saccharides with a DP greater than 10 are often referred to as oliogsaccharides. For example, inulin is often referred to as a nondigestible oligosaccharide, despite the fact that it is a polydisperse mixture with a DP ranging from 2 to 65.[38] Raffinose and stachyose, tri- and tetrasaccharides respectively composed of galactose, glucose, and fructose are found in beans. Other common oligosaccharides in foods are dextrins or starch hydrolysates. Starch hydrolysates are characterized based on their dextrose equivalent (DE), or the relative reducing power of all sugars in the mixture. Hydrolysates with a DE of 20 or below are referred to as maltodextrins and are mainly used for adding bulk to food products as well as their viscosity-modifying and film-forming properties. Oligosaccharides are also found in beer, the large majority being starch hydrolysates that result from the enzymatic activity of amylases during malting.[39]

Analysis of oligosaccharides released by partial acid hydrolysis or enzymatic attack of polysaccharide material is a technique often used to provide valuable information about molecular structure. For example, the (1→4) linkage attached to a (1→3) linked glucose unit in mixed linked (1→3)

(1→4)-β-D-glucan is preferentially cleaved by (1→3) (1→4)-β-D-glucan-4-glu-canohydrolase (lichenase). The oligosaccharides released from a lichenase digestion of β-glucan are primarily 3-O-β-cellobiosyl and 3-O-β-cellotriosyl-D-glucose; the relative proportion of which is indicative of the dominant polysaccharide structure. The analysis of these oligosaccharides is thus an important step in fully understanding the parent polysaccharide. Details on structural analysis of polysaccharides are given in Chapter 3.

Regardless of whether or not oligoaccharides are present in a sample as an inherent component of its makeup (e.g., beans) or owing to some hydro-lytic treatment resulting from direct sample manipulation (i.e., acid or enzy-matic hydrolysis), there is a necessity to identify and quantify them. Methods for analyzing oligosaccharides are similar to methods for analyzing monosaccharides. For example, the extraction of oligosaccharides from food products is accomplished as for monosaccharides, with hot 80% ethanol. They can be hydrolysed with acid or enzymes to their constituent monosac-charides and the hydrolysate subjected to analysis by chromatographic, chemical or enzymatic methods. In addition, size exclusion techniques such as high performance size exclusion chromatography or gel permeation chro-matography can be used to separate oligosaccharide mixtures by size.

2.4.1   Monosaccharide Composition

The monosaccharide composition of oligosaccharides can be determined by hydrolysis (acid or enzymatic) followed by a suitable method for identifying the released monosaccharides. Released monosaccharides may be identified using chromatographic techniques (HPLC, GC) provided suitable standards are used for comparing the retention times (see Section 2.3 for details on monosaccharide analysis).

2.4.2   Size Exclusion Chromatography (SEC)

Size exclusion chromatography is a chromatographic technique whereby molecules in a sample are separated based on their size. Molecules in an eluant stream (usually a buffer) are directed into a column filled with a gel packing of clearly defined pore size. The smaller molecules in a sample get held up in the pores, therefore spending more time in the column and eluting later than larger molecules, which essentially pass through the spaces between the pores and elute first. Separation is influenced by the size of the pores in the column packing. Effluant from the column is monitored using one or more detectors. Refractive index detectors are often used alone or in combination with light scattering and/or viscosmetric detectors.

Size exclusion chromatography has been used as a method for separating and characterizing mixtures of oligosaccharides based on size. Dextrins, or starch hydrolysates, can be rather complex mixtures of oligosaccharides with the size and proportion of linear and branched oligomers varying depending on the starting material (rice starch, corn starch, etc.) and hydrolysis method (enzymes, acid) even when the DE is the same. Dextrins of DE ranging from 4 to 25 were analyzed on a SEC system equipped with multi-angle light-scattering detector, enabling the acquisition of information with respect to molecular weight distribution.[40] In combination with high performance anion exchange chromatography (HPAEC), which provided information about the relative occurrence of individual oligosaccharides, SEC coupled with a molecular weight sensitive detector allowed a more complete dextrin sample profile than the DE value alone (which only reflects the number of reducing ends).

Conventional or low pressure size exclusion chromatography predates SEC. Low pressure systems employ large capacity columns (i.e., 1.6 × 70 cm and larger) singly or several in series that operate at a relatively low pressure compared to SEC and separate larger sample volumes. Therefore, they are often used for preparative or semi-preparative purposes, to clean samples (remove salts or other small molecular weight material from a sample) or isolate a specific fraction of interest in a sample prior to further analysis. Eluent concentration is often monitored with a detector (i.e., RI detector) so a chromatogram of sample distribution is obtained. Alternately, a eluent containing separated material is collected in tubes and the tube contents analyzed using another analytical technique, such as the phenol–sulfuric acid assay. By plotting the tube number vs. the carbohydrate concentration, a curve representing sample distribution can be obtained. For example, large and small molecular weight starch fractions in native and acid-modified starches from cereals, pulses, and tubers were separated on a Sepharose CL 4B gel (Pharmacia Fine Chemicals, Sweden) at 30ml/h with water as eluent.[41] Oligosaccharides arising from enzymatic degradation of carrageenan have also been separated successfully on a low pressure SEC system. Using Super-dex™ 30 (Pharmacia Fine Chemicals, Sweden) preparatory grade gel (a dextran-agarose composite), good separation and resolution of carrageenan oligosaccharides in large quantities with run times ranging from 16 hr[42] to several hours[43] depending on the sample type and the analytical conditions used. The structure of isolated oligosaccharides can then be determined using methods outlined in Chapter 3. Other examples of the use of gel-permeation chromatography for the purpose of purifying and/or separating oligosaccharides generated by the hydrolysis of polysaccharides abound, including galactoglucomannan oligosaccharides from kiwi fruit,[44] xyloglu-can oligosaccharides generated by the action of cellulase,[45] and oligosaccha-rides from olive xylogucan.[46]

Using gel-permeation chromatography is rooted in the fact that relatively large sample sizes may be applied to the column and therefore separated fractions can be collected in relatively large amounts. Additionally, it has been reported that conventional SEC separates some classes of oligosaccha-rides, such as a homologous series of maltodextrins, with better resolution than HPSEC.[47] One of the disadvantages of low pressure SEC systems is the time required for a single run, often requiring many hours or even days.

2.4.3   High Performance Anion Exchange Chromatography (HPAEC)

Using HPAEC for carbohydrate analysis is highlighted again when the separation of oligosaccharides is considered. The combination of column packing, eluent composition, and sensitive electrochemical detection enables the baseline resolution of a homologous series of oligosaccharides such as dextrins (starch hydrolysates) up to DP 40. Unlike monosaccharide analysis, where the mobile phase is typically sodium hydroxide alone, mobile phases for oligosaccharides contain sodium acetate as well to increase ionic strength and to ensure adequate pushing off of the oligosaccharides from the column. The Dionex Carbopac PA1 column has been used to separate dextrins up to DP 30[40] using a mobile phase combination of 80% A and 20%B at time zero with a linear gradient to 90% B and 10% A (where A = 100 mM NaOH and B = 100 mM NaOH containing 600 mM NaOAC). In the case of dextrins, sample preparation consists only of drying the sample, diluting an appropriate amount with water, and filtering (0.45 um filter) prior to injection. HPAEC has also been used as a preparative technique to separate the oligosaccharides in beer.[39] Again, using a NaOH/NaOAC buffer, an effluant was collected from the column in 20 second fractions and subjected to further analysis. Oligosaccharides (composed of glucose, galactose, and xylose) released by the action of cellulase on seed xyloglucan were separated on a CarboPak PA100 column (Dionex, Sunnyvale, CA) using gradient elution of sodium hydroxide and sodium acetate buffers.[45] Oligosaccharides such as isomaltose, kojibiose, gentiobiose, nigerose, and maltose from different varieties of honey have also been separated using HPAEC-PAD.[48]

Not limited to the separation of neutral oligosaccharides, oligogalacturonic acids from strawberry juice have been separated on a HPAEC system using a sodium hydroxide gradient and a CarboPac PA-100 column.[23] Oligogalacturonic acids resulting from pectin enzymatic depolymerisation were separated on a Mono-Q anion exchange column (Pharmacia, Upsala, Sweden) using gradient elution (Na_2SO_4 in phosphate buffer) and detected on a photodiode array detector. The anion exchange resin permitted good resolution of oligomers up to Dp 13.[49]

The main disadvantage of using HPAEC-PAD for analysing oligosaccharides is the absence of adequate standards for many oligosaccharide types (i.e., oligosaccharides from β-glucan, malto-oligosaccharides over DP 7), which prevents their quantification. In addition, pulsed amperometric detector response is not equivalent for all sample types and, in fact, detector response decreases with the increase of DP;[19] therefore purified standards are required in order to facilitate accurate quantification.

2.4.4   Enzymatic Analysis

Enzymatic hydrolysis of oliogsaccharides coupled with chromatographic separation and identification of released oligosaccharides is commonly used to quantify oligosaccharides in both simple and complex food systems. This method has successfully been applied to the analysis of fructo-oligosaccharides and inulin (AOAC 997.08). Inulin and oligofructose are fructans, (2→1) linked β-D-fructofuranosyl units with or without a terminal glucose,[50] found naturally in chicory, Jeruselem artichoke, and onions. Both inulin and oligofructose are polydisperse mixtures with DP values ranging from 2 to 60 and 2 to 10, respectively.[51] These fructans have received much attention in recent years based on their postulated positive nutritional benefits, including improved laxation and the stimulation of beneficial gut microflora, and their potential uses as functional food additives.[50] Because of inulins' larger DP it is less soluble than oligofructose and able to form micro-crystals in solution that impart a fat-like mouthfeel, enabling it to function as a fat mimetic. Oligofructose is more soluble than inulin and sweet tasting, making it appropriate for use as a humectant in baked goods and a binder in granola bars with improved nutritional functionality vs. corn syrup.

Both inulin and oligofructose in a food product may be quantified by subjecting a hot water extract to an enzymatic treatment that permits fructan determination by difference (sugar content before and after treatment) (AOAC 997.08).[51,52] Free fructose and sucrose are determined in the original sample, free glucose and glucose from starch are determined after amyloglucosidase treatment, and total glucose and total fructose are determined after fructozym hydrolysis. Glucose and fructose from fructans are then determined by difference. HPAEC-PAD is ideally suited for the analysis of these enzyme hydrolysates because it enables baseline resolution of all sugars of interest and facilitates straightforward quantification.

Similarly, a method has been developed for the extraction, detection, and quantification of inulin in meat products. Inulin is extracted from meat products[53] in hot aqueous solvent, treated with inulinase, and the released fructose quantified by HPLC with RI detection. The quantity of inulin and oligofructose is determined by subtracting the quantitiy of free fructose (determined from a blank run without enzyme) from the total fructose in a sample after enzyme treatment.

Alternately, fructans can be determined as glucose and fructose after enzyme hydrolysis using chemical/spectrophotometric methods[54] (AACC method 32-32). This method uses *p*-hydroxybenzoic acid hydrazide (PAHBAH) to measure fructans as reducing sugars after treatment with a fructanase. As with other methods, fructans are extracted in hot water and the extract is treated with a succession of enzymes to cleave sucrose into glucose and fructose and starch into glucose. The released monosaccharides are reduced to sugar alcohols to avoid interfering in the fructan assay. Fructanase is added to release fructose and glucose from the fructans and the quantity of reducing sugars present in the solution is then determined using the *p*-hydroxybenzoic acid hydrazide method where color generated by the reaction is measured at 410 nm. Advantages of this method vs. chromatographic methods include its relatively fast speed and that expensive chromatographic equipment is not required. The disadvantage of this method is that fructan content will be underestimated if fructans have been extensively depolymerized because the reducing ends will be reduced with other sugars prior to fructanase addition and therefore omitted from quantification when the PAHBAH is added.

2.5   Dietary Fiber Analysis

2.5.1   Definition of Dietary Fiber

The past few decades have seen much discussion and debate over the definition and measurement of dietary fiber. The definition, as it existed in the early 1970s, was a response to the observed physiological behavior of plant cell wall material in the human digestive system and the postulated benefits of this behavior.[55] This definition held that dietary fiber was plant cell wall material that was impervious to the action of digestive enzymes. This definition was later expanded to include all indigestible polysaccharides such as gums, mucilages, modified celluloses, oligosaccharides, and pectins.[56] The expanded definition arose because these additional substances behaved physiologically in a similar manner to compounds included in the original definition; they were edible and not digested and absorbed in the small intestine. Clearly the definition of dietary fiber is intricately linked with the physiological behavior of food compounds, but scientific methodology has tended to overshadow physiological function as the basis for definition.[57] The challenge over the past few years has been to arrive at a widely accepted (by nutrition professionals, regulators, and scientists) definition that reflects the physiological function of dietary fiber and to pursue methodologies that facilitate its quantitative analysis. The official AACC definition of dietary fiber[57] is as follows;

> Dietary fiber is the edible parts of plants or analogous carbohydrates that are resistant to digestion and absorption in the human small intestine with complete or partial fermentation in the large intestine. Dietary fiber includes polysaccharides, oligosaccharides, lignin, and associated plant substances. Dietary fibers promote beneficial physiological effects including laxation, and/or blood cholesterol attenuation, and/or blood glucose attenuation.

Other definitions have been put forth that differentiate between intact, intrinsic fiber in plants termed dietary fiber (i.e., that found in fruits or cereal grains) and which have been isolated or synthesized and added to a product, termed functional or added fiber.[58] Still others have argued that only plant cell wall material should be considered dietary fiber,[59,60] excluding such components as resistant starch and other noncell wall nonstarch polysaccharides. Despite the above, it is generally accepted that what constitutes dietary fiber are those compounds outlined in the AACC definition above.

Explicit in the definition above is the expectation of beneficial physiological consequences based on the ingestion of dietary fiber. Dietary fiber intake has been associated with reducing blood cholesterol levels, reducing the risk of coronary heart disease, attenuating postprandial blood glucose response, and improving laxation. Because dietary fiber is a term representing a fairly heterogeneous group of compounds, it is not expected that every type of dietary fiber will play a role in each of these beneficial physiological consequences. For example, insoluble fiber such a cellulose is associated with increasing fecal bulk and improving laxation,[61,62] while viscous dietary fiber, also termed soluble fiber, which includes β-glucan, psyllium gum, and pectin, is associated with lowering cholesterol[32,33,63,64] and moderating postprandial blood glucose levels.[31,65-67] There are several postulated mechanisms for the effects of dietary fiber. Soluble fiber is believed to bind cholesterol and bile acids, increasing their excretion.[68] Alternately, it has been suggested that short chain fatty acids produced by bacterial fermentation of fiber may beneficially affect lipid metabolism.[68,69] With respect to plasma glucose levels, it is thought that viscosity development in the gut caused by the presence of soluble fiber slows the movement of glucose into the blood stream, flattening the blood glucose peak that usually occurs after a meal. Included in the group termed viscous dietary fiber is β-glucan, psyllium gum, and guar gum. Despite this, the distinction between soluble and insoluble fiber in terms of physiological function is not clear, and in fact, there is support for the abandonment of the terms soluble and insoluble altogether.[58] For example, inulin and oligofructose are soluble dietary fibers that have a demonstrated ability to increase fecal weight[70] despite the fact they are soluble fibers. Additionally, many fiber sources contain both soluble and insoluble fiber and therefore provide the benefits associated with both types of fiber.

Given the nutritional importance of dietary fiber, there must be an adequate standardized method for determining the dietary fiber content in food products. This is important from a regulatory perspective, ensuring nutrition labels are reporting accurate data and that all those compounds behaving as a dietary fiber (are eaten, impervious to digestion in the small intestine, promote laxation, and/or attenuation of blood cholesterol and/or blood glucose levels) are captured and counted as dietary fiber in the analysis. For several reasons this has proven a less than simple task. Compounds that behave similarly physiologically do not necessarily have analogous solubility properties. For example, in what is probably the most used and widely recognized dietary fiber method (AOAC 985.29) there is an ethanol precipitation step (1 part sample solution, 4 parts ethanol) intended to separate those compounds considered digestible (glucose from starch hydrolysis, amino acids from protein hydrolysis, simple sugars) and therefore excluded from what is considered dietary fiber, and the indigestible residue. There are compounds, such as fructans which are partially soluble in 80% ethanol[71] and as such would be excluded from the analysis despite the fact research has suggested they are able to increase fecal bulk[72] and reduce serum lipid levels.[73,74] Polydextrose (a highly branched glucose polymer) and galacto-oligosaccharides have also historically been excluded from dietary fiber measurements based on their solubility in 80% ethanol. Realization of this fact necessitated the development of methods for quantifying fructo-oligosaccharides, inulin, polydextrose, and galacto-oligosaccharides.

2.5.2 Dietary Fiber Analysis

2.5.2.1 *Uppsala Method*

There are two fundamentally different approaches to analyzing dietary fiber. The Uppsala method[75] (AOAC 994.13; AACC 32-25) requires measuring the neutral sugars, uronic acids (pectic material), and Klason lignin (noncarbohydrate dietary fiber including native lignin, tannins, and proteinatous material) and summing these components to obtain a dietary fiber value (Figure 2.9). The sample is first subjected to an amylase and amyloglucosidase digestion to remove starch. Starch hydrolysates and low molecular weight sugars are separated from soluble fiber using an 80% ethanol precipitation, leaving a residue containing both soluble and insoluble fiber. Neutral sugars are determined after derivitisation as their alditol acetates by GC, uronic acids are assayed colorimetrically, and Klason lignin is determined gravimetrically.

2.5.2.2 *Enzymatic/Gravimetric Methods*

The second method is a measure by difference where dietary fiber residue is isolated, dried, weighed, and then this weight is adjusted for nondietary fiber material (i.e., protein and ash). Enzymatic-gravimetric methods such as official AOAC method 985-29 involve subjecting a sample to a succession of enzymes (α-amylase, amyloglucosidase, protease) to remove digestible material, an ethanol precipitation step to isolate nonstarch polysaccharide, and finally an ash and protein determination (Figure 2.10). The quantity of protein and ash is subtracted from the dry residue weight to obtain a total dietary fiber value. This method has been modified to permit the determination of total soluble and insoluble dietary fiber (AOAC 991.43).

With appropriate methods in place for the measurement of the majority of dietary fiber constituents, recent efforts have focused on establishing official methods for classes of compounds that behave physiologically like dietary fiber, but are soluble in 80% ethanol and hence excluded from commonly used official methods for measuring dietary fiber. These compounds include fructans (AOAC official method 997.08 or AOAC method 999.03), which are found in onions, leeks, chicory, and Jeruselum artichoke. Polydextrose, also soluble in 78% ethanol owing to its extensively branched structure, can be analyzed using AOAC method 2000.11. Fructans include both inulin and oligofructose, which rather than being single molecular species, are polydisperse mixtures of fructose polymers. Oligofructose has a DP range from

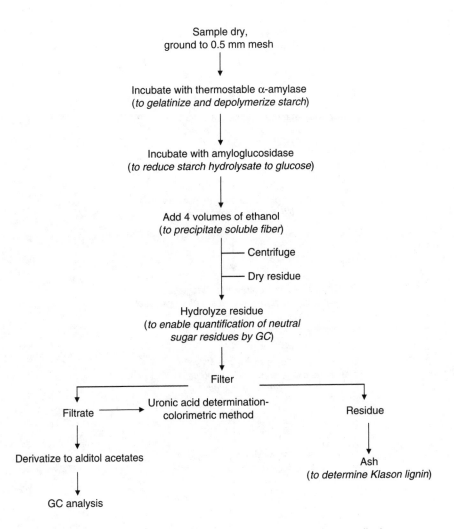

Total dietary fiber = neutral sugar residues + uronic acid residues + lignin

FIGURE 2.9
Total dietary fiber procedure — Uppsala method. (Adapted from AACC method 32–25.)

~2 to 10, while inulin typically ranges from ~2 to 60, but there is variation in this range depending on the inulin source. Because the DP of these compounds varies, their solubility in 80% ethanol varies, and therefore their inclusion in total dietary fiber (TDF) measurements is not complete.[76] In order to avoid overestimating the contribution of fructans by virtue of counting them twice, the inclusion of an inulinase in the TDF procedure will completely exclude fructans from being counted as a dietary fiber. Inulin and oligofructose can then be determined in a separate analytical procedure and their quantity as dietary fiber added to the TDF determined using

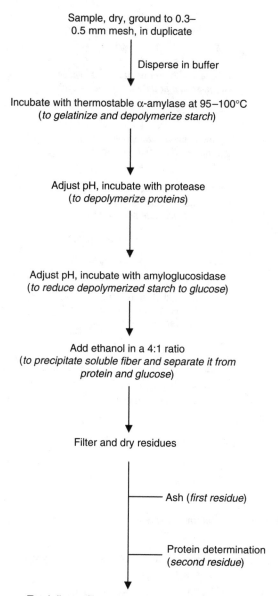

Sample, dry, ground to 0.3–
0.5 mm mesh, in duplicate

Disperse in buffer

Incubate with thermostable α-amylase at 95–100°C
(*to gelatinize and depolymerize starch*)

Adjust pH, incubate with protease
(*to depolymerize proteins*)

Adjust pH, incubate with amyloglucosidase
(*to reduce depolymerized starch to glucose*)

Add ethanol in a 4:1 ratio
(*to precipitate soluble fiber and separate it from
protein and glucose*)

Filter and dry residues

Ash (*first residue*)

Protein determination
(*second residue*)

Total dietary fiber = dry residue–ash-protein

FIGURE 2.10
Total dietary fiber procedure — enzymatic/gravimetric method. (Adapted from AACC method
32–05.)

conventional procedures.[77] Fructans are determined in food products as their
constituent monosaccharides after extraction in aqueous solvent and enzymatic hydrolysis. Released monosaccharides are determined using spectrophotometric or HPLC methods (see Section 2.4 for details).

2.5.3 Summary of Dietary Fiber Analysis

Given the nutritional implications of dietary fiber intake, its quantitative determination in food products is very important, but the measurement of dietary fiber has been hampered by controversies surrounding its very definition and the fact that dietary fiber encompasses a class of compounds with disparate properties. There are now several methods in place that permit the measurement of TDF and additional methods to specifically measure those compounds (i.e., polydextrose and fructans) excluded from conventional dietary fiber methods.

References

1. BeMiller, J. N. and Low, N. H. Carbohydrate analysis. Nielson, S. Suzanne, Ed. *Food Analysis*. 167–187. Gaithersburg, MD, Aspen Publishers. 1998.
2. Harris, P., Morrison, A., and Dacombe, C. A practical approach to polysaccharide analysis. Stephan, A. M., Ed. *Food Polysaccharides and Their Application*. 463–500. New York, Marcel Dekker Inc. 1995.
3. Pomeranz, Y. and Meloan, C. E. Carbohydrates. Pomeranz, Y. and Meloan, C. E., Eds. *Food Analysis: Theory and Practice*. 625. New York, Chapman & Hall. 1994.
4. Dubois, M., Gilles, K. A., Hamilton, J. K., Rebers, P. A., and Smith, F. Colorimetric method for determination of sugars and related substances. *Analytical Chemistry*, 28[3], 350–356. 1956.
5. Dische, Z. Color reactions of hexoses. Whistler, Roy L. and Wolfrom, M. L., Eds. *Methods in Carbohydrate Chemistry*. 1, 488–494. New York, Academic Press Inc. 1962.
6. Southgate, D. A. T. *Determination of Food Carbohydrates*. 75–84. Essex, England, Applied Science Publishers Ltd. 1976.
7. Laurentin, A. and Edwards, C. A. A microtiter modification of the anthrone-sulfuric acid colorimetric assay for glucose-based carbohydrates. *Analytical Biochemistry*, 315, 143–145. 2003.
8. Dische, Z. Color reactions of hexuronic acids. Whistler, R. L. and Wolfrom, M. L., Eds. *Methods in Carbohydrate Chemistry*, 1, 497–501. New York, Academic Press Inc. 1962.
9. Taylor, K. A. and Buchanan-Smith, J. G. A colorimetric method for the quantitation of uronic acids and a specific assay for galacturonic acid. *Analytical Biochemistry*, 201, 190–196. 1992.
10. Blumenkrantz, N. and Asboe-Hansen, G. New method for qunatitative determination of uronic acids. *Analytical Biochemistry*, 54, 484–489. 1973.
11. Kinter, P. K. and Van Buren, J. P. Carbohydrate interference and its correction in pectin analysis using the *m*-hydroxydiphenyl method. *Journal of Food Science*, 47, 756–764. 1982.
12. van den Hoogen, B. M., van Weeren, P. R., Lopes-Cardozo, M., van Golde, L. M. G., Barneveld, A. , and van de Lest, C. H. A. A microtiter plate assay for the determination of uronic acids. *Analytical Biochemistry*, 257, 107–111. 1998.
13. Meseguer, I., Aguilar, V., González, M. J., and Martínez, C. Extraction and colorimetric quantification of uronic acids of the pectic fraction in fruits and vegetables. *Journal of Food Composition and Analysis*, 11, 285–291. 1998.

14. Garleb, K. A., Bourquin, L. D., and Fahey, G. C. Jr. Neutral monosaccharide composition of various fibrous substrates: a comparison of hydrolytic procedures and use of anion-exchange high-performance liquid chromatography with pulsed amperometric detection of monosaccharides. *Journal of Agricultural and Food Chemistry*, 37, 1287–1293. 1989.

15. Pazur, J. H. Neutral polysachharides. Chaplin, M. F. and Kennedy, J. F., Eds. *Carbohydrate Analysis: a Practical Approach*. 55-96. 1986. Oxford, IRL Press.

16. Englyst, H. N. and Cummings, J. H. Simplified method for the measurement of total non-starch polysaccharides by gas-liquid chromatography of constituent sugars as aldelol acetates. *Analyst*, 109, 937–942. 1984.

17. Sloneker, J. H. Gas-liquid chromatography of alditol acetates. Whistler, R. L. and BeMiller, J. N., Eds. *Methods in Carbohydrate Chemistry*, 6, 20–24. New York, Academic Press, Inc. 1972.

18. Petkowicz, C. L. O., Sierakowski, M. R., Lea, J., Ganter, M. S., and Reicher, F. Galactomannans and arabinans from seeds of caesalpiniaceae. *Phytochemistry*, 49, 737–743. 1998.

19. Henshall, A. High performance anion exchange chromatography with pulsed amperometric detection (HPAEC-PAD): a powerful tool for the analysis of dietary fiber and complex carbohydrates. Cho, S. S., Prosky, L, and Dreher, M., Eds. *Complex Carbohydrates in Foods*, 267–289. New York, Marcel Dekker, Inc. 1999.

20. Andersen, R. and Sorensen, A. Separation and determination of alditols and sugars by high-pH anion-exchange chromatography with pulsed amperometric detection. *Journal of Chromatography A*, 897, 195–204. 2000.

21. Cataldi, T. R. I., Margiotta, G., and Zambonin, C. G. Determination of sugars and alditols in food samples by HPAEC with integrated pulsed amperometric detection using alkaline eluents containing barium or strontium ions. *Food Chemistry*, 62, 109–115. 1998.

22. De Ruiter, G. A., Schols, H. A., Voragen, A. G. J., and Rombouts, F. M. Carbohydrate analysis of water-soluble uonic acid containing polysaccharides with high-performance anion-exchange chromatography using methanolysis combined with TFA hydrolysis is superior to four other methods. *Analytical Biochemistry*, 207, 176–185. 1992.

23. Versari, A., Biesenbruch, S., Barbanti, D., Farnell, P. J., and Galassi, S. HPAEC-PAD analysis of oligogalacturonic acids in strawberry juice. *Food Chemistry*, 66, 257–261. 1999.

24. McCleary, B. V. and Codd, R. Measurement of (1-3)(1-4)-β-D-glucan in barley and oats: a streamlined enzymic procedure. *Journal of the Science of Food and Agriculture*, 55, 303–312. 1991.

25. McCleary, B. V., Gibson, T. S., and Mugford, M. D. C. Measurement of total starch in cereal products by amyloglucosidase-α-amylase method: Collaborative study. *Journal of the Association of Official Analytical Chemistry*, 80, 571–579. 1997.

26. McCleary, B. V. Galactomannan quantitation in guar varieties and seed fractions. *Lebensmittel-Wissenschaft and Technologie*, 14, 188–191. 1981.

27. McCleary, B. V. Enzymatic determination of galactomannans. BeMiller, J. N., Ed. *Methods in Carbohydrate Chemistry*. X, 85–90. New York, John Wiley & Sons, Inc. 1994.

28. Hough, L. and Jones, J. K. N. Enzymic methods for determination of D-glucose. Whistler, R. L. and Wolfrom, M. L., Eds. *Methods in Carbohydrate Chemistry.* 1, 400–404. New York, Academic Press. 1962.

29. Sturgeon, R. J. Enzymatic determination of D-galactose. BeMiller, J. N., Ed. *Methods in Carbohydrate Chemistry.* X, 15–16. New York, John Wiley & Sons, Inc. 1994.

30. Mustranta, A. and Östman, C. Enzymatic determination of lactose and galactose in foods: NMKL collaborative methods performance study. *Journal of AOAC International* 80, 584–590. 1997.

31. Wood, P. J., Braaten, J. T., Scott, F. W., Riedel, K. D., Wolynetz, M. S., and Collins, M. W. Effect of dose and modifications of viscous properties of oat gum on plasma glucose and insulin following an oral glucose load. *British Journal of Nutrition,* 72, 731–743. 1994.

32. Braaten, J. T., Wood, P. J., Scott, F. W., Wolynetz, M. S., Lowe, M. K., Bradley-White, P., and Collins, M. W. Oat β-glucan reduces blood cholesterol concentration in hypercholesterolemic subjects. *European Journal of Clinical Nutrition,* 48, 465–472. 1994.

33. Maki, K. C., Davidson, M. H., Ingram, K. A., Veith, P. E., Bell, M. L., and Gugger, E. Lipid response to consumption of a beta-glucan containing ready-to-eat cereal in children and adolescents with mild-to-moderate primary hypercholesterolemia. *Nutrition Research,* 23, 1527–1535. 2003.

34. McCleary, B. V. and Nurthen, E. J. Measurement of (1-3)(1-4)-β-D-glucan in malt, wort and beer. *Journal of the Institute of Brewing,* 92, 168–173. 1986.

35. Schorsch, C. , Garnier, C., and Doublier, J. L. Viscoelastic properties of xanthan/galactomannan mixtures: comparison of guar gum with locust bean gum. *Carbohydrate Polymers,* 34, 165–175. 1997.

36. McCleary, B. V., Clark, A. H., Dea, I. C. M., and Rees, D. A. The fine structure of carob and guar galactomannans. *Carbohydrate Research,* 139, 237–260. 1985.

37. Anonymous, IUB-IUPAC, and Joint Commission on Biochemical Nomenclature. Abbreviated Terminology of Oligosaccharide Chains. *Journal of Biological Chemistry,* 257, 3347–3351. 1982.

38. Van Loo, J., Cummings, J., Delzenne, N., Englyst, H., Franck, A., Hopkins, M., Kok, N., Macfarlane, G., Newton, D., Quigley, M., Roberfroid, M., van Vliet, T., and van den Heuvel, E. Functional food properties of non-digestible oligosaccharides: a consensus report from the ENDO project (DGXII-CT94-1095). *British Journal of Nutrition,* 81, 121–132. 1999.

39. Vinogradov, E. and Bock, K. Structural determination of some new oligosaccharides and anlysis of the branching pattern of isomaltooligosaccharides from beer. *Carbohydrate Research,* 309, 57–64. 1998.

40. White, D. R., Hudson, P., and Adamson, J. T. Dextrin characterization by high-performance anion-exchange chromatography-pulsed amperometric detection and size-exclusion chromatography-multi-angle light scattering-refractive index detection. *Journal of Chromatography. A,* 997, 79–85. 2003.

41. Singh, V. and Ali, S. Z. Acid degradation of starch. The effect of acid and starch type. *Carbohydrate Polymers,* 41, 191–195. 2000.

42. Caram-Lelham, N., Sundelöf, L. O., and Andersson, T. Preparative separation of oligosaccharides from κ–carrageenan, sodium hyaluronate, and dextran by Superdex™ 30 prep. grade. *Carbohydrate Research,* 273, 71–76. 1995.

43. Knutsen, S. H., Sletmoen, M., Kristensen, T., Barbeyron, T., Kloareg, B., and Potin, P. A rapid method for the separation and analysis of carrageenan oligosaccharides released by iota- and kappa-carrageenase. *Carbohydrate Research*, 331, 101–106. 2001.
44. Schröder, R., Nicolas, P., Vincent, S. J. F., Fischer, M., Reymond, S., and Redgwell, R. J. Purification and characterisation of a galactoglucomannan from kiwifruit (*Actinidia deliciosa*). *Carbohydrate Research*, 331, 291–306. 2001.
45. Tiné, M. A. S., de Lima, D. U., and Buckeridge, M. S. Galactose branching modulates the action of cellulase on seed storage xyloglucan. *Carbohydrate Polymers*, 52, 135–141. 2003.
46. Vierhuis, E., York, W. S., Kolli, V. S. K., Vincken, J. P. , Schols, H. A., Van Alebeek, G. J. W. M., and Voragen, A. G. J. Structural analyses of two arabinose containing oligosaccharides derived from olive fruit xyloglucan: XXSG and XLSG. *Carbohydrate Research*, 332, 285–297. 2001.
47. Churns, S. C. Recent progress in carbohydrate separation by high-performance liquid chromatography based on size exclusion. *Journal of Chromatography A*, 720, 151–166. 1996.
48. Weston, R. J. and Brocklebank, L. K. The oligosaccharide composition of some New Zealand honeys. *Food Chemistry*, 64, 33–37. 1999.
49. Deconinck, T. J.-M., Ciza, A., Sinnaeve, G. M., Laloux, J. T., and Thonart, P. High-performance anion-exchange chromatograpy-DAD as a tool for the identification and quantification of oligogalacturonic acids in pectin depolymerisation. *Carbohydrate Research*, 329, 907–911. 2000.
50. Niness, K. R. Inulin and oligofructose: What are they? *Journal of Nutrition*, 129, 1402S–1406S. 1999.
51. Prosky, L. and Hoebregs, H. Methods to determine food inulin and oligofructose. *Journal of Nutrition*, 129, 1418S–1423S. 1999.
52. Quemener, B. , Thibault, J., and Coussement, P. Determination of inulin and oligofructose in food products, and integration in the AOAC method for measurement of total dietary fiber. *Lebensmittel-Wissenschaft un-Technologie*, 27, 125–132. 1994.
53. Vendrell-Pascuas, S., Castellote-Bargalló, A. I. , and López-Sabater, M. C. Determination of inulin in meat products by high-performance liquid chromatography with refractive index detection. *Journal of Chromatography A*, 881, 591–597. 2000.
54. McCleary, B. V., Murphy, A., and Mugford, D. C. Measurement of oligofructan and fructan polysaccharides in foodstuffs by an enzymic/spectrophotometric method: Collaborative study. *Journal of the Association of Official Analytical Chemists*, 83, 356–364. 2000.
55. Trowell, H. C. Definition of dietary fiber. *The Lancet*, 303, 503. 1974.
56. Trowell, H. C., Southgate, D. A. T., Wolever, T. M. S., Leeds, A. R., Gassull, M. A., and Jenkins, D. J. A. Dietary fiber redefined. *The Lancet*, 307, 967. 1976.
57. Annonymous. The definition of dietary fiber. *Cereal Foods World* 46, 112–126. 2001.
58. Slavin, J. Impact of the proposed definition of dietary fiber on nutrient databases. *Journal of Food Composition and Analysis*, 16, 287–291. 2003.
59. Englyst, H. N. and Hudson, G. J. The classification and measurement of dietary carbohydrates. *Food Chemistry*, 57, 15–21. 1996.
60. Goodlad, R. A. and Englyst, H. N. Redefining dietary fiber: potentially a recipe for disaster. *The Lancet*, 358. 2001.

61. Stephen, A. M., Dahl, W. J., Johns, D. M., and Englyst, H. N. Effect of oat hull fiber on human colonic function and serum lipds. *Cereal Chemistry*, 74, 379–383. 1997.

62. Spiller, R. C. Pharmacology of dietary fiber. *Pharmacology and Therapeutics*, 62, 407–427. 1994.

63. Brown, L., Rosner, B., Willet, W., and Sacks, F. M. Cholesterol lowering effects of dietary fiber: a meta-analysis. *American Journal of Clinical Nutrition*, 69, 30–42. 1999.

64. Jenkins, D. J. A., Kendall, C. W. C., Vuksan, V., Vidgen, E., Parker, T., Faulkner, D., Mehling, C., Garsetti, M., Testolin, G., Cunnane, S. C., Ryan, M. A., and Corey, P. N. Soluble fiber intake at a dose approved by the US Food and Drug Administration for a claim of health benefits: serum lipid risk factors for cardiovascular disease assessed in a randomized controlled crossover trial. *American Journal of Clinical Nutrition*, 75, 834–839. 2002.

65. Bourdon, I., Yokoyama, W., Davis, P., Hudson, C., Backus, R., Richter, D., Knuckles, B., and Schneeman, B. O. Postprandial lipid, glucose, insulin, and cholecystokinin responses in men fed barley pasta enriched with β-glucan. *American Journal of Clinical Nutrition*, 69, 55–63. 1999.

66. Lu, Z. X., Walker, K. Z., Muir, J. G., Mascara, T., and O'Dea, K. Arabinoxylan fiber, a byproduct of wheat flour processing, reduces the postprandial glucose response in normoglycemic subjects. *American Journal of Clinical Nutrition*, 71, 1123–1128. 2000.

67. Yokoyama, W. , Hudson, C. A., Knuckles, B., Chiu, M. M., Sayre, R. N., Turnlund, J. R., and Scheeman, B. O. Effect of barley β-glucan in durum wheat pasta on human glycemic response. *Cereal Chemistry*, 74, 293–296. 1997.

68. Jenkins, J. A., Kendall, C. W. C., and Ransom, T. P. P. Dietary fiber, the evolution of the human diet and coronary heart disease. *Nutrition Research* 18, 633–652. 1998.

69. Bartnikowska, E. The role of dietary fiber in the prevention of lipid metabolism disorders. Sungsoo Cho, S., Prosky, L., and Dreher, M., Eds. *Complex Carbohydrates in Foods*. 53–62. New York, Marcel Dekker, Inc. 1999.

70. Flamm, G., Glinsmann, W., Kritchevsky, D., Prosky, L., and Roberfroid, M. Inulin and oligofructose as dietary fiber: a review of the evidence. *Critical Reviews in Food Science and Nutrition*, 41, 353–362. 2001.

71. Ku, Y., Jansen, O., Oles, C. J., Lazar, E. Z., and Rader, J. I. Precipitation of inulins and oligoglucoses by ethanol and other solvents. *Food Chemistry*, 81, 125–132. 2003.

72. Jenkins, D. J. A., Kendall, C. W. C., and Vuksan, V. Inulin, oligofructose and intestinal function. *Journal of Nutrition*, 129, 1431S–1433S. 1999.

73. Brighenti, F., Casiraghi, M. C., Canzi, E., and Ferrari, A. Effect of consumption of a ready-to-eat brakfast cereal containing inulin on the intestinal milieu and blood lipids in healthy male volunteers. *European Journal of Clinical Nutrition*, 53, 726–733. 1999.

74. Causey, J. L., Feirtag, J. M., Gallaher, D. D., Tungland, B. C., and Slavin, J. L. Effect of dietary inulin on serum lipids, blood glucose and the gastrointetinal environment in hypercholesterolemic men. *Nutrition Research*, 20, 191–201. 2000.

75. Theander, O., man, P., Westerlund, A. R., and Pettersson, D. Total dietary fiber determined as neutral sugar residues, uronic acid residues, and klason lignin (the uppsala method): Collaborative study. *Journal of AOAC International*, 78, 1030–1044. 1995.

76. Coussement, P. Inulin and oligofructose as dietary fiber: analytical, nutritional and legal aspects. Sungsoo Cho, S., Prosky, L., and Dreher, M., Eds. *Complex Carbohydrates in Foods*. 203–212. New York, Marcel Dekker, Inc. 1999.
77. Dysseler, P., Hoffem, D., Fockedey, J., Quemener, B., Thibault, J. F., and Coussement, P. Determination of inulin and oligofructose in food products (modified AOAC dietary fiber method). Sungsoo Cho, S., Prosky, L., and Dreher, M., Eds *Complex Carbohydrates in Foods*. 213–227. New York, Marcel Dekker, Inc. 1999.

3

Structural Analysis of Polysaccharides

Steve W. Cui

CONTENTS

3.1   Introduction: Strategy for Polysaccharide Structural Analysis

The structure analysis of polysaccharides requires specialized techniques, which differ significantly from those methods used for small molecules and other biopolymers. To understand the primary structure of a polysaccharide, the following information is essential:

- Monosaccharide composition: nature and molar ratios of the monosaccharide building blocks
- Linkage patterns: linkage positions between the glycosidic linkages and branches
- Ring size: distinction of furanosidic and pyranosidic rings
- Anomeric configuration: α- or β-configuration of the glycosidic linkage
- Sequences of monosaccharide residues and repeating units;
- Substitutions: position and nature of OH–modifications, such as O–phosphorylation, acetylation, O–sulfation, etc.
- Molecular weight and molecular weight distribution (covered in Chapter 4)

Since structural analysis of polysaccharides is a complex and demanding task, a good strategy is necessary before starting any experiments. Figure 3.1 summarizes the necessary steps frequently used for elucidating a detailed structure of a polysaccharide.

A polysaccharide extracted from plant materials or food products is usually purified before being subjected to structural analysis (see Chapter 1 for details). The first step of characterizing a polysaccharide is the determination of its purity, which is reflected by its chemical composition, including total sugar content, levels of uronic acid, proteins, ash, and moisture. Colorimetric methods described in Chapter 2 are suitable for estimating the contents of

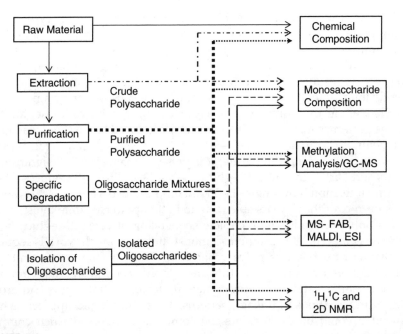

FIGURE 3.1
Strategy and methods for structure analysis of polysaccharides.

total sugars and uronic acids. The second step is the determination of monosaccharide composition, which will unveil structural information such as the number of monosaccharides present in the polysaccharide and how many of each sugar unit. Monosaccharide composition is determined by analyzing monosaccharides using high performance liquid chromatography (HPLC) or gas-liquid chromatography (GC) after a complete acid hydrolysis (see Chapter 2 for details).

The ring size and glycosidic linkage positions of sugar units in a polysaccharide is established by methylation analysis and/or cleavage reduction. The anomeric configuration is conventionally determined by chromium trioxide oxidation,[1] and this method can be combined with modern techniques, such as fast atom bombardment (FAB) Mass Spectrometry to afford more structural information.[2] Two dimensional (2D) Nuclear Magnetic Resonance (NMR) spectroscopy has proved to be the method of choice for this purpose because of the characteristic chemical shifts and coupling constants of the anomeric signals (both ^{1}H and ^{13}C) and wide availability of the instrument. The use of available and well characterized highly specific and purified enzymes can also give information leading to linkage position and configuration of the hydrolyzed sugar residues.

The relative position of each individual sugar, including its linkage derivatives, unfolds the sequence of polysaccharide chains. Sequencing of polysaccharides is difficult to achieve because of the heterogeneous nature of the polysaccharide structure, high molecular weight, and polydispersity of the polymer chains. Naturally occurring polysaccharides usually have structural regularity, and some have repeating units. Cellulose is a simple polysaccharide that has only one single sugar repeating unit (i.e., 1,4-linked-β-D-glucosyl residue). However, in most cases, the repeating units are frequently interrupted by other sugar units or have substitution groups and/or branches along the polymer chains. For example, with cellulose, another linkage pattern, e.g., a (1→3)-linkage, may be inserted to the cellulose molecular chain to form a (1→3) (1→4)-mix-linked β-D-glucan. This type of polysaccharide is found in nature; it is present in cereals, most significantly in oats and barley. In cereal β-D-glucans, the (1→4)-linked-β-D-glucosyl residues are interrupted by single (1→3)-linked-β-D-glucosyl residues. Although it has been over thirty years since the first attempt to elucidate the structure of this polysaccharide, the complete sequencing of cereal β-D-glucans has yet to be achieved. This example demonstrates how much work is required to solve the structure of a polysaccharide that only has one sugar unit with just two different linkages. In nature, many polysaccharides contain multiple monosaccharides and some may contain up to six neutral sugars and uronic acids with variations in linkage patterns. Thus, one should appreciate how difficult it could be to elucidate the complete structure of such complex polysaccharides.

Thanks to the rapid development of modern technologies during the last three decades, many techniques, including fast atom bombardment mass spectrometry (FAB-MS), matrix-assistant laser desorption ionization (MALDI-MS) and electrospray ionization (ESI-MS) spectrometry and one and two- (multi)-dimensional NMR spectroscopy have been developed. These modern techniques and methodologies have been shown to be extremely powerful for solving the structural problems of polysaccharides. In addition, numerous highly specific and purified enzymes have become readily available. All of these advances in science and technology have made the structural analysis of polysaccharide much easier. The primary goal of this chapter is to present the current methodologies used for structural analysis of polysaccharides.

3.2   Determination of Linkage Pattern: Methylation Analysis, Reductive Cleavage, and Peroxidation

3.2.1   Methylation Analysis

Methylation analysis has been used to determine the structure of carbohydrate for over a century and it is still the most powerful method in carbohydrate structural analysis.[3-7] Current methylation analysis consists of two steps:

- Chemical derivatization
- Gas-liquid chromatograph–mass spectroscopy (GC-MS)

3.2.1.1 Methylation Reaction

The derivatization of a polysaccharide for methylation analysis includes conversion of all free hydroxyl groups into methoxyls followed by acid hydrolysis. Acidic hydrolysis of the resulting poly-methyl-ethers only cleaves the inter-glycosidic linkages and leaves the methyl-ether bonds intact. The hydrolyzed monomers are reduced and acetylated to give volatile products, i.e., partially methylated alditol acetate (PMAA), which can be identified and quantitatively determined by gas-liquid chromatography equipped with a mass spectroscopic detector (GC-MS). Figure 3.2 depicts the procedures and chemical reactions involved in methylation analysis. The substitution pattern of the O-acetyl groups on the PMAA reflects the linkage patterns and ring sizes of the corresponding sugar in the original polymer. However, this method gives no information on sequences or the anomeric configuration of the glycosidic linkages. In addition, this method cannot distinguish whether an alditol is derived from a 4–O–linked aldopyranose or the corresponding 5–O–linked aldofuranose. These drawbacks can be overcome by a method called reductive cleavage, which is described in Section 3.2.2.

Experimentally, the reaction of converting the hydroxyl groups into methoxyls requires an alkaline environment and methyl group provider. Silver oxide–methyl iodide[3] and sodium hydroxide–methyl sulphate[4] were used in the past. These methods were then replaced by dimsyl sodium (sodium methylsulfinymethanide) and methyl iodide.[5] More recently a simpler procedure using dry powdered sodium hydroxide and methyl iodide has been adapted.[6] This method has been modified by using a sodium hydroxide suspension in dry dimethyl sulfoxide (DMSO).[7] A procedure adapted from Ciucanu's method is prepresented in Figure 3.3.

A prerequisite for the methylation reaction is complete solubilization of polysaccharides in DMSO. This solubilization can be achieved by constant stirring and/or ultrasonic treatment at elevated temperatures (up to 70°C). Incomplete solubilization will cause undermethylation, which subsequently leads to incorrect conclusions about the structure. The undermethylation is usually caused by incomplete methylation of the insoluble portion of the polymer. The methylated polysaccharide is recovered by partition between water and methylene chloride or by dialysis. The subsequent hydrolysis of methylated polysaccharide is carried out in mineral acids. Trifluoroacetic acid (TFA) is frequently used since it can be easily removed by evaporation. Typically, a methylated polymer is hydrolyzed in 4.0 M TFA at 100°C or 120°C for 4 to 6 hours and the TFA is removed by evaporation under a stream of nitrogen. Being an inert gas, nitrogen can help to prevent undesirable chemical reactions, such as oxidation. After hydrolysis, the hydrolysate

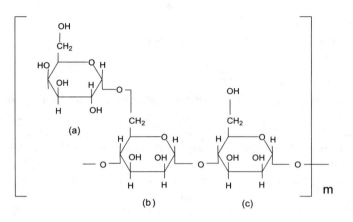

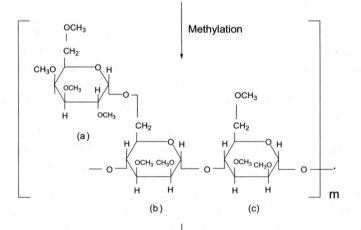

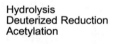

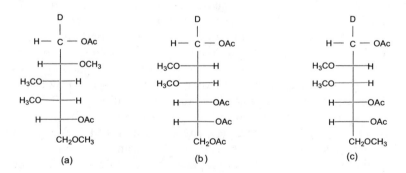

FIGURE 3.2
Illustrations of chemical reaction in methylation analysis.

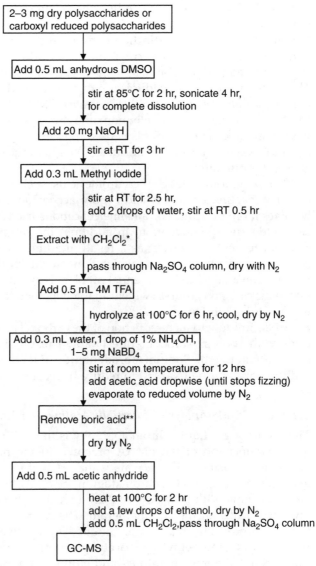

```
┌─────────────────────────────────┐
│ 2–3 mg dry polysaccharides or   │
│ carboxyl reduced polysaccharides│
└─────────────────────────────────┘
                │
┌─────────────────────────────────┐
│ Add 0.5 mL anhydrous DMSO       │
└─────────────────────────────────┘
                │  stir at 85°C for 2 hr, sonicate 4 hr,
                │  for complete dissolution
        ┌─────────────────────┐
        │ Add 20 mg NaOH      │
        └─────────────────────┘
                │  stir at RT for 3 hr
      ┌───────────────────────────┐
      │ Add 0.3 mL Methyl iodide  │
      └───────────────────────────┘
                │  stir at RT for 2.5 hr,
                │  add 2 drops of water, stir at RT 0.5 hr
        ┌─────────────────────────┐
        │ Extract with CH₂Cl₂*    │
        └─────────────────────────┘
                │  pass through Na₂SO₄ column, dry with N₂
        ┌──────────────────────┐
        │ Add 0.5 mL 4M TFA    │
        └──────────────────────┘
                │  hydrolyze at 100°C for 6 hr, cool, dry by N₂
   ┌────────────────────────────────────┐
   │ Add 0.3 mL water,1 drop of 1% NH₄OH,│
   │          1–5 mg NaBD₄              │
   └────────────────────────────────────┘
                │  stir at room temperature for 12 hrs
                │  add acetic acid dropwise (until stops fizzing)
                │  evaporate to reduced volume by N₂
        ┌─────────────────────┐
        │ Remove boric acid** │
        └─────────────────────┘
                │  dry by N₂
     ┌───────────────────────────┐
     │ Add 0.5 mL acetic anhydride│
     └───────────────────────────┘
                │  heat at 100°C for 2 hr
                │  add a few drops of ethanol, dry by N₂
                │  add 0.5 mL CH₂Cl₂,pass through Na₂SO₄ column
            ┌──────────┐
            │  GC-MS   │
            └──────────┘
```

* transfer the mixture to a 5 ml vial using glass pipette, wash reaction vial with 1 ml CH₂Cl₂ three times; use deionized water (3–5 ml) to wash the CH₂Cl₂ solutions three times

** add 0.5 ml 5% acetic acid in methanol, evaporate, add 0.5 ml methanol, evaporate (three times each).

FIGURE 3.3
A widely adapted methylation procedure. (Modified from Ciucanu and Kerek, 1984.)[6]

should be brought to neutral pH if mineral acids, rather than TFA, are used. For example, barium carbonate can be used to neutralize sulfuric acid and the barium sulphate precipitate can be removed by centrifugation or filtration. The monosaccharides obtained by hydrolysis are reduced to alditols by treating the hydrolysate with sodium borodeuteride under alkaline conditions. Sodium borodeuteride introduces a deuterium atom onto the C1 position and facilitates the distinction between the C1 and C6 carbons. Excess borodeuteride is converted to boric acid by adding acetic acid, which is co-evaporated with methanol as its methyl ester under a stream of nitrogen. The partially methylated alditols are then acetylated with acetic anhydride to give partially methylated alditol acetates (PMAA), which are analyzed by GC-MS. The congruence of retention time and mass spectrum of each PMAA with those of known standards is used to identify the monosaccharide unit and its linkage pattern while the area or height of the chromatographic peak is used for quantification.

When polysaccharides contain uronic acids, such as rhamnogalacturonans of plant pectins or other acidic polysaccharides, methylation analysis becomes more difficult. The alkaline conditions used for methylation analysis could cause β-elimination. In addition, the uronic acids are generally resistant to acid hydrolysis; as a result, the linkage information of uronic acid and the neutral sugars to which they are attached could be lost during methylation analysis. This potential problem is avoided by a chemical reduction of the carboxyl group. A widely adapted method for uronic acid reduction is summarized in Figure 3.4.[8-10]

Carbodiimide activated reduction of the carboxyl groups of glycosyluronic acids with sodium borodeuteride ($NaBD_4$) results in an easily identified sugar (deuterized). Sodium borodeuteride ($NaBD_4$) can be replaced by sodium borohydride ($NaBH_4$) to reduce the cost of reagents.

3.2.1.2   GC-MS Analysis of Partially Methylated Alditol Acetates (PMAA)

The capillary gas-liquid chromatography is the best method for separation and quantification of the PMAA prepared in the previous section.[11] The advantage of using alditol acetate is that each aldose sugar derivative will give only one peak on the chromatogram. Special attention should be given when working with polymers containing ketoses because their derivatives give two peaks on the chromatogram. The retention times of PMAAs are highly reproducible on a specific column, and relative retention times (relative to 2,3,4,6-tetramethyl-1,5-diacetyl glucitol) are often used in the literature. The relative retention times and elution order of some of the PMAAs will depend on the coating material used in the GC column. Thus comparisons with literature values should only be made for same columns running under identical conditions. Figure 3.5 illustrates a typical gas-liquid chromatogram of PMAAs derived from galactomannans.

The electron-impact fragmentation patterns of the mass spectra of PPMAs are well documented for all linkage patterns and of all known sugars. The substitution pattern of PMAA can be readily determined based on the following rules:

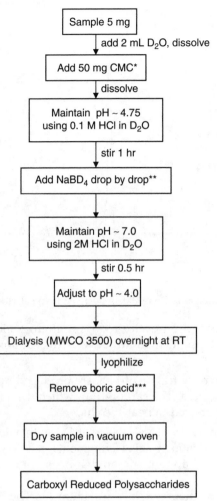

FIGURE 3.4
Reduction of carboxyl groups for Ciucanu and Kerek methylation analysis.

- Rule 1: Primary fragments are formed by cleavage of the alditol backbone.

- Rule 2: The charge always resides on the fragment with a methoxy-bearing carbon atom adjacent to the cleavage point.

- Rule 3: Fragmentation between two adjacent methoxy-bearing carbon atoms is favored over fragmentation between a methoxy-bearing

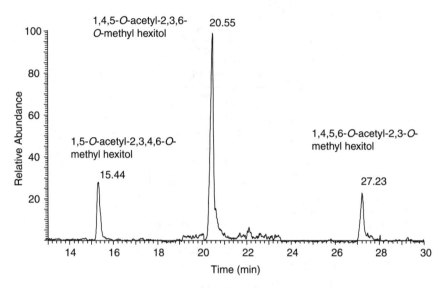

FIGURE 3.5
A typical gas-liquid chromatogram of PMAAs derived from galactomannans.

carbon atom and an acetoxy carbon atom, which itself is highly favored over fragmentation between two acetoxy-bearing carbon atoms.

- Rule 4: Secondary fragment-ions are produced by the loss of methanol or acetic acid. The carbon from which the substituted group is cleaved is preferred to bear the charge.
- Rule 5: When the PPMA is labeled at C-1 with a deuterium atom, the charge-to-mass ratio (m/z) of a fragment ion that contains C-1 is even, whereas m/z of a fragment ion that does not contain C-1 is odd.

These rules are further explained in Figure 3.6. The symmetry introduced by converting sugars into alditols is avoided by introducing a deuterium at the C1 position during the reduction reaction. The deuterium atom in a PMAA gives diagnostic fragments in the mass spectrum. However, some stereoisomeric partially methylated alditol acetates give very similar or identical mass spectra, which make it impossible to distinguish them on the basis of mass spectra only. For example, glucose, galactose, and mannose are steroisomers that will give very similar mass spectra for the same substitution patterns. Fortunately, the relative retention times (in GC) of the partially methylated alditol acetates derived from these three sugars are significantly different, which makes the identification of these three sugars and their different substitution patterns much easier. In addition, other supplementary structural information, such as monosaccharide composition, NMR spectrum, etc., is extremely helpful for identifying and confirming the sugar units and their linkage patterns.

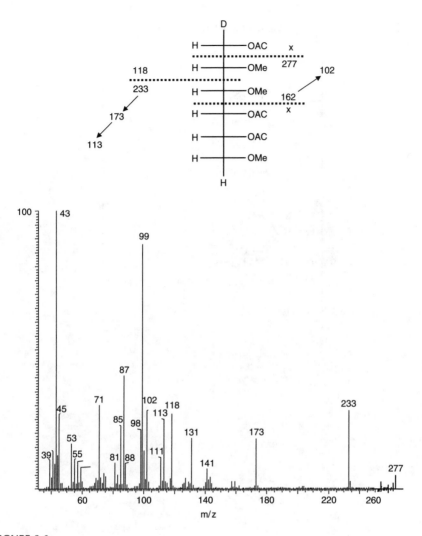

FIGURE 3.6
MS fragmentation patterns and Mass Spectrum of a PMAA (1,4,5-tri-O-acetyl-2,3,6-tri-O-methyl-hexitol).

3.2.2 Reductive Cleavage Analysis

One of the major drawbacks of the above described methylation analysis is that some of the structural information, such as anomeric conformation, is lost during the conversion of sugars into alditols. Methylation analysis also cannot distinguish if an alditol was from a 4–O–linked aldopyranose or from its corresponding 5–O–linked aldofuranose because both sugar units will give the same alditol according to the standard methylation procedure (Figure 3.7). A protocol called reductive cleavage can solve such problems.[12]

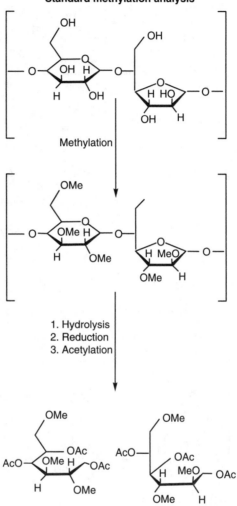

FIGURE 3.7
Demonstration of a 4–*O*–linked aldopyranose and its corresponding 5–*O*–linked aldofuranose gives the same PMAA.

In the cleavage reduction, methylated polysaccharide is depolymerized by triethylsilylane (TES) and TMS-*O*-triflate (trimethylsilyl methanesulfonate) to give partially methylated anhydroalditols. The partially methylated anhydroalditols are acetylated only at the second linkage position of the sugar ring because the anomeric position is deoxygenated (Figure 3.8). As a result, different reaction products will be obtained depending on whether furanosides or pyranosides are cleaved as illustrated in Figure 3.8.

Reductive cleavage of permethylated galactomannans with TMS-*O*-mesylate/BF3-etherate will afford four derivatives (Figure 3.9).[13] The 1,4-linked

Reductive cleavage method

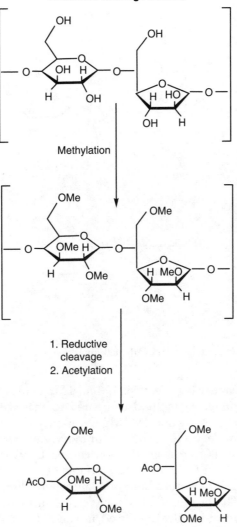

FIGURE 3.8
Demonstration of a 4–O–linked aldopyranose and its corresponding 5–O–linked aldofuranose gives different products.

D-mannopyranosyl residue gives a 4-O-acetyl-1,5-anhydro-2,3,6-tri-O-methyl-D-mannitol(1) whereas the 1,4,6-linked D-mannopyranosyl residue, i.e., the branching sugar unit, gives 4,6-di-O-acetyl-1,5-anhydro-2,3-di-O-methyl D-mannitol (2). The terminal D-galactopyranosyl residues give 1,5-anhydro-2,3,4,6-tetra-O-methyl-D-galactitol (3); similarly, the nonreducing end D-mannopyranosyl residue gives a small amount of 1,5-anhydro-2,3,4,6-tetra-O-methyl-D-mannitol(4),[13] which can be identified by GC-MS.

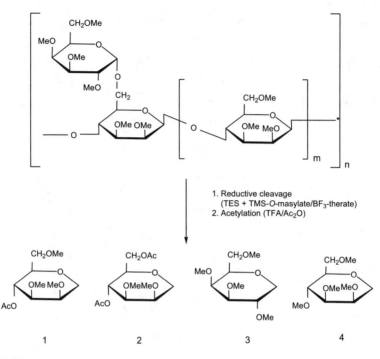

1. Reductive cleavage
   (TES + TMS-*O*-masylate/BF$_3$-therate)
2. Acetylation (TFA/Ac$_2$O)

FIGURE 3.9
Cleavage reduction of galactomannans. (Adapted from Kiwitt-Haschemie et al., 1996.)[13]

Reductive cleavage using triethysilane and TMS-*O*-mesylate/BF3-etherate as catalyst is a preferred method compared to that with TMS-*O*-triflate, because the cleavage reduction by TMS-*O*-triflate could give isomerization products (for example, via formation of furanosyl rings) and possible incomplete cleavage.[13] This ring isomerization caused by TMS-*O*-triflate occurs more frequently when there are trace amounts of water present; the extent of isomerization can be reduced but not completely eliminated by addition of solid CaH$_2$.[14] The results from reductive cleavage correspond well with the results by hydrolytic cleavage used in standard methylation analysis. Therefore, reductive cleavage not only can be used to confirm the ring size of monosaccharides, but also as an alternative method for the standard methylation analysis for providing linkage information of the monosaccharide building blocks.

3.2.3 Peroxidation

Polysaccharides containing free hydroxyl groups have the potential to react with oxidation reagents as described in Chapter 1. The oxidation reaction could be used to elucidate structural information of polysaccharides. For example, vicinyl-glycols (two neighboring hydroxyl groups) can react with

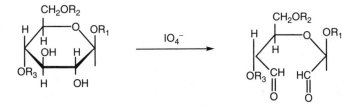

FIGURE 3.10
Periodate oxidation of a sugar unit forms two aldehydic groups with the cleavage of the carbon chain.

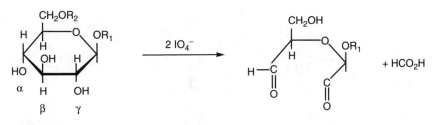

FIGURE 3.11
Periodate oxidation of a sugar unit consumes two molecular proportion of periodate and forms two aldehydic groups and one formic acid.

periodic acid or its salts to form two aldehydic groups upon the cleavage of the carbon chain, as shown in Figure 3.10.

This reaction will quantitatively consume one molar equivalent of periodate. In the case of $\alpha\beta\gamma$-triols, a double cleavage of the carbon chain will occur on both sides of the β position. This reaction consumes two molar equivalents of periodate and forms two aldehydic groups and one formic acid, as illustrated in Figure 3.11.

For polysaccharides, sugar units with different linkage patterns will vary significantly in the way they react with periodates. For example, nonreducing end sugar residue and/or 1→6-linked nonterminal residues have three adjacent hydroxyl groups; double cleavages will occur and the reaction consumes two molar equivalents of periodate and gives one molecular equivalent of formic acid (Figure 3.11). Nonterminal units, such as 1,2- or 1,4-linked residues will consume one equivalent of periodate without formation of formic acid. Sugar units that do not have adjacent hydroxyl groups, such as 1,3-linked residues or branched at C-2 or C-4 positions, will not be affected by this reaction. A quantitative determination of periodate consumed and the formic acid formed, combined with the information on the sugar units surviving the oxidation reaction, will provide clues to the nature of the glycosidic linkage and other structural features of the polysaccharides.

Periodate oxidation can be used to estimate the degree of polymerization of linear 1→4-linked polysaccharides. Each 1→4-linked polymer chain will release three formic acid equivalents after the oxidation reaction: one from the nonreducing end and two from the reducing end (Figure 3.12).

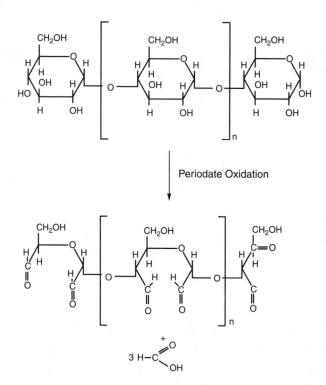

FIGURE 3.12
Periodate oxidation of 1→4-linked polysaccharides for determination of DP.

Periodate oxidation can also be used to estimate the ratio of terminal to nonterminal sugar residues of a branched polysaccharide. Two situations should be noted: (1) in highly branched polysaccharides, such as glycogen and gum arabic, the formic acid produced from the reducing end becomes insignificant and (2) in polysaccharides containing 1→6-linkages within chains, this method becomes invalid because the 1→6-linkages also liberate formic acid by periodate oxidation.[15]

Experimentally, the polysaccharide is oxidized in a dilute solution of sodium periodate at lower temperatures (e.g., 4°C). The amount of formic acid produced and periodate consumed are determined at time intervals. A constant value of two consecutive measurements of formic acid and/or periodate indicates the end of the reaction. The periodate concentration can be measured by titrimetric or spectrophotometric methods. Formic acid can be determined by direct titration with standard alkali or indirectly by the liberation of iodine from a solution of potassium iodide and iodate. For detailed experimental procedures and method of calculation, the readers are referred to an early review article.[15] The examination of the structural features of the surviving sugar units by FAB–MS, MALDI-MS and NMR spectroscopy will afford valuable linkage and branching information about the polysaccharides.

3.3 Specific Degradation of Polysaccharides

Since polysaccharides usually have extremely large molecular size and broad distribution of molecular weight (from a few thousand to up to several million Da), except for simple polysaccharides such as cellulose, the complete sequencing of a complex polysaccharide is practically impossible. In order to obtain sequence information of a complex polysaccharide, the polymer is frequently degraded into oligosaccharide repeating units or building blocks (if they exist) in a controlled manner. The resulting oligosaccharides mixture is fractionated and each oligosaccharide fully characterized. The knowledge of the complete structure of the oligosaccharides and the specific method used to cleave the polymer will eventually lead to the sequence of the polysaccharides. However, depending on how the oligomers are arranged in the polymer, the information obtained by this method may not always provide significant sequence information for the overall polysaccharide. This problem could be resolved or partially resolved by using more selective or specific degradation methods. The procedures described in this chapter may ultimately lead to the complete sequence of a polysaccharide structure if it contains a highly regular repeating unit. For those polysaccharides containing less regular or ordered oligosaccharide units, these methods can only afford information on the major structural features of the polysaccharides.

3.3.1 Partial Degradation by Acid Hydrolysis

3.3.1.1 Controlled Acid Hydrolysis

Partial degradation of polysaccharides by acid hydrolysis is based on the fact that some glycosidic linkages are more labile to acid than others. For example, furanosyl rings and deoxy sugars are usually considered as weak glycosidic linkages that can be easily hydrolyzed by acid. Hydrolysis of the glycosidic linkages of 6-deoxyhexoses is about 5 times faster than their corresponding hexoses.[16] If a polysaccharide contains only a limited number of acid-labile glycosidic linkages, a partial hydrolysis will afford a mixture of monosaccharides and oligosaccharides. Detailed characterization of these products or any residual of the polysaccharide backbone will provide meaningful information about the structure of the polysaccharides.

Partial acid hydrolysis of fully methylated polysaccharides often furnishes useful information on the positions at which the oligosaccharides were linked in the original polysaccharides. For example, a polysaccharide consisting of D-galactopyranose and D-galactofuranose residues can be selectively hydrolyzed under mild acidic conditions after methylation, as shown in Figure 3.13. The D-galactofuranosyl linkage is preferably hydrolyzed under mild acid conditions and the resultant products are reduced with borodeuteride and remethylated with trideuteriomethyl iodide (Figure 3.13). A trideuteriomethyl group at O-3 in the D-galactopyranose residue can be determined by mass spectrometry (MS). Likewise the locations of the three trideuteriomethyl groups on the reducing end unit can also be identified by MS. The combination of the structural information on the disaccharide deriv- ative and the mild acid hydrolysis method provides meaningful information leading to the structure of the repeating unit.[16]

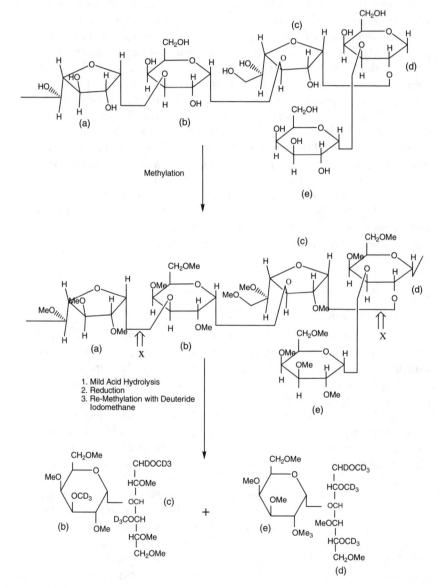

FIGURE 3.13
Acid hydrolysis of methylated polysaccharides.

There are two types of glycosidic linkages that are highly resistant to acid hydrolysis, e.g., those of uronic acids and 2-amino-2-deoxyhexose residues.[16] Polysaccharides containing isolated uronic acid or uronic acid blocks can be treated with fairly strong acids to afford aldobiouronic acids or oligo-uronic acids. Structural characterization of the hydrolyzed mono- and/or oligomer will help to elucidate the sequence of the original polysaccharide structure. The same principles can be applied to polysaccharides containing 2-amino-2-deoxyaldose units.

3.3.1.2 Methanolysis

Methanolysis breaks the glycosidic linkages of permethylated polysaccharides by introducing a methyl group and consequently forms methyl glycosides. Glycosidic bonds differ in their susceptibility to methanolysis. The rate of reaction is dependent on the anomeric configuration, position of the glycosidic linkage, and the identity of the monosaccharide. Therefore, monitoring products of a time course methanolysis of permethylated polysaccharides is useful for determining sequences, branching patterns, and location of substituents.[17] The permethylated polysaccharides are gradually degraded by acid catalyzed methanolysis and gives methyl glycosides at the released reducing terminal. A free hydroxyl group will be formed from each released glycosidic oxygen and each hydrolyzed substituent.[17] Sequence and branching information can be derived from the number of free hydroxyl groups produced. If hydrolytic removal of one or more residues occurs without the generation of a free hydroxyl group, the removed residues must be at the reducing end of the intact oligosaccharide (Figure 3.14a). In a case where the methanolysis process produced two free hydroxyl groups, two different branches must have been simultaneously hydrolyzed (Figure 3.14b).

Methanolysis reagent is prepared by introducing dry HCl gas into methanol or deuteromethanol until the solution becomes hot (approximately 1 M of HCl).[17] After cooling, an aliquot of this reagent is added to the permethylated polysaccharides (see Section 3.2 for methylation procedure) in a small reaction vial. The sample is then incubated at 40°C. Aliquots of the samples are taken after a few minutes and analyzed immediately by FAB-MS. Further aliquots are taken at suitable time intervals. The temperature can be increased to 60°C if the hydrolysis is too slow (feedback from FAB-MS results). Fresh reagent should be added after each aliquot is removed to prevent the samples from drying.

3.3.1.3 Acetolysis

Acetolysis is a process of heating polysaccharides in a mixture of acetic anhydride, acetic acid, and sulphuric acid in the ratio of 10:10:1.[16] This reaction will cause peracetylation and cleavage of selected glycosidic bonds. For example, 1,6-linkages are relatively stable to acid hydrolysis, however, they are preferentially cleaved by acetolysis and afford oligosaccharide

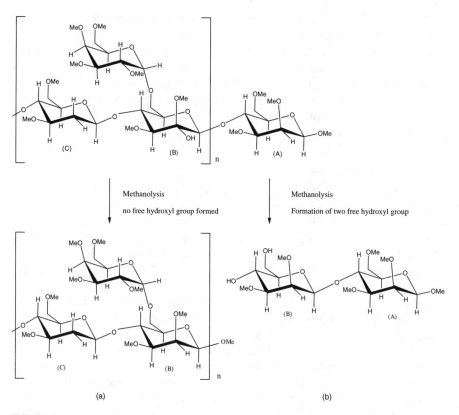

FIGURE 3.14
Methanolysis of polysaccharides or oligosaccharides. (a) Methanolysis of permethylated polysaccharides with no free hydroxyl group formed. (b) Release of two free hydroxyl groups. (Adapted from Lindberg, Lonngren, and Svensson, 1975.)[16]

peracetates.[16,18] The complete characterization of oligosaccharides derivatives produced by acetolysis will shed light on the structural features of the native polysaccharides. Please note that 1→2 and 1→3 linkages are comparatively resistant to acetolysis. Glycosidic linkages in an α-configuration are more susceptible to acetolysis than those in a β-configuration.[19,20]

3.3.2　Selective Degradation by Oxidation

3.3.2.1　Smith Degradation

As described in the previous section, the oxidation of a simple glycoside with periodate acid or its salts yields a dialdehyde and formic acid. The dialdehyde could be reduced with borohydrides to give the corresponding alcohols. The end product is a true acetal and sensitive to acid. These acetals are ready to be hydrolyzed even with dilute mineral acid at room temperature. In contrast, the surviving sugar units in glycosidic linkages are resistant

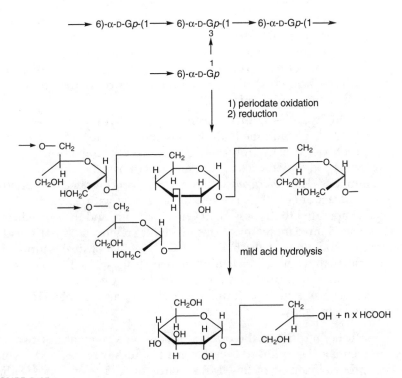

FIGURE 3.15
Smith Degadation of dextran. (Adapted from Lindberg, Lonngren, and Svensson, 1975.)[16]

to weak acid hydrolysis. The combination of periodate oxidation, reduction, and mild acid hydrolysis is known as the Smith Degradation. The consequent characterization of the resulting surviving monosaccharides and/or oligosaccharides will shed light on the fine structure of the original polysaccharide.

For example, dextran contains periodate-stable units: 1→6-linked α-D-glucose that are linked through position 3 and flanked by 1→6 linked α-D-glucose units (Figure 3.15). The product resulting from the controlled degradation by oxidation, reduction and mild acid hydrolysis processes is 3-*O*-α-D-glucopyranosyl-D-glycerol.[16]

3.3.2.2 Procedures for Smith Degradation

Controlled degradation of β-D-glucan is an example from literature:[21]

- Periodate oxidation: β-D-glucan (0.5 g) is treated with 0.04 M sodium periodate (500 mL) at 5°C for 135 hr. The periodate uptake is 0.58 mole per D-glucose residue and becomes constant after 90 hrs. A known quantity of ethylene glycol (0.98 g) is added to destroy the excess periodate, and after 30 min, the solution is dialyzed against distilled water to remove inorganic salts. The periodate and iodate could also be removed by precipitation with lead acetate.

- Reduction with sodium borohydride: 1.1 gram of sodium borohydride is added to the dialyzed solution (500 mL) to reduce the aldehyde group. The reduction reaction is continued for about 10 hrs and the excess borohydride is destroyed by dropwise addition of hydrochloric acid (1 N). The neutral solution is concentrated under vacuum at 40°C to 150 mL.

- Mild acid hydrolysis: the concentration of hydrochloric acid is adjusted to 0.5 N and the hydrolysis is continued for 8 hrs at room temperature. The hydrolyzate is deionized using both anion-exchange and cation-exchange resins to remove the chloride and sodium ions. The deionized solution is then concentrated and dried under a vacuum at 40°C. The residue is treated with methanol and then evaporated to dryness at 40°C in a vacuum, and the procedure is repeated three more times. This procedure removes the boric acid by co-evaporation of the boric acid with methanol in the form of volatile methyl borate (see methylation analysis section). The resultant residue can be analyzed for erythritol and a series of erythrityl glycosides using high performance liquid chromatography (HPLC) mass spectroscopy.

It is worth noting that in the original literature, a sequence of two and three consecutive (1→3)-linked β-D-glucosyl residues was reported[21] which was later proven incorrect by 2D NMR spectroscopy.[22] The evidence leading to the presence of consecutive (1→3)-linked β-D-glucosyl residues was the detection of 2-O-β-laminaribiosyl-D-erythritol (O-β-D-glucopyranosyl-(1→3)-O-β-D-glucopyranosyl-(1→2)-D-erythriltol) and 2-O-β-laminaritriosyl-D-erythritol. This false identification was possibly prompted by incomplete oxidation and inaccurate identification of the tri- and bisaccharide-erythritols.

The formation of erythritol suggests that the polysaccharide contains an adjacent (1→4)-linked D-glucose unit, which was later confirmed by 2D NMR spectroscopy and specific enzyme hydrolysis.[22,23] This controlled degradation may lead to the formation of a glycolaldehyde acetal of 2-O-β-D-glucopyranosyl-D-erythritol, which should be considered when determining the amount of D-glucopyranosyl-D-erythritol produced by the reaction (Figure 3.16c).

3.3.2.3   Oxidation with Chromium Trioxide

All fully acetylated aldopyranosides in β-anomeric configuration can be easily oxidized by chromium trioxide in acetic acid.[1] By contrast, the anomer having an axially attached aglycon (α-configuration) is oxidized only slowly. This preferred oxidation reaction can be used to distinguish the anomeric configuration of sugar residues in oligosaccharides. For example, two oligosaccharides, a cellobiose and a maltose, are first reduced to alditols, followed by acetylation. The acetylated esters are treated with chromium trioxide in acetic acid. Any residue that survives the oxidation reaction is α-D-linked (maltose), as illustrated in Figure 3.17.[16]

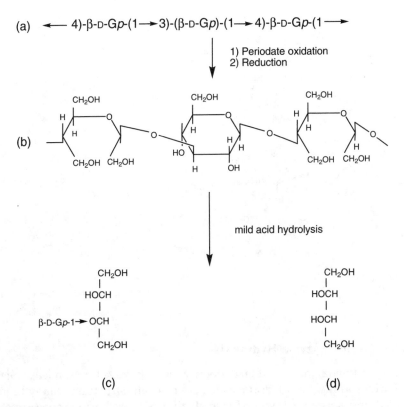

(a) ◄──── 4)-β-D-G*p*-(1──►3)-(β-D-G*p*)-(1──►4)-β-D-G*p*-(1──►

1) Periodate oxidation
2) Reduction

(b)

mild acid hydrolysis

$$CH_2OH$$
$$|$$
$$HOCH$$
$$|$$
β-D-G*p*-1──►OCH
$$|$$
$$CH_2OH$$

(c)

$$CH_2OH$$
$$|$$
$$HOCH$$
$$|$$
$$HOCH$$
$$|$$
$$CH_2OH$$

(d)

FIGURE 3.16
Smith Degradation of (1→3)(1→4)-linked-β-D-glucan. (Adapted from Goldstein, Hay, Lewis, and Smith, 1965.)[21]

When this method is applied to acetylated polysaccharides, the sequence of sugar residues may be determined by comparing the results of an original methylation analysis with that of the oxidized sample. The ester linkages formed during the chromium trioxide oxidation are cleaved and replaced by methyl groups in the subsequent methylation step.[5] Recently, the chromium trioxide oxidation procedure combined with FAB-MS has been successfully applied to assign anomeric configurations of pyranose sugars in oligosaccharides.[2] However, acetylated furanosides will be oxidized by the reagent independent of their anomeric configuration. The difference in reactivity between the anomeric forms of dideoxy sugars is very small. In addition, some problems may be encountered if the acetylation of the polysaccharide is not complete: residues containing free hydroxyl groups will also be oxidized with chromium trioxide regardless of their anomeric configuration.[16]

Other specific degradation methods are also available in the literature, such as β-elimination, deamination (for amino containing polysaccharides), etc.[16] Readers are encouraged to read the review by Lindberg and co-workers[16] and to be alert to more recent publications.

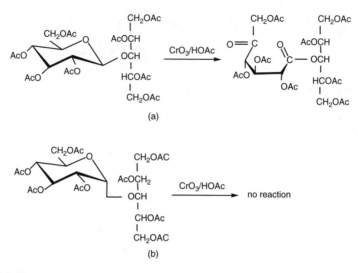

(a)

(b)

FIGURE 3.17

CrO_3 oxidation of (a) cellobiose and (b) maltose. (Adapted from Lindberg, Lonngren, and Svensson, 1975.)[16]

3.3.3   Specific Enzyme Hydrolysis

Specific endo-enzymes are able to identify major linkage types, release oligo-saccharides in high yield, and cleave acid resistant structures. Many highly purified and specific enzymes for analyzing polysaccharide structures are well characterized and commercially available.[24] In this section only two examples are presented to demonstrate the principles and procedures of specific enzyme hydrolysis of polysaccharides.

The first example is enzymic analysis of the fine structure of galactomannans. Galactomannans consist of a linear $(1\rightarrow4)$-β-D-mannopyranose backbone with side groups (α-D-galactopyranosyl units) that are $(1\rightarrow6)$ linked to the mannosyl backbone. The molar ratio of galactose to mannose varies with origins, typically in the range of 1.0:1.0~1.1, 1.0:1.6–1.8, 1.0:3.0, and 1.0:3.9~4.0 for fenugreek, guar, tara, and locust bean galactomannans, respectively. Galactomannans are probably the mostly studied gums, and numerous methods have been described in the literature for analyzing their fine structure.[25] One of the most significant methods is using highly purified and well characterized β-mannanases (EC. 3.2.1.78).[25] β-Mannanases can specifically hydrolyze the mannan backbone chain; however, enzymes from different sources will cause variations in the oligosaccharides produced. For example, β-mannanase isolated from *Aspergillus niger* is very effective in depolymerizing even highly substituted galactomannans; by contrast, β-mannanase isolated from germinated guar seed requires longer consecutively unsubstituted D-mannosyl units to cleave the D-mannan chain. As

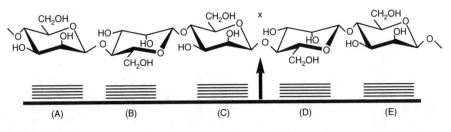

FIGURE 3.18
Action mode of endo-1,4-β-D-mannanase (EC.3.2.1.78) for hydrolysis of galactomannans. (Adapted from McCleary et al., 1985.)[26]

illustrated in Figure 3.18, guar seed β-mannanase requires five D-mannosyl units (A–E) to be active, and it can not hydrolyze the polymer at point x if B, C, or D is substituted by D-galactosyl unit. In contrast, rapid hydrolysis of the galactomannan chain by *A. niger* β-mannanase only requires four D-mannosyl units (B–E), and the presence of a D-galactosyl unit at sugar unit C has no effect on the hydrolysis at point x. However, substitutions on sugar units B or D will prevent the hydrolysis at point x.[26] The produced oligosaccharides can be fractionated by size exclusion chromatography (e.g., BioGel P-2) or analyzed by high performance ion exchange chromatography (see Chapter 2). The structures of the oligosaccharides produced are dictated strictly by the action pattern and binding requirements of the particular β-mannanase used. Hence, a complete analysis of the structures and quantities of the oligosaccharides released by a specific enzyme will furnish valuable information on the fine structure of the original galactomannan.

The second example is enzymic hydrolysis of $(1{\rightarrow}3)(1{\rightarrow}4)$-mixed-linked β-D-glucans. $(1{\rightarrow}3)(1{\rightarrow}4)$-β-D-Glucans are cell wall polysaccharides of cereal endosperm and aleurone cells. Current interest in β-D-glucans is largely due to its beneficial physiological effect on human health. Of the numerous methods for studying the structure of β-D-glucan, a highly purified and specific $(1{\rightarrow}3)(1{\rightarrow}4)$-β-D-glucan endohydrolase (lichenase, E.C. 3.2.1.73) has played a major role in elucidating the structure of this polymer. Lichenase specifically cleaves the $(1{\rightarrow}4)$-linkage of the 3-O-substituted glucose unit in β-D-glucans. The action mode of this enzyme is depicted in Figure 3.19.[27]

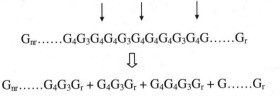

FIGURE 3.19
Action mode of lichenase (1,3-1,4-β-D-glucanase, EC 3.2.1.73) for hydrolyzing $(1{\rightarrow}3)(1{\rightarrow}4)$-linked-β-D-glucan. (Adapted from Stone, 1994.)[27]

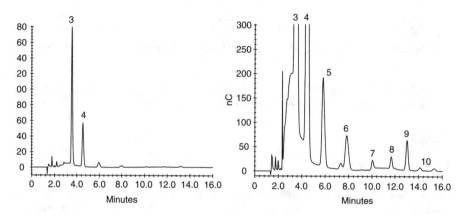

FIGURE 3.20

Oilgosaccharide profiles of cereal β-D-glucans after lichenase (EC 3.2.1.73) digestion; number labeled on the peaks represent degree of polymerization (DP).

Oligosaccharides released by lichenase hydrolysis can be analyzed by gel chromatography[27] or by a high performance anion exchange chromatography equipped with pulsed amperometric detector (PAD)[28] or by capillary electrophorosis.[29]

The most popular method is the high performance anion exchange chromatography equipped with pulsed amperometric detector (PAD) developed by Wood and his co-workers.[28] In this method, aliquots of the digest are diluted with water up to 20 fold to analyze the relative amount of trisaccharide and tetrasaccharide. Higher concentration is required to analyze oligosaccharide of degree of polymerization (DP) greater than 4 due to their low levels in β-D-glucan. Typical chromatographic profiles of the oligosaccharides released by lichenase are shown in Figure 3.20.

The relative amounts of oligosaccharides produced by lichenase treatment constitute a fingerprint of the structure of a β-D-glucan; this is expressed by the ratio of tri- to tetrasaccharides.[30] The sum of the two oligosaccharides constitutes 92 to 93% of cereal β-D-glucans. A comparison of oligosaccharide composition of three β-D-glucans is presented in Table 3.1. This example demonstrates that specific enzyme hydrolysis is a powerful method to analyze the structure of polysaccharides provided that highly purified and specific enzymes are available.

3.4   Mass Spectroscopy and Sequencing of Oligosaccharides

Mass spectroscopy is a technique for determining the masses of electrically charged molecules or particles. It is one of the key techniques in structural

TABLE 3.1

Oligosaccharides Released from Cereal β-D-Glucans after Lichenase Hydrolysis

| β-D-Glucan Source | Peak Area Percent % | | | | Molar Ratio |
	Tri	Tetra	Tri+Tetra	Penta to Nona	Tri/Tetra
Wheat bran	72.3	21.0	93.3	6.7	4.5
Barley	63.7	28.5	92.2	7.8	3.3
Oat	58.3	33.5	91.9	8.1	2.2

Note: Tri, tetra, penta to nona are numbers of glucosyl units in oligosaccharides released by lichenase hydrolysis.

Source: Adapted from Cui and Wood, 2000.[30]

analysis of polysaccharides. The application of modern mass spectroscopy in structural analysis of complex carbohydrates has been driven by the rapid advances in life science during the last two decades of the 20th century. Understanding of fine structures and sequences of glycans and glycan containing components, including glycoproteins and glycolipids, is critical in order to appreciate the function of cells. Conventional electron impact ionization mass spectroscopy (EI-MS) and chemical ionization (CI) only work for vaporized compounds. Newer ionization techniques, such as fast atom bombardment (FAB), electrospray ionization (ESI), and more recently, matrix-assisted laser desorption ionization (MALDI), have been developed for analyzing nonvolatile biomolecules, including oligosaccharides and some small molecular weight polysaccharides, proteins, glycoproteins, and glycolipids.[19,31] These new techniques offer broader analytical versatility, higher sensitivity, and more precise results. Successful applications of these techniques in determining the structures of complex polysaccharides, glycoproteins, and glycolipids have been demonstrated by a number of world class carbohydrate research laboratories.[31-33] More information on the theory, instrumentation, and detailed methodology of mass spectrometry and its application in carbohydrate structural analysis can be found in several recent reviews.[7,19,31,34,35] Readers are also recommended to visit the website provided by mass spectrometry www server, Department of Chemistry, Cambridge University (http://www-methods.ch.cam.ac.uk/meth/ms/theory/), where a wealth of concise information about mass spectroscopy can be found.

Food scientists interested in carbohydrate structural analysis should take advantage of these proven useful techniques to solve the structural problems of complex polysaccharides encountered in food materials. These would include nonstarch polysaccharides from cereals, mucilage, and cell wall materials from oil seeds, pulse crops, and legumes, pectins from fruits and vegetables, and new hydrocolloid gums from novel sources. This section introduces several selected mass spectrometric techniques that have been frequently used to provide molecular weight profiles and sequences of oligosaccharides, low molecular weight polysaccharides, and glycoproteins.

3.4.1   FAB-MS

3.4.1.1   Basic Procedures and Principles of FAB-MS

Fast atom bombardment mass spectrometry was first reported in 1982 to determine the molecular weight of oligosaccharides.[36] In a FAB-MS experiment, carbohydrate samples are first mixed with a nonvolatile organic solvent (matrix) that can dissolve the analyte. The carbohydrate–matrix mixture is then inserted into the mass spectrometer and bombarded by a fast atom beam, typically of Ar or Xe atoms at bombardment energies of 4 to 10 KeV. This process will generate both positive and negative ions, which are selectively extracted and analyzed by the mass spectrometer.

Purified oligosaccharides or glycoproteins can be analyzed directly; however, derivatization of oligosaccharides will significantly increase the sensitivity. The fragmentation pathways of derivatives are more predictable than those native samples, which can lead to correct assignment of a polysaccharide structure.[17]

Matrix selection and sample preparation are important for a successful FAB-MS experiment. A good matrix can dissolve the sample and keep a fresh homogeneous surface for FAB-MS. More importantly, the matrix can also extend the lifetime of the generated signal and therefore enhance the sensitivity of the method. Thioglycerol, 3-nitrobenzyl alcohol (3-NBA), and triethylene glycol monobutyl ether are commonly used matrixes for analyzing oligosaccharides.

3.4.1.2   FAB-MS Fragmentation Pathways

FAB is a soft ionization technique and is suitable for low volatility molecules. It produces large responses for the pseudo-molecular ions, e.g., $[M+H]^+$ or $[M-H]^-$ and also gives fragment ions containing a wealth of structural information about the oligosaccharides. There are two major fragmentation pathways for carbohydrate, namely A-type cleavage and β-cleavage.[17,34] The possible fragmentation pathways are depicted in Figure 3.21 and are explained as follows:

- A-type cleavage starts from the nonreducing side of a glycosidic bond. It occurs mostly in permethyl and peracetyl derivatives; if a sugar bears an amino group, the cleavage is preferred at this residue.

- β-Cleavage occurs frequently in native oligosaccharide samples. It also involves glycosidic breaking; however, no charge is produced at the cleavage site because a hydrogen transfer occurs after the bond is broken. Charges of fragments from β-cleavage are derived from protonation or cationization.

- Double cleavages could occur simultaneously in different parts of a molecular ion. It could be a combination of A-type cleavage and

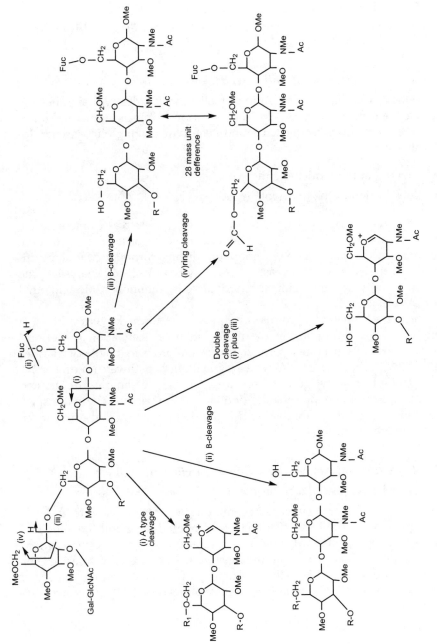

FIGURE 3.21
Fast atom bombardment mass spectrum (FAB-MS) fragmentation pathways. (Adapted from Dell, et al., 1994.)[17]

β-cleavage or two β-cleavages. The double cleavage usually has a lower abundance compared to single cleavage.

- Other fragmentation pathways, such as ring cleavages or loss of substituents (e.g., methanol or water), may also exist, but the abundance is usually very low.

3.4.1.3   Derivatization of Oligosaccharides

As mentioned earlier, derivatization of oligosaccharides will significantly improve the sensitivity of FAB-MS and afford more precise structural information. Two methods have been found useful for making oligosaccharides derivatives:[17]

- Tagging at the reducing end
- Substitution of the hydroxyl groups

The original intention of tagging an oligosaccharide was to increase the sensitivity in FAB-MS. New tagging agents have been developed to serve dual purposes — one is as a chromophore during chromatographic purification and the other is to increase the sensitivity in FAB-MS analysis.[17] The most commonly used reducing end derivative is *p*-aminobenzoic acid ethyl ester (ABEE).

Protection of functional groups is frequently performed by introducing a methyl or acetyl group using the methylation or acetylation procedures described earlier. Once the hydroxyl groups are protected, ogilosaccharides will follow a characteristic A-type fragmentation pathway (Figure 3.21). When methyl or acetyl groups are present in native samples, deuterated reagents are recommended for the derivatization reaction. The mass spectrum can be used to easily distinguish the deuterated methyl group from the native methyl groups.

3.4.1.4   Structural Analysis of Oligosaccharides by FAB-MS*

3.4.1.4.1   Molecular and Fragment Ions Calculation

The accurate and average masses of common sugar residues and their derivatives are given in Table 3.2. The masses of nonreducing and reducing end moieties are represented in Table 3.3.[17] Accurate masses are used to assign ^{12}C peaks in resolved spectra while the average masses are used to assign the center of an unresolved ion cluster. The rules for calculating the masses of the molecular and fragment ions were summarized by Dell and co-workers as follows:[17]

---

* Most of the materials described in this section are adapted from a review article by Dell et al., 1994.[17]

TABLE 3.2

Residue Masses of Sugars Commonly Found in Glycoconjugates

Monosaccharides	Native		Permethylated		Deuteromethylated		Peracetylated		Deuteroacetylated	
	Accurate Mass	Average Mass	Accurate Mass	Average Mass	Accurate Mass	Average Mass	Accurate Mass	Average Mass	Accurate Mass	Average Mass
Pentose (Pent)	132.0432	132.1161	160.0736	160.1699	166.1112	166.2069	216.0634	216.1907	222.1010	222.2277
Deoxyhexose (deoxyHex)	146.0579	146.1430	174.0892	174.1968	180.1269	180.2337	230.0790	230.2176	236.1167	236.2545
Hexose (Hex)	162.0528	162.1424	204.0998	204.223	213.1563	213.2785	288.0845	288.2542	297.1410	297.3097
Hexuronic acid (HexA)	176.0321	176.1259	218.0790	218.2066	227.1355	227.2620	260.0532	260.2005	266.0909	266.2375
Heptose (Hept)	192.0634	192.1687	248.1260	248.2762	260.2013	260.3501	360.1056	360.3178	372.1810	372.3917
N-Acetylhexosamine (HexNAc)	203.0794	203.1950	245.1263	245.2756	254.1828	254.3311	287.1005	287.2695	293.1382	293.3065
2-Keto-3-deoxyoctonate (KDO)	220.0583	220.1791	276.1209	276.2866	288.1962	288.3605	346.0900	346.2909	355.1465	355.3464
Muramic acid (Mur)	275.1005	275.2585	317.1475	317.3392	326.2039	326.3946	317.1111	317.2958	320.1299	320.3143
N-Acetylneuraminic acid (NeuAc)	291.0954	291.2579	361.1737	361.3923	376.2678	376.4847	417.1271	417.3698	426.1836	426.4252
N-Glycolyneuraminic acid (NeuGc)	307.0903	307.2573	391.1842	391.4186	409.2972	409.5295	475.1326	475.4064	487.2079	487.4804

Source: Adapted from Dell et al., 1994.[17]

TABLE 3.3

Masses of Nonreducing and Reducing End Moieties

Terminal Group	Native		Permethylated		Deuteromethylated		Peracetylated		Deuteroacetylated	
	Accurate Mass	Average Mass	Accurate Mass	Average Mass	Accurate Mass	Average Mass	Accurate Mass	Average Mass	Accurate Mass	Average Mass
Nonreducing end	1.0078	1.0079	15.0235	15.0348	18.0423	18.0533	43.0184	43.0452	46.0372	46.0372
Free reducing end	17.0027	17.0073	31.0184	31.0342	34.0372	34.0527	59.0133	59.0446	62.0321	62.0631
Reduced reducing end	19.0184	19.0232	47.0497	47.077	53.0874	53.1139	103.0395	103.0978	109.0772	109.1347
Sum of terminal masses (including the proton) for molecules that are not reduced	19.0184	19.0232	47.0497	47.0769	53.0873	53.1139	103.0395	103.0977	109.0772	109.1347
Sum of terminal masses (including the proton) for reduced molecules	21.0340	21.0391	63.081	63.1197	72.1375	72.1751	147.0657	147.1509	156.1222	156.2063

Source: Adapted from Dell et al., 1994.[17]

- The [M+H]$^+$ value is obtained by adding the increments derived by protonatation of both reducing and nonreducing ends (Table 3.3) to the sum of masses of the residues. The molecular weight of a derived oligosaccharide is the sum of the mass increments for each residue and for each natural substituent plus the mass of one protecting group at the nonreducing end and the reducing end increment.

- The masses of an A-type fragment ion are calculated by adding the nonreducing end increment to the sum of the residue masses.

- The mass of an ion derived from one or more β-cleavages is the same as that of a quasimolecular ion of the same sugar composition; however, if the oligosaccharides were derivatized, the incremental mass of one functional group is subtracted for each β-cleavage event that occurred.

- The mass of an ion derived from a double cleavage resulting from one A-type and one β-cleavage is calculated by adding one hydrogen to the sum of the residue masses.

By applying the above rules, combined with proper analytical strategy, the following structural information could be obtained from FAB-MS for oligosaccharides:

- *Composition:* The compositional assignments of an oligosaccharide could be achieved from the molecular ions, i.e., [M+H]$^+$ or [M–H]$^-$; these include number and type of sugar units, substitute functional groups, and aglycone.

- *Sequencing:* Sequencing information is derived from fragment ions produced by the intact material and from quasimolecular and fragment ions after enzymatic or chemical manipulation. For example, a time course methanolysis of permethylated oligosaccharides can be used to determine sequences, branching patterns, and location of substituents. Gradual degradation that occurs during the methanolysis of permethylated sample can be monitored by FAB-MS which will give the sequence information of the native polysaccharides. During methanolysis, methyl glycosides are produced at the released reducing end. Free hydroxyl groups are formed by each released glycosidic oxygen and each hydrolyzed substituent. The number and position of free hydroxyl groups present in each methanolysis fragment are used to derive sequence and branching information of the oligosaccharides (see Section 3.3.1 for details on methanolysis of permethylated polysaccharides).

- *Linkage assignment:* Although methylation analysis has been used as the primary method for deducing linkage information of polysaccharides, FAB-MS can provide useful linkage information when combined with some degradation procedures. For example, FAB-MS analysis of Smith Degradation products after proper derivatization can be used to identify the residues that were susceptible to periodate cleavage, therefore affording information about linkages of the intact oligosaccharide. In addition, ring cleavage in FAB-MS could also provide linkage information of the oligosaccharides.

- *Anomeric configuration:* Anomeric configuration of sugars is usually determined by specific enzyme digestion or by NMR spectroscopy. FAB-MS can be used in combination with the chromium trioxidation procedure,[1] to provide anomeric configuration information about a sugar residue. As discussed in previous section, α- and β-linkages of acetylated oligosaccharides have different stabilities under chromium trioxidation: the β-linkages are rapidly oxidized and the progress of oxidation can be monitored by FAB-MS. The oxidation of a glycosidic bond will give an increase of 14 mass units (amu). The sites of the oxidation can be established from fragmentation patterns in FAB-MS. The oxidized residues are β-linkage; in contrast, the unoxidized residues are likely in α-linkage. Nevertheless, confirmation of α- or β-linkages by other methods is highly recommended.

3.4.1.4.2 Applications of FAB-MS in the Structural Analysis of Oligosaccharides

3.4.1.4.2.1 Negative FAB-MS of Native Oligosaccharides — Native oligosaccharides usually have weaker signals in positive ion FAB mass spectra; however, relatively strong signals can be obtained in negative ion FAB. For example, a xyloglucan oligosaccharide isolated from tamarind seed was reduced to alditol to give a structure shown in Figure 3.22. A negative ion FAB mass spectrum of the alditol was recorded using 1-amino-2,3-dihydroxy-propane (aminoglycerol) as the matrix. As shown in Figure 3.23, the spectrum

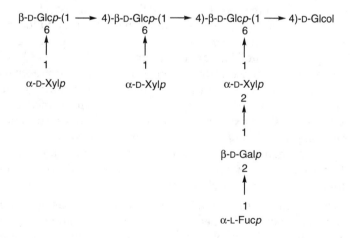

FIGURE 3.22
Structure of a xyloglucan alditol. (Adapted from York, et al., 1990.)[37]

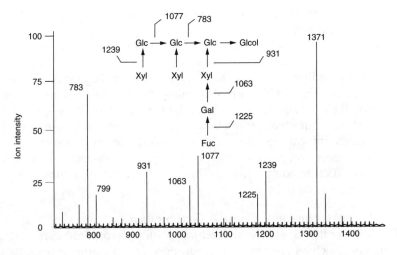

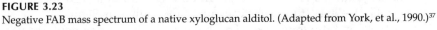

FIGURE 3.23
Negative FAB mass spectrum of a native xyloglucan alditol. (Adapted from York, et al., 1990.)[37]

has an abundant [M–H]⁻ ion. A-type cleaves were evident as explained in the spectrum.[37] The low intensity ion at m/z 799 ([XylGlc₃Glcol]) was derived from a double cleavage by loss of two side chains from the [M–H]⁻ ion.

3.4.1.4.2.2   Positive FAB-MS of Derivatised Oligosaccharides — The positive ion FAB mass spectrum of acetylated XG oligoglycosyl alditol (same structure as shown in Figure 3.22) was recorded using thioglycerol as the matrix, and the spectrum is shown in Figure 3.24. The positive ion FAB mass spectrum

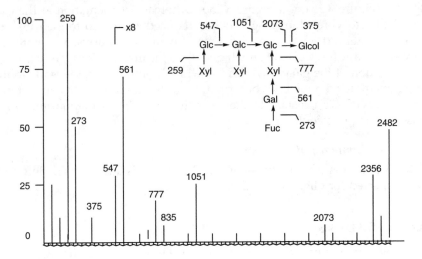

FIGURE 3.24
Positive ion FAB mass spectrum of an acetylated xyloglucan alditol. (Adapted from York, et al., 1990.)[37]

gives a intensive molecular ion $[M+NH_4]^+$ (m/z 2482) and its fragments (m/z 2398, 2356). Characteristic A-type (nonreducing end) fragment ions (m/z 2073, 1051, 777, 561, 547, 273 and 259) along with low intensity signal of $[M+H]^+$ (m/z 2423) are clearly evident. The molecular ion $[M+NH_4]^+$ is confirmed by adding NaOAc to the thioglycerol matrix to give a [M+Na]+ ion (m/z 2487). The positive ion FAB-mass spectrum also gives low abundance double cleavage ions, such as m/z 835, which corresponds to a $[Hex_2Pent]^+$ structure that cannot be formed by a single fragmentation.[37] By combining the information derived from both the negative ion spectrum of the native sample and positive ion spectrum of acetylated sample, structural information such as composition and sequence of the xyloglucan oligosaccharide can be obtained.[37]

3.4.1.4.2.3 Assignment of Anomeric Configuration — The combination of FAB-MS with other analytical procedures could afford additional structural information, such as anomeric configuration. For example, a trisaccharide Fucα1→2Gal1→3Glc (fucosyllactose, structure shown in Figure 3.25) contains both α- and β-pyranosides. This oligosaccharide is first deuteroacetylated, followed by oxidation with chromium trioxide. Positive FAB mass spectra of deuteroacetylated fucosyllactose before and after chromium trioxide oxidation are shown in Figure 3.25A. After oxidation, deuteroacetylated fucosyllactose yielded molecular ions at m/z 939 and 970 (Figure 3.25b); the later is 14 amu higher than the molecular ion before oxidation (m/z 956, Figure 3.25a); this confirms that only one glycosidic bond has been oxidized. The molecular ion of m/z 939 is the product of two oxidation reactions, as shown in Figure 3.25c. The terminal nonoxidized fucosyl residue gives a very abundant A-type ion at m/z 282 in both oxidized and un-oxidized spectra (Figure 3.25a and Figure 3.25b), whereas fragmentation at the oxidized galactosyl residue is significantly reduced compared to their un-oxidized counterpart ions. A small degree of oxidation is associated with the fucosyl residue (m/z 296, 14 amu higher than the m/z 282 ion). The evidence for oxidation of the galactosyl residue with chromium trioxide leads to the assignment of a β-anomeric conformation for the sugar unit. In contrast, the fucosyl residue is resistant to the chromium trioxide oxidation, therefore, has a α-anomeric configuration.[2]

3.4.1.4.3 Summary of FAB-MS

Table 3.4 summarizes the types of FAB-MS analysis and probable information expected for oligosaccharides.

3.4.2 MALDI-MS

The development of the laser desorption (LD) ionization technique during the last half century laid the foundation for mass spectroscopic analysis of nonvolatile polar biological and organic macromolecules and polymers. The

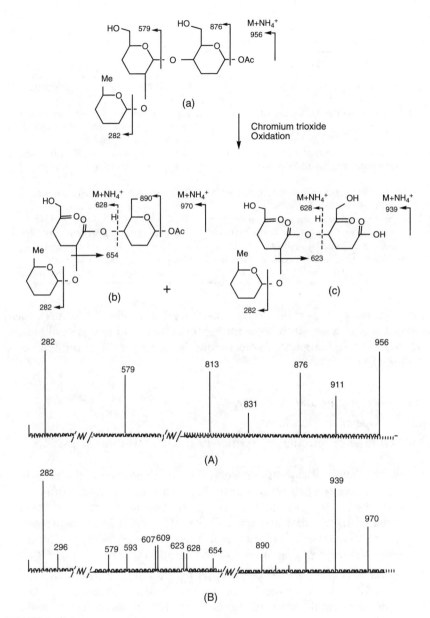

FIGURE 3.25
FAB-MS of deuteroacetylated fucosyllactose; (a) and (A): before oxidation; (b) and (B): after oxidation; (c) double oxidation. (Adapted from Khoo and Dell, 1990.)[2]

recent development of MALDI combined with the time of flight (TOF) separation system is the method of choice for structural characterization of oligosaccharides, including sequencing, branching and linkage and profiling of oligosaccharides.[31,38,39] Advantages of MALDI over other ionization methods

TABLE 3.4

FAB-MS Methods and Derived Information

	FAB-MS Experiment Strategy	Probable Data and information
1	Both positive and negative FAB-MS	Composition; sequence (depending on sample)
2	Time course methanolysis, monitored by FAB-MS	Types of functional groups and the residues to which they are attached; some sequence data
3	Perdeuteroacetylated and analyzed by both positive and negative FAB-MS	Composition and sequence information, locations of functional groups (e.g., acetyl)
4	Permethylated and analyzed by positive, sometimes negative FAB-MS (when sulphate or other negative ions present)	Composition information, sequence data, linkage information
5	Time course methanolysis of permethylated sample, monitored by FAB-MS	Sequence data and branching patterns
6	FAB-MS of samples after periodate oxidation or chromium trioxide oxidation	Sequence data and branch patterns, and possible anomeric configuration data

include its wide mass range, high sensitivity (fmol-pmol) and mass measurement accuracy. In addition, sample preparation and experiment are relatively easy to operate and the technique is also tolerant of buffers, salts, and detergents.

3.4.2.1   *Principles and Procedures*

3.4.2.1.1   Operation Procedures

A MALDI mass spectrometry experiment usually has three steps:

- *Sample Preparation:* sample preparation involves mixing the analyte with matrix (see below) and formation of a solid solution or suspension. The analyte molecules are required to be monomerly dispersed in the matrix so that they are completely isolated from each other. A homogenous solid solution is formed by evaporating any liquid solvent(s) used in preparation of the solution before analysis.

- *Matrix Excitation:* the second step is to irradiate a portion (usually about 100 μm in diameter) of the sample/matrix mixture with a UV or IR laser light. Some of the laser energy incident on the solid solution is absorbed by the matrix, causing rapid vibrational excitation and ionization.

- *Analyte Ionization and Detection:* some of the charges in the photo-excited matrix are passed to the analyte molecules which form clusters of single analyte molecules surrounded by neutral and excited matrix clusters. The matrix molecules are evaporated to give an excited analyte molecule, hence leading to the formation of the typical

[M+X]$^+$ type of ions (where X= H, Li, Na, K, etc.) which are subsequently detected by a mass detector. Negative ions can also be formed from reactions involving deprotonation of the analyte, which leads to the formation of [M–H]$^-$ molecular ions. Either positive or negative ions will be extracted from the ion source for TOF focusing and then are analyzed by the detector.

3.4.2.1.2 *Matrix Selection*

Matrixes used in MALDI experiments must have the following characteristics:

- Ability to embed and isolate analytes
- Soluble in solvents compatible with analyte
- Vacuum stable
- Absorb the laser wavelength
- Cause co-desorption of the analyte upon laser irradiation and promote analyte ionization

Aromatic acids with a chromophore are typical matrixes for ultraviolet lasers. The mid-infrared laser wavelengths are also possible for MALDI, however, different matrix compounds are required.

3.4.2.2 **Fragmentation Pathways**

3.4.2.2.1 *Systematic Nomenclature for Oligosaccharide Fragmentation*

Oligosaccharides ionized by various techniques are governed predominantly by glycosidic cleavages between the monosaccharide rings and by cross ring cleavages. A systematic nomenclature for carbohydrate fragmentations was originally proposed by Professor Costello's group for FAB-MS/MS spectra, and now has been adapted throughout the mass spectrometry field.[40] The systematic nomenclature is illustrated in Figure 3.26 and is explained below:

- Fragment ions that contain nonreducing terminus are represented by capital letters starting from the beginning of the alphabet (A, B, C, ...).
- Fragment ions that contain the reducing end of the oligosaccharide or the aglycon are represented by letters from the end of the alphabet (X, Y, Z, ...).
- Subscripts indicate the cleaved ions.
- A and X ions are produced by cleavage across the glycosidic ring, and are labelled by assigning each ring bond a number and counting clockwise. Examples for two cross ring cleavage ions are demonstrated in Figure 3.26.

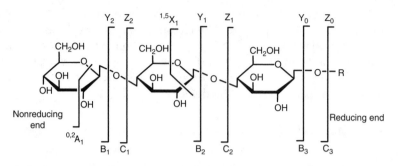

FIGURE 3.26
Systematic nomenclature for mass spectrum of oligosaccharides. (Adapted from Zaia, 2004 and Domon and Costello, 1988.)[31,40]

- Ions produced from the cleavage of successive residues are labelled as Am, Bm, and Cm, with m=1 for the nonreducing end and Xn, Yn, and Zn, with n=1 for the reducing end residue.
- Y_0 and Z_0 refer to the fragmentation of the bond linked to the aglycone if there is any.

3.4.2.2.2 *Fragmentation Characteristics*

According to the systematic nomenclature, the following rules generally apply to the fragmentation of oligosaccharides:

- The glycosidic cleavages, Bn, Cn, Yn, Zn which are prominent at low energy, are the most common fragmentations; they reveal details on the sequence and branching of the constituent monosaccharides.
- The cross ring cleavages, An and Xn, are often induced by higher energy collisions; these cleavages afford information on the linkages.
- Permethylated derivatives produce the most informative spectra; the cleavage is also more specific and the signals are more abundant. In addition, the extent of multiple bond rupture is low in this case. Permethylation of oligosaccharides could increase the sensitivity by 10x. Ions are often observed as sodium adducts $[M+Na]^+$.
- Analysis of underivatized carbohydrates results in more complex spectra; it is usually performed when sample quantity is limited.

3.4.2.3 *Examples of Applications*

- Example 1: A positive MALDI-CID spectrum of a high mannose N-linked glycan $(Man)_6(GlcNAc)_2$ isolated from ribonuclease, is presented in Figure 3.27. The major fragmentations are shown on the structural fomula.[35] The major ions are 1419.5 [M + Na], 1198.5 (B_4),

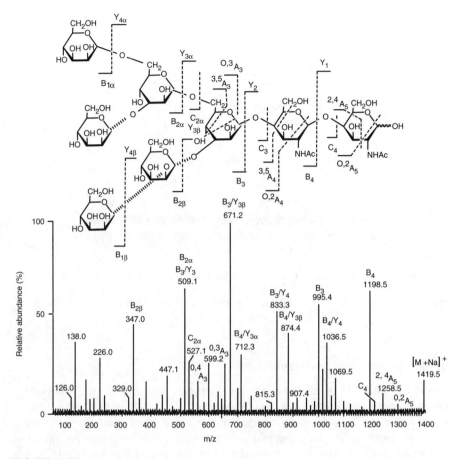

FIGURE 3.27
MALDI-TOF mass spectrum of an oligosaccharide (MAN)$_6$(GlcNAc)$_2$ and fragmentations. (From Harvey, 2003. With permission.)[35]

995.4 (B$_3$), and double cleavage ions including 833.3 (B$_3$/Y$_4$, 671.2 (B3/Y$_{3\beta}$). α and β representing two major branches in the disaccharides.

- Example 2: An oligosaccharide mixture generated by endoglucanase treatment of a xyloglucan polysaccharide was subject to MALDI-TOF MS analysis. The MALDI-TOF mass spectrum of the oligosaccharides was recorded in reflection mode and positive ion detection using 2,5-dihydroxybenzoic acid as matrix, and the mass spectrum is shown in Figure 3.28. Based on molecular weight, ions at m/z 1085, 1247, 1393, 1409 and 1555 can be attributed to the molecular ion [M+Na]$^+$ of Hex$_4$Pent$_3$ (four hexose and 3 pentose residues), Hex$_5$Pent$_3$, Hex$_5$Pent$_3$dHex (dHex, deoxy-hexose), Hex$_6$Pent$_3$, and Hex$_6$Pent$_3$dHex, respectively.[41] This experiment demonstrades that MALDI-TOF-MS can be used for profiling the molecular weight of oligosaccharide mixtures.

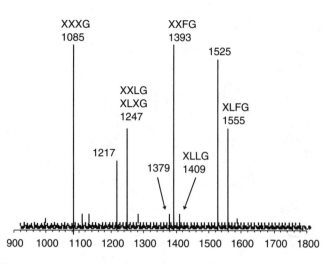

FIGURE 3.28

MALDI-TOF mass spectrum of a mixture oligosaccharides generated by endoglucanase treatment of the xyloglucan-rich extract isolated from the cell wall of *A. spinosa* leaves. Peaks labeled with symbols represent varieties of oligosaccharides identified. (Adapted from Ray et al., 2004.)[41]

3.4.3   Summary of Mass Spectroscopy

Structural analysis of oligosaccharides includes determination of monosaccharide composition, linkage and branching patterns (the hydroxyl groups involved in the linkage of one residue with another) and sequence. No single mass spectrometric (MS) technique can provide all of these parameters, however, a combination of several techniques could help to deduce the complete structure of oligosaccharides, hence the structural features of polysaccharides.

3.5   NMR Spectroscopy

NMR spectroscopy has become the most powerful and noninvasive physico-chemical technique for determining polysaccharide structures. It can provide detailed structural information of carbohydrates, including identification of monosaccharide composition, elucidation of α- or β-anomeric configurations, establishment of linkage patterns, and sequences of the sugar units in oligosaccharides and/or polysaccharides.

The principle of NMR spectroscopy is based on the magnetic properties of some nuclei. Depending upon the atomic number and mass number of a nucleus, there is an associated angular momentum spin number. For a particular spin number, an isotopic nucleus may give rise to a magnetic field that can absorb energy from a pulsed radio frequency in a strong magnetic field and subsequently the energy can be released when the radio frequency is removed. The release of energy will simultaneously give a weak signal, which represents the structural information about the individual nucleus and its surroundings. The released signal is detected, analyzed, and expressed as chemical shift (measured in ppm) and spin coupling. The most useful nuclei in carbohydrate research are ^{1}H and ^{13}C. Every polysaccharide has a unique spectrum in both ^{1}H and ^{13}C NMR spectroscopy. In other words, a NMR spectrum contains all of the structural information about the interested oligosaccharides or polysaccharides. Unfortunately, the signals of carbohydrates in NMR spectra are frequently crowded in a narrow region, especially for the ^{1}H NMR spectrum, mostly between 3 to 5 ppm. As a result, the interpretation of ^{1}H NMR spectra becomes difficult if a polysaccharide contains many similar sugar residues. The most recent development in two and multi-dimensional NMR techniques has significantly improved the resolution and sensitivity of NMR spectroscopy. For example, homonuclear correlated spectra are extremely useful for assigning ^{1}H resonances while the complete assignment of ^{13}C- resonances is achieved by ^{1}H-^{13}C heteronuclear correlation. Long range correlation techniques, such as nuclear Overhauser enhancement (NOE) and heteronuclear multiple bond correlation (HMBC), are most useful in providing sequence information of polysaccharides.

In this section, several frequently used one and two dimensional NMR techniques are introduced for elucidation of the structure of polysaccharides.

3.5.1 ^{1}H NMR Spectrum

A NMR spectrum is presented in chemical shifts (δ, ppm) relative to internal references (e.g., TMS, tetramethylsilane). In the proton spectrum, all chemical shifts derived from carbohydrates, including mono-, oligo- and polysaccharides, are in the range of 1 to 6 ppm. The anomeric protons from each monosaccharide give recognizable signals depending on their α- or β-configurations. For example, most of the α-anomeric protons will appear in the region of 5 to 6 ppm while most of the β-anomeric protons will appear in the 4 to 5 ppm range. The chemical shifts of different protons in carbohydrates are assigned in Figure 3.29.

^{1}H NMR signals are much more sensitive than ^{13}C signals due to their natural abundance. As a result high ^{1}H NMR signals can be used for quantitative purposes in some applications. However, most of the proton signals fall within a 2 ppm chemical shift range (3 to 5 ppm), which results in substantial overlap of the signals. Therefore, using one dimensional proton NMR alone to solve a structural problem of a complex polysaccharide or oligosaccharide is very difficult.

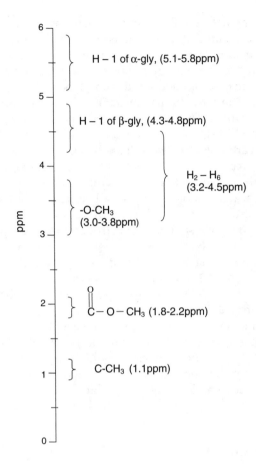

FIGURE 3.29
Illustration of chemical shifts of carbohydrates in ^{1}H NMR spectroscopy.

3.5.2   ^{13}C NMR Spectrum

Although ^{13}C-NMR has a much weaker signal, it has significant advantages over ^{1}H-NMR spectroscopy in the analysis of polysaccharides because the chemical shifts in ^{13}C-NMR spectrum are spread out over a broader range. This broad distribution of signals helps to overcome the severe overlapping problems associated with the proton spectrum. In a ^{13}C spectrum, signals from the anomeric carbons appear in the 90 to 110 ppm whereas the nonanomeric carbons are between 60 to 85 ppm. For polysaccharides with de-oxygen sugars, the –CH$_3$ signals appear in a much higher field (15 to 20 ppm). Of the two types of anomeric protons, signals derived from α-anomeric carbons mostly appear in the region of 95 to 103 ppm region whereas most of the β-anomeric carbons will appear between 101 to 105 ppm. For polysaccharides containing uronic acids, signals from the carboxyl carbons will appear in a much lower field, i.e., 170 to 180 ppm. The signals of carbon atoms having primary hydroxyl groups, such as C-6 in pyranoses and C-5 in furanoses, will appear at a higher field of 60 to 64 ppm, whereas the signals of carbon atoms with secondary hydroxyl groups (C2,3,4 in pyranoses and C2,3 in furanoses) will appear in the region of 65 to 87 ppm. For alkylated carbon atoms (C5 in pyranoses and C4 in furanoses) the signal will shift 5 to 10 ppm to the lower field. Substitution at a sugar ring by another sugar residue results in chemical shift changes to lower field by 4 to 10 ppm.[42] This phenomenon is called glycosylation shifts. The glycosylation shift also influences its adjacent carbons which tend to shift upfields by a small amount (0.9 to 4.6 ppm).[42] Other carbon resonances remain virtually unaffected. The illustration of chemical shifts of carbohydrates in a ^{13}C NMR spectrum is depicted in Figure 3.30.

To assign a NMR spectrum from an unknown polysaccharide, the first step is to compare the obtained spectrum with literature values. Further assignment of the spectrum can be completed by using 2D NMR techniques (see following section). Comparison with simulated NMR spectra will also be useful provided knowledge on composition and linkage pattern has been acquired by chemical analysis.

3.5.3 Two Dimensional NMR Spectroscopy and Structural Analysis of Polysaccharides

3.5.3.1 Assignment of ^{13}C and [1]H-Resonances

3.5.3.1.1 Homonuclear and Total Correlated Spectroscopy (COSY and TOCSY)

COSY is a ^{1}H homonuclear shift correlation spectrum which contains information on spin coupling networks within a constituent residue through the observation of cross peaks off the diagonal (Figure 3.31a). The main advantage of COSY over traditional J-resolved experiment is the disentanglement of the overlapping multiplets. The strategy of assigning a COSY spectrum is to find one unmistakably characteristic signal from which to begin the tracing of a spin system or network. An anomeric proton is often chosen as the starting point because it is connected to a carbon bearing two oxygen atoms, which is probably the most down field ^{1}H signal. For a typical aldohexopyranosyl ring, the coupling network follows the order of: H-1 couples with H-2; H-2 couples with H-1 and H-3; H-3 couples with H-2 and H-4; H-4 couples with H-3 and H-5; H-5 couples with H-4 and H-6; H-6 couples with H-5 and H-6. This scalar connectivity is shown in Figure 3.31b.

A COSY spectrum of (1→3)(1→4)-mix-linked β-D-glucan isolated from wheat bran is shown in Figure 3.32. The scalar connectivity of one of the three major sugar residues are marked in the spectrum. Note that $H_{3,4}$ and $H_{4,5}$ were unresolvable due to similar chemical shifts.

Total correlated spectroscopy (TOCSY), also known as homonuclear Hartmann-Hahn spectroscopy (HOHAHA), correlates protons that are in the same spin system and yields both long range and short range correlations. It is useful for establishing the scalar connectivity or J-network if the proton signals are from within a spin system, especially when the multiplets overlap

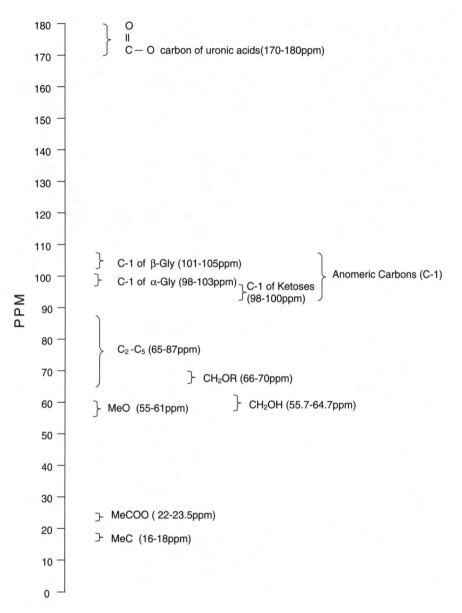

FIGURE 3.30
Illustration of chemical shifts of carbohydrates in ^{13}C NMR spectroscopy.

or there is extensive second order coupling. A J-network is defined as a group of protons that are serially linked via ^{1}H-^{1}H J coupling (scalar coupling).[42] For example, all protons in a single sugar residue belong to the same J-network. A TOCSY spectrum of the same wheat β-D-glucan is shown in Figure 3.33. J-network is identified by signals on a same horizontal or vertical

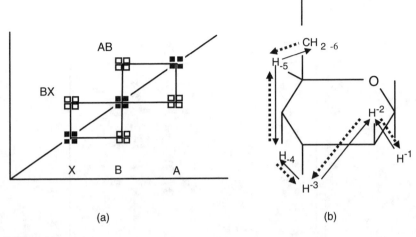

(a) (b)

FIGURE 3.31
Demonstration of scalar connectivity in a correlated spectroscopy (COSY).

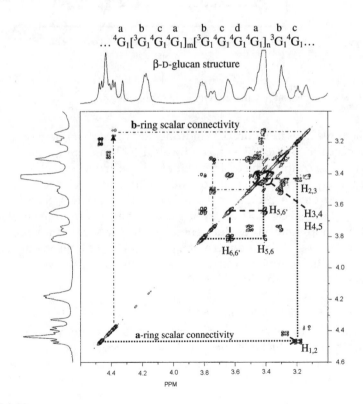

FIGURE 3.32
COSY spectrum of wheat β-D-glucan and scalar connectivity. (From Cui et al., 2000.)[22]

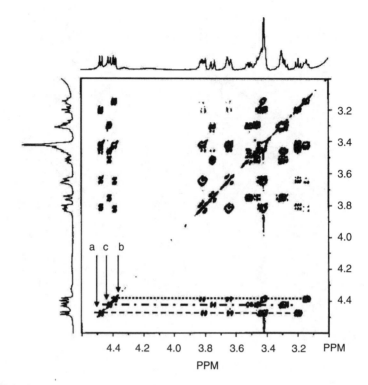

FIGURE 3.33

TOCSY spectrum of wheat β-D-glucan. (From Cui et al., 2000.)[22]

line. The three lines in Figure 3.33 indicate there are three J-networks present in the spectrum. TOCSY is a very powerful tool to distinguish J-networks and is used to confirm the assignment of ¹H spectrum.

When interpreting the spectra of both COSY and TOCSY, care should be taken if two vicinal protons have no or small couplings. Such examples include H-4 and H-5 ($J_{4,5}$=2–3 Hz) of galactopyranosyl residue and H-1 and H-2 of mannopyranosyl residue. These small or no couplings prevent the detection of cross peaks and therefore, prohibits the establishment of a complete set of the spinning system. In other words, a J-network cannot be established for such residues.[42]

3.5.3.1.2 *¹³C-¹H Heteronuclear Correlated Spectroscopy (HETCOR)*

Identification of the relative position of protons to one another in a molecular structure helps to elucidate the structure; so does the position of a given proton relative to its corresponding carbon. ¹³C-¹H heteronuclear correlated spectroscopy (HETCOR) allows one to match the protons with the corresponding carbons in a molecule. In such a spectrum, cross peaks arise from connectivity between a ¹³C nucleus and its corresponding directly linked

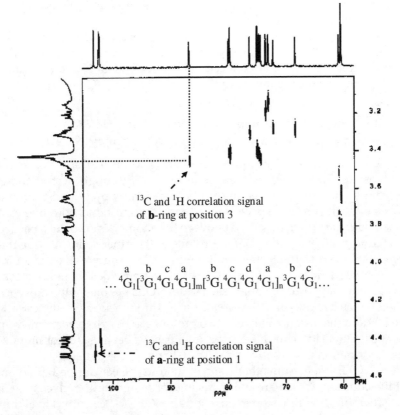

FIGURE 3.34
^{13}C-^1H heteronuclear correlated spectrocscopy of wheat β-D-glucan.

proton. An example of a $^{13}C/H^1$ HETCOR spectrum of (1→3)(1→4)-mix-linked β-D-glucan isolated from wheat bran is shown in Figure 3.34. The 1D spectra of β-D-glucan (projected on the x,y axes) appeared crowded. Much clearer signals can be identified from the 2D spectrum which enables the one to one connection of each of the protons and their corresponding carbon signals. HETCOR spectrum is particularly useful for assigning one set of resonances (frequently ^1H resonances) if the other set of data is available (or partially available) in the literature. Proton resonances of most of the polysaccharides are overlapped, which makes the complete assignments of all ^1H signals difficult. In contrast, ^{13}C resonances are spread out and the assignments of most of the monosaccharide signals in oligosaccharides or polysaccharides are readily available in the literature.[42,43] As a result, the chemical shifts of protons attached to each carbon can be deduced from the greater resolving power of the ^{13}C spectrum. However in a HETCOR experiment, the low abundant ^{13}C is detected. The low sensitivity of ^{13}C NMR method presents some problem if the amount of the sample is limited. There is a method similar to HETCOR, but with much higher sensitivity: i.e., ¹H detected ¹H-¹³C chemical shift correlated spectroscopy. In this method, the abundant ¹H is detected, which gives a much higher sensitivity. This will allow one to trace the scalar connectivity between ¹H and ¹³C atoms through indirect detection of the low natural abundant nuclei ¹³C via ¹H nuclei.[42]

3.5.3.2   Sequence Determination by NOESY and HMBC

3.5.3.2.1   Nuclear Overhauser Effect Spectroscopy (NOESY)

The NOE is an incoherent process in which two nuclear spins cross relax. This cross relaxation causes changes in one spin through perturbations of the other spin. Nuclear Overhauser effect spectroscopy (NOESY) provides information on through space rather than through bond couplings. NOE connectivities are often observed between the anomeric proton of a particular sugar residue (A in Figure 3.35) to protons of the other sugar residue that is glycosidic linked to the former (B in Figure 3.35). The presence of an inter-residue NOE defines the glycosidic linkage and provides sequence information of a polysaccharide or oligosaccharide. NOESY is one of the most useful techniques as it allows one to correlate nuclei through space (distance smaller than 5Å) and the distance between two protons can be extracted by measuring cross peak intensity. Thus NOE experiments have been useful for measuring the conformation of carbohydrates (see Chapter 5 for details).

In a NOESY spectrum, there are numerous peaks derived from inner residue coupling which are uninteresting for assigning the glycosidic linkages. Therefore, the first step in analyzing a NOESY spectrum is to eliminate those uninteresting peaks by comparing the NOESY spectrum against COSY or TOCSY spectra, and focusing on the remaining peaks.

3.5.3.2.2   Heteronuclear Multiple Bond Connectivity (HMBC)

The HMBC experiment detects long range coupling between proton and carbon (two or three bonds away) with great sensitivity. Once all the carbon-13 resonances are assigned, especially the anomeric and glycosidic linked carbons, unambiguous glycosidic linkages and sequences of the sugar residues can be established through the long range C-H correlation, as shown in Figure 3.35b. This technique is very valuable for detecting indirectly quaternary carbons coupled to protons and is especially useful if direct carbon-13 is impossible to obtain due to a low amount of material available. This very useful sequence provides information about the skeleton of a molecule. It is also very useful in carbohydrate research as a sequence analysis tool that provides unique information concerning connectivities across glycosidic linkages. Following is an example of using HMBC to establish the glycosidic linkages and sequence of wheat β-D-glucan. As shown in Figure 3.36,

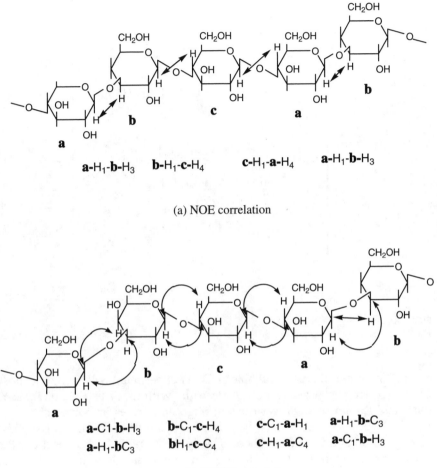

a-H₁-**b**-H₃   **b**-H₁-**c**-H₄   **c**-H₁-**a**-H₄   a-H₁-**b**-H₃

(a) NOE correlation

a-C1-**b**-H₃   **b**-C₁-**c**-H₄   **c**-C₁-**a**-H₁   a-H₁-**b**-C₃

a-H₁-**b**C₃   **b**H₁-**c**-C₄   **c**-H₁-**a**-C₄   a-C₁-**b**-H₃

(b) HMBC correlation

FIGURE 3.35
Illustration of NOE and HBMC effect and connectivity of wheat β-D-glucan.

the proton at position 1 of the a-ring is correlated to the C-3 of b-ring. Likewise, the H1 of b-ring is correlaed to C-4 of the c-ring and the H1 of c-ring is correlaed to C-4 of the a-ring. A complete assignment of the ¹H and ¹³C spectra is summarized in Table 3.5.

3.5.3.3 Summary of NMR Spectroscopy

NMR spectroscopy has been demonstrated to be the most powerful tool in structural analysis of polysaccharides. However, the first step approach is always to have a simple proton and ¹³C NMR spectra that can be compared

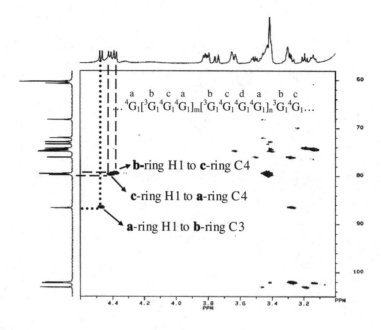

FIGURE 3.36
HMBC spectrum of wheat β-D-glucan.

against any literature data available. A COSY experiment will overcome most of the overlap problems in the 1D spectrum; with the assistance of TOCSY, the identification of sugar residues in oligosaccharides or polysaccharides, i.e., the J-network, can be established. The HECTOR is useful for assigning all the proton and C-13 resonances. The application of 2D NOE and HMBC allow one to determine the sequences and inter-residue linkage positions. Of course, there are many other powerful 2D NMR techniques for structural analysis of complex polysaccharides, and new techniques are still being developed. For further details in the area, please follow the suggested references and be alert to publications in the current literature.[42,44]

TABLE 3.5

Complete Assignment of ^{13}C and ^{1}H NMR Spectra of Wheat β-D-Glucan Based on Shift-Correlated Spectroscopy (COSY and TOCSY) (Figure 3.32 and Figure 3.33), Heteronuclear Correlation (Figure 3.34) and Heteronuclear Multi-bond Correlation (HMBC) (Figure 3.36) Spectrocopies

Glucose Residue	Assigned C, H Position	^{13}C Resonance (ppm)	^{1}H Resonance (ppm)
(1→4)-linked β-D-glucose	1	102.95	4.47
(Residue **a** in Figure 3.35)	2	71.85	3.19
	3	74.03	3.43
	4	79.31	3.42
	5	74.55	3.42
	6	60.12	3.64
			3.81
(1→3)-linked β-D-glucose	1	102.07	4.38
(Residue **b** in Figure 3.35)	2	72.70	3.14
	3	86.43	3.42
	4	68.02	3.29
	5	75.90	3.30
	6	60.50	3.51
			3.74
(1→4)-linked β-D-glucose	1	101.94	4.42
(Residue **c** in Figure 3.35)	2	73.16	3.28
	3	74.28	3.43
	4	79.53	3.42
	5	74.65	3.42
	6	60.12	3.64
			3.81

References

1. Angyal, S. J. and James, K. Oxidation of carbohydrates with chromium trioxide in acetic acid. *Australian Journal of Chemistry*, 23, 1209–1215. 1970.
2. Khoo K.H. and Dell, A. Assignment of anomeric configurations of pyranose sugars in oligosaccharides using a sensitive FAB-MS strategy. *Glycobiology*, 1, 83–91. 1990.
3. Purdie, T. and Irvin, J. C. Purdie methylation. *Journal of the Chemical Society*, 83, 1021. 1903.
4. Haworth, W. N. Haworth methylation. *Journal of the Chemical Society*, 107, 13. 1915.
5. Hakomori, S. A rapid permethylation of glycolipids and polysaccharides catalyzed by methylsulfinyl carbanion in dimethylsulfoxide. *Journal of Biochemistry* (Tokyo), 55, 205–208. 1964.
6. Ciucanu, I. and Kerek, F. A simple and rapid method for the permethylation of carbohydrates. *Carbohydrate Research*, 131, 209–217. 1984.

7. Dell, A. Preparation and desorption mass spectrometry of permethyl and per-acetyl derivatives of oligosaccharides. *Methods in Enzymology*, 193, 647–660. 1990.

8. Taylor, R. L. and Conrad, H. E. Stoichiometric depolymerization of polyuronides and glycosaminoglycuronans to monosaccharides following reduction of their carbodiimide-activated carboxyl groups. *Biochemistry*, 11, 1381–1388. 1972.

9. York, W. S., Darvill, A. G., McNeil, M., Stevenson, T. T., and Albersheim, P. Isolation and characterization of plant cell walls and cell wall components. *Methods in Enzymology*, 118, 3–40. 1986.

10. Singthong, J., Cui, S. W., Ningsanond, S., and Goff, H. D. Structural characterization, degree of esterification and some gelling properties of Kruio Ma Noy pectin. *Carbohydrate Polymers*, 58, 391–400. 2004.

11. Sweet, D. P., Albersheinm, P., and Shapiro, R. Partially ethylated alditol acetates as derivatives for elucidation of the glycosyl linkage composition of polysaccharides. *Carbohydrate Research*, 40, 199–216. 1975.

12. Jun, J. and Gary, G. R. A new catalyst for reductive cleavage of methylated glycans. *Carbohydrate Research*, 163, 247–261. 1987.

13. Kiwitt-Haschemie, K., Renger, A., and Steinhart, H. A comparison between reductive-cleavage and standard methylation analysis for determining structural features of galactomannans. *Carbohydrate Polymers*, 30, 31–35. 1996.

14. Bennek, J. A., Rolf, D., and Gray, G. R. *Journal of Carbohydrate Chemistry*, 2, 385–393. 1983.

15. Hay, G. W., Lewis, B. A., and Smith, F. Determination of the average chain length of polysaccharides. *Methods in Carbohydrate Chemistry*, 5, 377–380. 1965.

16. Lindberg, B. L., Lönngren, J., and Svensson, S. Specific degradation of polysaccharides. *Advances in Carbohydrate Chemistry and Biochemistry*, 31, 185–240. 1975.

17. Dell, A., Reason, A. J., Khoo, K., Panico, M., McDowell, R. A., and Morris, H. R. Mass spectrometry of carbohydrate-containing biopolymers. *Methods in Enzymology*, 230, 108–132. 1994.

18. Guthrie, R. D. and McCarthy, J. F. Acetolysis. *Advances in Carbohydrate Chemistry*, 11–23. 1967.

19. Harvey, D. J. Review: identification of protein-bound carbohydrates by mass spectrometry. *Proteomics*, 1, 311–328. 2001.

20. Pazur, J. H. Neutral polyscharrides. In *Carbohydrate Analysis, A Practical Approach*, Chaplin, M. Z. and Kennedy, J. F., Eds. IRL Press, Oxford, 55–96. 1986.

21. Goldstein, I. J., Hay, G. W., Lewis, B. A., and Smith, F. Controlled degradation of polysaccharides by periodate oxidaton, reduction, and hydrolysis. *Methods in Carbohydrate Chemistry*, 5, 361–370. 1965.

22. Cui, W., Wood, P. J., Blackwell, B., and Nikiforuk, J. Physicochemical properties and structural characterization by 2 dimensional NMR spectroscopy of wheat β-D-glucan — comparison with other cereal β-D-glucans. *Carbohydrate Polymers*, 41, 249–258. 2000.

23. Wood, P. J., Weisz, J., and Blackwell, B. A. Structural studies of (1-3)(1-4)-β-D-glucans by ^{13}C-NMR and by rapid analysis of cellulose-like regions using high-performance anion-exchange chromatography of oligosaccharides released by lichenase. *Cereal Chemistry*, 71, 301–307. 1994.

24. BeMiller, J. N. *Methods in Carbohydrate Chemistry*. 10. 1994.

25. McCleary. B.V. Enzymic analysis of the fine structure of galactomannans. *Methods in Carbohydrate Chemistry*, 10, 175–182. 1994.

26. McCleary, B., Clark, A., Dea, I., and Rees, D. A. The fine-structures of carob and guar galactomannans. *Carbohydrate Research*, 139, 237–260. 1985.
27. Stone, B. A. Enzymic determination of (1→3)(1→4)-β-D-glucan. *Methods in Carbohydrate Chemistry*, 10, 91–104. 1994.
28. Wood, P. J., Erfle, J. D., Teather, R. M., Weisz, J., and Miller, S. S. Comparison of (1-3)(1-4)-beta-d-glucan-4-glucanohydrolases (EC-3.2.1.73) from fibrobacter-succinogenes and from bacillus-subtilis and use of high-performance anion-exchange chromatography in product characterization. *Journal of Cereal Science* 19, 65–75. 1994.
29. Johansson, L., Virkki, L., Maunu, S., Lehto, M., Ekholm, P., and Varo, P. Structural characterization of water solyble β-glucan of oat bran. *Carbohydrate Polymers*, 42, 143–148. 2000.
30. Cui, W. and Wood, P. J. Relationships between structural features, molecular weight and rheological properties of cereal β-D-glucans. In *Hydrocolloids-Part 1*, Nishinari, K., Ed. Elsevier Science, Amsterdam, 159–168. 2000.
31. Zaia, J. Mass spectrometry of oligosaccharides. *Mass Spectrometry Reviews*, 23, 161–227. 2004.
32. Dell, A. and Morris, H. R. Fast-atom-bombardment mass-spectrometry for carbohydrate-structure determination. *Carbohydrate Research*, 115, 41–52. 1983.
33. York, W. S., Impallomeni, G., Hisamatsu, M., Albersheim, P., and Darvill, A. G. Eleven newly chracterized xyloglycan oligoglycosyl alditos: the specific effects of sidechain structure and location on ^{1}H NMR chemical shifts. *Carbohydrate Research*, 267, 79–104. 1995.
34. Dell, A. FAB-mass spectrometry of carbohydrates. *Advanced Carbohydrate Chemistry Biochemistry*, 45, 19–72. 1987.
35. Harvey, D. J. Matrix-assisted laser desorption/ionization mass spectrometry of carbohydrates and glycoconjugates. *International Journal of Mass Spectrometry*, 226, 1–35. 2003.
36. Barber, M., Bordoli, R. S., Sedgwick, R. N., and Tyler, A. N. Fast atom bombardment mass spectrometry. *Analytical Chemistry*, 54, 645A–657A. 1982.
37. York, W. S., Halbeek, H., Darvill, A. G., and Albersheim, P. Structural analysis of xyloglucan oligosaccharides by [1]H-n.m.r. spectroscopy and fastatom-bombardment mass spectrometry. *Carbohydrate Research*, 200, 9–31. 1990.
38. Harvey, D. J., Bateman, R. H., and Green, M. R. High-energy collision-induced fragmentation of complex oligosaccharides ionized by matrix-assisted laser desorption/ionization mass spectrometry. *Journal of Mass Spectrometry*, 32, 167–187. 1997.
39. Zhao, Y., Kent, S. B., and Chait, B. T. Rapid, sensitive structure analysis of oligosaccharides. *Proceedings of The National Academy of Sciences of The United States of America*, 94, 1629–1633. 1997.
40. Domon, B. and Costello, C. E. A systematic nomenclature for carbohydrate fragmentations in FAB-MS/MS spectra of glucoconjugates. *Glycoconjugate Journal*, 5, 397–409. 1988.
41. Ray, S., Loutelier-Bourhis, C., Lange, C., Condamine, E., Driouich, A., and Lerouge, P. Structural investigation of hemicellulosic polysaccharides from *Argania spinosa*: characterisation of a novel xyloglucan motif. *Carbohydrate Research*, 339, 201–208. 2004.
42. Agrawal, P. K. Determination of the chemical structure of complex polysaccharides by heteronuclear NMR Spectroscopy. *Phytochemistry*, 31, 3307. 1992.

43. Bock, K., Pedersen, S., and Pedersen, H. Carbon-13 nuclear magnetic resonance data for oligosaccharides. *Advances in Carbohydrate Chemistry and Biochemistry,* 42, 193–226. 1984.
44. Dabrowski, J. Two-dimensional and related NMR methods in structural analyses of oligosaccharides and polysaccharides. In *Two-Dimensional NMR Spectroscopy: Applications for Chemists and Biochemists, Second Edition,* Croasmun, W.R. and Carlson, R.M. Eds. John Wiley & Sons, Hoboken, NJ. 741–780. 1994.

4

Understanding the Physical Properties of Food Polysaccharides

Qi Wang and Steve W. Cui

CONTENTS

4.1   Introduction

Most food polysaccharides are classified as water soluble polysaccharides as they readily dissolve or disperse in hot or cold aqueous solutions. Water soluble polysaccharides are either naturally present or purposely added to food systems to control functional properties and provide desired textures of food products. The most important functional properties of food polysaccharides are water binding capacity and enhancing viscosity. As polysaccharides can dramatically increase the solution viscosity at a relatively low concentration, they are often used as a viscofier in liquid and semi-solid foods. They are also used to stabilize food products such as emulsion, foam, and frozen dairy products. Some water soluble polysaccharides may form gels that have been used for controlling many textural properties of semi-solid foods. Polysaccharide gels are three dimensional, liquid water containing networks showing solid-like behavior with characteristic properties such as strength, hardness, and brittleness. All these functional properties exhibited by food polysaccharides are dependent on the structure, molecular weight, and concentration of the polysaccharides present. This chapter will focus on the physical properties of food polysaccharides and associated characterization methods.

4.2   Molecular Weight

4.2.1   Molecular Weight Distribution and Averages

Molecular weight is a fundamental characteristic of polysaccharides and its determination is important in relation to many physical properties of these materials. Polysaccharides are polydisperse in molecular weight, meaning that each polysaccharide contains chains of different numbers of monosaccharide

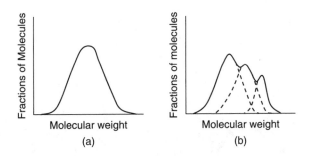

FIGURE 4.1
Typical types of molecular weight distribution for natural polysaccharides. (a) Quasi-continuous distribution. (b) Polydisperse mixture.

units giving a distribution of molecular weight. The distribution of molecular weight for a given polysaccharide varies, depending on the factors such as pathway and environments of synthesis and the extraction conditions used to isolate the polysaccharide. Figure 4.1 shows two typical types of molecular weight distribution for natural polysaccharides: quasi-continuous distribution and polydisperse mixture. A precise description of molecular weight of a polysaccharide requires information on both the average molecular weight and molecular weight distribution. The former is described by different molecular weight averages, which correspond to measurements of specific factors; the latter is usually described by the mode of distribution and polydispersity.

Four types of statistically described molecular weight averages are in common use. They are number average molecular weight (M_n), weight average molecular weight (M_w), z-average molecular weight (M_z), and viscosity average molecular weight (M_v). These averages are based on statistical approaches that can be described mathematically. The mathematical description of these averages in terms of the numbers of molecules n_i or concentrations c_i having molecular weight M_i are:

$$M_n = \frac{\sum_{i=1}^{\infty} M_i n_i}{\sum_{i=1}^{\infty} n_i} = \frac{\sum_{i=1}^{\infty} c_i}{\sum_{i=1}^{\infty} c_i / M_i} \tag{4.1}$$

$$M_w = \frac{\sum_{i=1}^{\infty} M_i^2 n_i}{\sum_{i=1}^{\infty} M_i n_i} = \frac{\sum_{i=1}^{\infty} M_i c_i}{\sum_{i=1}^{\infty} c_i} \tag{4.2}$$

$$M_z = \frac{\sum\limits_{i=1}^{\infty} M_i^3 n_i}{\sum\limits_{i=1}^{\infty} M_i^2 n_i} = \frac{\sum\limits_{i=1}^{\infty} M_i^2 c_i}{\sum\limits_{i=1}^{\infty} M_i c_i} \qquad (4.3)$$

$$M_v = \left(\frac{\sum\limits_{i=1}^{\infty} M_i^{1+\alpha} n_i}{\sum\limits_{i=1}^{\infty} M_i n_i} \right)^{1/\alpha} = \left(\frac{\sum\limits_{i=1}^{\infty} M_i^{\alpha} c_i}{\sum\limits_{i=1}^{\infty} c_i} \right)^{1/\alpha} \qquad (4.4)$$

In Equation 4.4, α is the Mark-Houwink exponent (see Equation 4.22). The number average molecular weight corresponds to a measure of chain length of polysaccharides. Physically, it can be measured by a technique that counts molecules, such as membrane osmometry and end group analysis. It is highly sensitive to small molecules present in the mixture. Most of the thermodynamic properties are related to the number of particles present and thus are dependent on M_n. M_w corresponds to a measure of the size of the polymer chain. It emphasizes the heavier molecules to a greater extent than does M_n. Experimentally, M_w can be measured by a technique that measures the property relative to the size of the molecules, such as light scattering and sedimentation equilibrium. Bulk properties such as viscosity are particularly affected by M_w. The z-average molecular weight is a third power average, which can be determined by sedimentation techniques. Properties such as melt elasticity are more closely related to M_z. The viscosity average molecular weight is obtained by the measurement of intrinsic viscosity, which can be related to the molecular weight, for example, through the Mark-Houwink relationship (Section 4.3.3.2). As illustrated in Figure 4.2, for heterogeneous polysaccharides, $M_z > M_w > M_v > M_n$.[1]

A convenient measure of the range of molecular weight present in a distribution is the ratio M_w/M_n, called the polydispersity index. The most probable distribution for polydisperse polymers produced by condensation syntheses has a polydispersity index of 2.0. The polydispersity index of natural polysaccharides is usually in the range of 1.5 ~ 2.0. Using fractionation techniques, such as size exclusion chromatography or fractional precipitation, polysaccharide fractions of low polydispersity index (close to 1) can be obtained. Figure 4.3 is an example showing that an oat β-glucan sample with a wide range of molecular weight distribution is fractionated into several sub-fractions of low polydispersity index using gradient ammonium sulphate precipitation.[2]

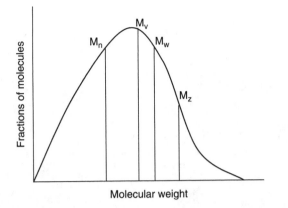

FIGURE 4.2
Molecular weight distributions and average molecular weight for a heterogeneous molecular weight system. (Adapted from Carraher, 2000.)[1]

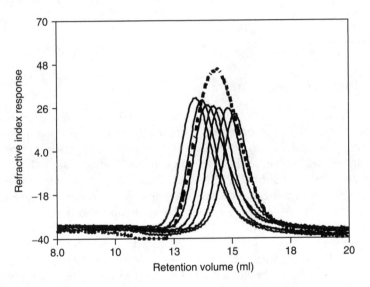

FIGURE 4.3
HPSEC profiles for a $(1\rightarrow3)(1\rightarrow4)$-$\beta$-D-glucan (dashed line) and the sub-fractions (solid line) obtained by gradient ammonium sulphate precipitation. The average molecular weight of each fraction decreases gradually from left to right. (Adapted from Wang et al., 2003.)[2]

4.2.2 Determination of Molecular Weight

Typical techniques for the determination of molecular weight (MW) of polysaccharides and the corresponding information obtained from each technique are listed in Table 4.1. These can be grouped into three categories:

TABLE 4.1

Typical Molecular Weight Determination Methods Applicable to Polysaccharides

Method	Type of MW Average	Applicable MW Range	Other Information
Light scattering (LS)	M_w		Shape
Membrane osmometry	M_n	<500,000	
Centrifugation			
Sedimentation equilibrium	M_z		
Sedimentation velocity	Gives a real MW only for monodisperse systems		
Small angle x-ray scattering	M_w		
Size exclusion chromatography (SEC)	Calibrated		MW distribution
Viscometry	Calibrated		
End-group analysis	M_n	<20,000	
Coupled SEC-LS			M_w, M_n, MW distribution, shape
Viscometry and sedimentation	Calibrated		
Sedimentation velocity and dynamic LS	M_w		

Source: Adapted from Carraher, 2000.[1]

absolute, relative, and combined techniques. Absolute techniques for MW determination include membrane osmometry, static light scattering, and sedimentation equilibrium. These techniques require no assumptions about molecular conformation and do not require calibration using standards of known molecular weight. Relative techniques require either knowledge/ assumptions concerning molecular conformation or calibration using standards of known MW. Viscometry and size exclusion chromatography (SEC) methods belong to this group. A combined technique uses information from two or more methods, such as, sedimentation velocity combined with dynamic light scattering, sedimentation velocity combined with intrinsic viscosity, and SEC combined with online (or offline) static light scattering, or sedimentation equilibrium. In this section, a number of most commonly used techniques for determination of molecular weight of polysaccharides are discussed.

4.2.2.1   *Membrane Osmometry*

Osmotic pressure is one of the common colligative properties that can be conveniently measured for polysaccharides. Suppose a polymer solution and the corresponding solvent are placed in two chambers that are separated by a semipermeable membrane (Figure 4.4), from the perspective of thermodynamics, the polymer solution exerts osmotic pressure across the membrane boundary due to the chemical potential difference of the pure solvent

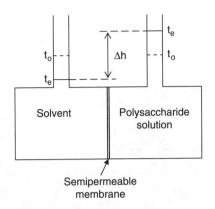

FIGURE 4.4
Schematic diagram of a classic osmometer, where t_0 represents the initial liquid level and t_e is the liquid levels when the equilibrium is reached.

and the solvent in polymer solution. This thermodynamic driving force pushes the solvent flow into the polymer containing solution, because the membrane limits the passage of polymer molecules while allowing the solvent and some small molecules to pass (depending on pore sizes of the membrane selected). When the chemical potential of diffusible solvent molecules is equal on both sides of the membrane, equilibrium is restored.

For a thermodynamically ideal solution, such as a very dilute solution, the osmotic pressure π is related to number average molecular weight M_n and polymer concentration c through the Van't Hoff's equation:[3]

$$\frac{\pi}{c} = \frac{RT}{M_n} \tag{4.5}$$

where π/c is called reduced osmotic pressure. R is the gas constant and T is the absolute temperature. In polymer solutions, deviations from idealized system are usually pronounced and the relationship may be expressed in terms of the virial equation:

$$\frac{\pi}{c} = RT \left(\frac{1}{M_n} + A_2 c + A_3 c^2 + \cdots \right) \tag{4.6}$$

where A_2 and A_3 are second and third virial coefficients, respectively. According to the above equation, M_n may be determined by a plot of π/c vs. c extrapolated to zero concentration (Figure 4.5). If the plot is a straight line, the intercept gives RT/M_n, and the slope of the plot yields RTA_2. If the plot is not linear, a curve fitting approach needs to be used to determine the intercept and virial coefficients, although virial coefficients beyond A_3 are usually of no physical interest.

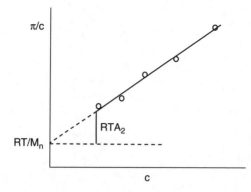

FIGURE 4.5

Plot of π/c vs. concentration c used to determine number average molecular weight M_n and second virial coefficient A_2.

Experimentally, a classic osmometry measures the difference in height (Δh) of the liquids at equilibrium (Figure 4.4), from which the osmotic pressure is calculated as:

$$\pi = \Delta h \rho g \qquad (4.7)$$

where ρ is the density of the polysaccharide solution and g is the gravity. For neutral polysaccharides, the osmotic pressure measurement can be made in water solution. For polysaccharides with charged groups, measurements have to be made in salt solutions such as 0.1 M – 1 M NaCl or LiI solutions, in order to suppress the charge effects. The osmotic pressure method is less sensitive to high molecular weight polysaccharides. In practice, this method is only useful for polysaccharides having MW less than 500,000 g/mol.[4] The lower limit of molecular weight that can be measured is determined by the molecular weight cut off of the membrane used.

4.2.2.2   Static Light Scattering

The amount of light scattered by a polymer solution is related to the mass of the solute molecules. Hence, the measurement of the intensity of scattered light by a dilute polymer solution enables us to determine the molecular weight of polymers. Static light scattering affords a convenient method for deriving several molecular parameters simultaneously, including weight average molecular weight, virial coefficients and radius of gyration R_g (see Chapter 5 for definition). The normalized intensity of scattered light $R(q)$ as a function of scattering wave vector (q) and polysaccharide concentration (c) is given as:[5]

$$\frac{Kc}{R(q)} = \frac{1}{M_w P(q)} + 2A_2 c \qquad (4.8)$$

where K is a contrast constant, $P(q)$ is called particle scattering factor. For a random coil, $P(q)$ can be expressed in terms of q and R_g:

$$P(q) = 1 - \frac{q^2 R_g^2}{3} + \dots \tag{4.9}$$

$$q = \frac{4\pi}{\lambda} \sin\left(\frac{\theta}{2}\right) \tag{4.10}$$

where λ is the wavelength and θ is the scattering angle. In the determination of M_w in dust-free dilute solutions, the intensity of scattered light from sample solutions is measured at different concentrations and different angles. The scattering data is usually analysed using the so called Zimm plot approach. This includes a double extrapolation to zero concentration and zero angle. A typical Zimm plot is constructed by plotting $Kc/R(q,c)$ vs. $q^2 + kc$ (Figure 4.6). An arbitrary constant (k) is selected to separate the angle dependent curves from different concentrations. The two limiting curves at $c = 0$ and $\theta = 0$ intersect the ordinate at the same point. This point gives the reciprocal molecular weight $1/M_w$. R_g is obtained from the slope of the curve at $c = 0$. The initial slope of the curve at $\theta = 0$ is $2A_2$. Apart from the Zimm plot approach, the experiment data can also be analyzed by other associated plots such as Debye plot, Berry plot, and Gunnier plot.[6]

A fairly large inaccuracy and irreproducibility are usually associated with the light scattering technique when it is applied to polysaccharides. Most of the problems arise from the difficulty of preparing optically clear solutions

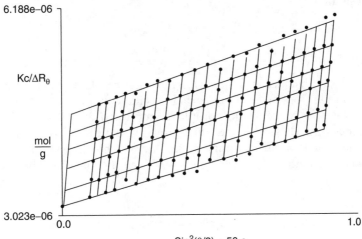

FIGURE 4.6
Zimm plot for a $(1\rightarrow3)(1\rightarrow4)$-$\beta$-D-glucan in 0.5 M NaOH aqueous solution.

that are free of dust and molecular aggregates. Polysaccharides tend to form large macromolecular aggregates in aqueous solution. In some cases, it appears impossible to separate aggregates from the individual molecules through usual methods such as centrifugation and filtration; consequently, only apparent particle weight is measured. Extreme caution has to be taken in interpreting light scattering data. Poor reproducibility is often an indication of the existence of aggregates. Extensive efforts have been made to eliminate aggregates by selection of appropriate solvents, or by chemically modifying polysaccharides to reduce hydrogen bonding. It is useful to carry out experiments using more than one solvent, or to confirm the results by other independent methods. Detailed procedures for preparation and clarification of polymer solutions are given by Tabor (1972)[7] and Harding et al. (1991).[4]

4.2.2.3   *Ultracentrifugation (Sedimentation)*

Sedimentation is a frictional property that describes the transport of mass from the surface toward the bottom. In ultracentrifuge sedimentation, macromolecules in solution are forced to sediment in a special centrifuge cell by a centrifugal force much greater than gravitational force. There are two basic sedimentation methods, sedimentation equilibrium and sedimentation velocity. In the equilibrium technique, diffusion plays an equally important role as sedimentation. Under appropriate conditions, for a macromolecule at a given molecular weight, the force of sedimentation equals the force of diffusion. Consequently, there is no net transport of the macromolecule (sedimentation equilibrium). The basic equation describing the distribution of polymer concentration $J(r)$ at sedimentation equilibrium is given for an ideal system as:[4]

$$M_w(r) = \frac{2RT}{(1-\upsilon\rho)\omega^2} \frac{d\ln J(r)}{dr^2} \tag{4.11}$$

where r is the distance of a given point in the cell from the center of the rotor, ω is the rotor speed, υ the partial specific volume, and ρ the solution density. Polymer concentration profile is monitored, usually by a Rayleigh interference optical system, and transformed into plots of $\ln J(r)$ vs. r^2, from which the (point) weight average molecular weight $M_w(r)$ can be obtained according to Equation 4.11. The average M_w can then be calculated as:

$$M_w = \frac{J(b)-J(a)}{J_0(b^2-a^2)} \frac{2RT}{\omega^2(1-\upsilon\rho)} \tag{4.12}$$

where a and b are the distance from the center of the rotor to the cell meniscus and cell bottom respectively, and J_0 is the initial loading concentration. For polysaccharides, such an equilibrium distribution is generally achieved in

24 to 48 hours depending on the nature of the polymer and the experimental conditions.

Sedimentation equilibrium method can be used to measure a wide range of molecular weight compared to light scattering and osmotic pressure methods. However, the experiment is time consuming and data interpretation is often complicated by the thermodynamic nonideality of polysaccharide solution. The technique is not popularly used in polysaccharide characterization.

In sedimentation velocity method, the ultracentrifuge is operated at extremely high speed so that diffusion is negligible compared to sedimentation. The rate of sedimentation during ultracentrifugation is measured by monitoring the boundary movement using an optical device which detects the sharp change in refractive index at the boundary. The rate of sedimentation is defined by the sedimentation coefficient (S) and is inversely proportional to the square of the angular velocity of rotation (ω) and the distance from the center of rotation to the point of observation in the cell (r):

$$S = \frac{1}{\omega^2 r}\frac{dr}{dt} \qquad (4.13)$$

Sedimentation coefficient is related to molecular weight M, solution density (ρ), specific volume of the polymer (υ), and frictional coefficient (f) through the Svedberg's equation:

$$S = \frac{M(1-\upsilon\rho)}{f} \qquad (4.14)$$

Compared to sedimentation equilibrium, sedimentation velocity measurement can be completed in a short period of time. It is particularly useful for monodisperse systems, providing qualitative data on molecular weight. Quantitative data can be obtained by combining sedimentation velocity with other methods such as dynamic light scattering. For polydisperse polysaccharide samples, sedimentation velocity can only provide some qualitative information on the average molecular weight and molecular weight distribution.

4.2.2.4 *Viscometry*

The viscosity of a polymer solution is directly related to the relative size and shape of the polymer molecules. Viscometry is one of the most widely used methods for the characterization of polysaccharide molecular weight because it requires minimal instrumentation. The experiments can be easily carried out and data interpretation is simple and straightforward. It simply requires the measurement of relative viscosity η_r and polymer concentrations (for a series of dilute solutions). From these data, intrinsic viscosity [η] can be calculated (see Section 4.3.3.1 for details). The molecular weight of the polysaccharides is then calculated via the Mark-Houwink relationship

(Equation 4.22). The Mark-Houwink equation contains two constants, k and α, which can be determined by one of the absolute methods discussed above, using a series of known molecular weight, ideally monodisperse substances. It is customary to distinguish the type of k and α values determined by different methods. For example, if osmometry was employed to determine molecular weight, the k and α values and subsequent molecular weight are designated as number average values. Thus, viscometry does not yield absolute molecular weight values, it only gives a relative measure of a polymer's molecular weight. It is worth noting that any factors that may change chain extension will lead to changes in k and α values; examples are degree of branching and the distribution of certain repeating units. Ideally, the chemical composition and structure of the material under investigation should resemble those of the calibration substances.

4.2.2.5   Size Exclusion Chromatography

Size exclusion chromatography (SEC), also called gel permeation chromatography (GPC) or gel filtration chromatography, is a form of chromatography based on separation by molecular size rather than chemical properties. In particular, high performance size exclusion chromatography (HP SEC) technique has been widely used for determination of MW and MWD of polysaccharides. When a polymer solution is pushed through a SEC column, polymer chains are separated according to differences in hydrodynamic volume by the column packing material. The column is constructed as such to allow access of smaller molecules and exclude the larger ones. Thus, the retention volume (or time) of a fraction provides a measure of the molecular size. The resulting chromatogram represents a molecular size distribution. SEC does not yield an absolute molecular weight. It requires calibration using standards of known molecular weight by converting the retention volume to a molecular weight for a given column set. The calibration can be achieved by several approaches including the most popular peak position method and universal calibration method. In the peak position approach, narrow fraction standards of known MW, such as pullulan and dextran are often used to calibrate the column. The peak retention volumes of several standards are recorded, from which a calibration curve of log MW vs. retention volume is plotted and used for determination of unknown molecular weight. Similar to viscometry, the difference in structure between the calibration standards and the test sample may lead to over- or under-estimation of molecular weight. In this case, values should be reported as pullulan or dextran equivalent molecular weight if pullulan or dextran were used as MW standards.

To overcome the inconsistency caused by using improper standards, a universal calibration approach was suggested based on the observation that the product of intrinsic viscosity [η] and MW is proportional to hydrodynamic volume of a polymer.[8] For different polysaccharides, a plot of log [η]MW vs. retention volume usually emerges to a common line, i.e., the

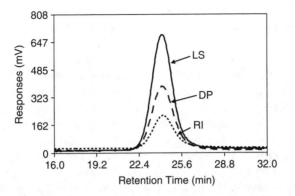

FIGURE 4.7
Typical HP SEC chromatograph of $(1\rightarrow3)(1\rightarrow4)$-$\beta$-D-glucan in aqueous solution obtained from a multi-detection system. RI-refractive index detector, DP-differential viscometry and LS-light scattering detector.

so-called universal calibration curve. The universal calibration curve for a particular column set is obtained using narrow fraction standards of known molecular weight. The retention volume of a test sample is determined using the same column and chromatography conditions, and the peak MW is read from the calibration curve, provided the intrinsic viscosity is known.

In the last two decades or so, methods for determination of molecular weight have been facilitated by connecting GPC with several detectors including a laser light scattering detector or a sedimentation equilibrium detector for direct measurement of molecular weight.[4] This allows the measurements of molecular weight and concentration for each elution fraction simultaneously. Thus, apart from the average molecular weights, information on the molecular weight distribution and molecular conformation can also be obtained. Figure 4.7 depicts a typical chromatogram obtained from a commercial HP SEC system coupled with three detectors: a refractive index detector for concentration determination, a differential viscometer for viscosity measurement, and a light scattering detector for direct molecular weight determination.[9] In a single measurement, parameters such as weight average MW, radius of gyration, intrinsic viscosity, and polydispersity index can be extracted simultaneously.

4.3 Properties of Solutions and Dispersions

4.3.1 Polysaccharides — Water Interactions

4.3.1.1 Solubility

Polysaccharides display a wide range of solubility, some readily dissolve in cold water and some are only soluble in hot water. Insoluble ones cannot be dissolved even in boiling water. Molecular structure and molecular weight are the two primary factors that determine the solubility of polysaccharides. Generally, neutral polysaccharides are less soluble than the charged ones (polyelectrolytes). Polysaccharides with highly regular conformation that can form crystalline or partial crystalline structures are usually insoluble in water. For example, linear polysaccharides with high regularity in structure, such as (1→4) linked β-D-glucans (cellulose) or β-D-mannans are essentially insoluble in aqueous medium. Solubility increases with reduced regularity of molecular structure. The irregularity of molecular structure prevents the formation of a closely packed structure and allows many polysaccharides to readily hydrate and dissolve when water is available. Branching or substitution of polysaccharide chains may reduce the possibility of intermolecular association, hence increases the solubility. Highly branched polysaccharides usually have good water solubility. For example, by introducing single β-D-galactopyranosyl substituents (1→6) linked to mannan backbone, the resulting galactomannans are fairly soluble in water. Generally, any structures that contain especially flexible units such as (1→6) linkages lead to easier solubility because of a large favorable entropy of the solution.

Polysaccharides contain a large number of hydroxyl groups, oxygen atoms, and other groups that may interact with water molecules mainly through hydrogen bonding. These interactions control the mobility of water in food systems, affect ice crystal formation and growth, and thus may exert a particular influence on the textural and functional properties of foods. Unlike small solutes, polysaccharides do not depress the freezing point of water significantly because they are substances of high molecular weight. However, the presence of polysaccharide molecules may modify the structure of water that is bound to it so that this water will be difficult to freeze. In an aqueous solution of polysaccharide, water can be divided into bound water and unbound water. Unbound water freezes at the same temperature as normal water. Bound freezable water freezes at a lower temperature than normal water. Polysaccharides may also provide cryostabilization through restricting ice crystal growth by adsorption to nuclei or active crystal growth sites.

4.3.1.2 *Factors Affecting Solubility and Dissolution Rate*

Dissolution is an entropically driven process. A homogeneous solution is obtained only when the Gibbs free energy is negative. For polysaccharides, the increase of entropy of dissolution is limited by the conformational constraints of polysaccharide chains. Many polysaccharides only form colloidal dispersions in aqueous medium that are not in thermodynamic equilibrium.

Polysaccharides used in food applications are usually sold as powders of different particle sizes. The sizes of polysaccharide powders are also polydisperse, and the average particle size and the range of particle size distribution vary with source of materials and preparation methods. Figure 4.8

FIGURE 4.8
Scanning electron microscope images of guar gum powders. (From Wang et al., 2003. With permission.)[10]

shows a scanning electron microscope image of commercial guar gum powders.[10] The average particle sizes of many commercial polysaccharide powders are in the range of 50 to 150 μm. The dissolution rate of polysaccharide powders generally increases with reduction in particle size.

Molecular weight is an important factor in determining the dissolution rate. For samples of similar particle size, dissolution rate usually decreases with increasing molecular weight. This can be explained by the fact that the disentanglement from the particle surface and subsequent diffusion to bulk solution of large molecules takes a longer time compared to that of small molecules. The dissolution rate of guar gum powders was found to be inversely related to the molecular weight of galactomannans.[10] Since interactions between polysaccharides and water depend mostly on hydrogen bonding, temperature and pressure also influence the solubility and dissolution kinetics. High temperature and pressure generally promote the dissolution of polysaccharides.[11] The initial moisture content may also affect the dissolution of certain polysaccharides, but in various ways. In general, dissolution rate increases with the level of residual solvent in the solid polymer.[12,13] However, in some cases, the residual of solvent may lead to formation of ordered structures, and therefore reduce the dissolution rate.

4.3.2   Concentration Regime

According to de Gennes (1979),[14] polymer solutions are classified into three types according to their concentration: dilute solutions, semidilute solutions, and concentrated solutions. In a dilute solution each polymer molecule and the solvent associated with it occupies a discrete hydrodynamic domain within the solution. The isolated molecules provide their individual contribution to the bulk properties of the system almost independently. Studies of dilute solution allow us to probe fundamental molecular properties of polysaccharide such as conformation, molecular weight, molecular weight distribution, and interaction properties. As the concentration increases to a critical point, individual molecular domains begin to interpenetrate or overlap each other. This concentration is called the overlap concentration or critical concentration (c*). When the polymer concentration (c) exceeds the c*, polymer chains start to interact with each other and form entangled networks; the solution is said to be in the semi-dilute region. The flow behavior of semi-dilute solutions is of great importance to food applications of polysaccharides. When c >> c*, the solution is considered in the concentrated region, in which polymer chain is described to be in a collapsed state. In the context of food polysaccharides study, only dilute and semi-dilute solutions are of great interest.

4.3.3   Dilute Solutions

4.3.3.1   Definitions of Viscosity Terms

Viscosity (η) is a measure of the resistance to flow of a fluid. It is defined as the ratio of applied shearing stress (τ) to rate of shear strain ($d\gamma/dt$) (or simply as shear rate $\dot{\gamma}$):

$$\eta = \frac{\tau}{d\gamma/dt} = \frac{\tau}{\dot{\gamma}} \tag{4.15}$$

According to the Newton's law, for an ideal viscous liquid, the applied stress is proportional to the rate of shear strain but is independent of the strain; in other words, the viscosity η is a constant. Such fluid obeys Equation 4.15, is called a Newtonian fluid. Those polymer solutions or dispersions that do not obey Equation 4.15 are called non-Newtonian fluids where the viscosity is shear-rate dependent. When the solution viscosity decreases with increasing shear rate, the solution is said to have a shear thinning flow behavior; by contrast, if the solution viscosity increases with increasing shear rate, the solution has a shear thickening flow behavior. The term fluidity is often used in the food industry, which is the reciprocal of viscosity.

When polysaccharide is dissolved in a solvent (water or other aqueous solutions), the viscosity of the solution (η) is always higher than that of the solvent (η_s). The ratio of solution viscosity to solvent viscosity is called relative viscosity (η_{rel}).

$$\eta_{rel} = \frac{\eta}{\eta_s} \tag{4.16}$$

Another associated term, specific viscosity η_{sp}, describes the fractional increase in viscosity upon addition of polysaccharide:

$$\eta_{sp} = \frac{\eta - \eta_s}{\eta_s} = \eta_{rel} - 1 \qquad (4.17)$$

For most polysaccharides, especially the random coil type, or highly branched globular type, dilute solutions under shear show essentially Newtonian flow behavior, as such, the viscosity is independent of shear rate. However, shear thinning flow behavior is observed for dilute solutions of some rigid polysaccharides, such as xanthan and some β-glucans.[15] This is a result of the progressive orientation of the stiff molecules in the increasing field of shear.

For dilute polymer solutions, η_{sp} and η_{rel} usually increase with polymer concentration which can be described by the Huggins and the Kramer equations:

$$\eta_{sp} = [\eta]c + K'[\eta]^2 c^2 \qquad (4.18)$$

$$\ln (\eta_{rel}) = [\eta]c + (K' - 0.5)[\eta]^2 c^2 \qquad (4.19)$$

where K' is the Huggins coefficient. The parameter $[\eta]$ is called intrinsic viscosity which is defined as dividing the specific viscosity by polymer concentration and extrapolated to zero concentration:

$$[\eta] = \lim_{c \to 0}\left(\frac{\eta_{sp}}{c} \right) \qquad (4.20)$$

From Equation 4.19, dividing $\ln \eta_{rel}$ by c and extrapolating to zero concentration also yields the intrinsic viscosity:

$$[\eta] = \lim_{c \to 0}\left(\frac{\ln \eta_{rel}}{c} \right) \qquad (4.21)$$

The unit of intrinsic viscosity is volume per unit weight, cm^3/g or dl/g. Thus intrinsic viscosity is not a viscosity but a measure of the hydrodynamic volume occupied by the isolated polymer chains in a given solvent. It depends primarily on the molecular structure (linear vs. branched, rigid vs. flexible) and molecular weight of polysaccharides as well as on the solvent quality. However, unlike reduced or specific viscosity, intrinsic viscosity does not depend on polymer concentration.

Equation 4.18 and Equation 4.19 form the theoretical basis for the determination of $[\eta]$. The η_r of dilute solutions of several concentrations are determined and plotted in the forms of η_{sp}/c or $\ln(\eta_{rel})/c$ vs. c. Both lines should ideally extrapolate to the same point at zero concentration as illustrated in

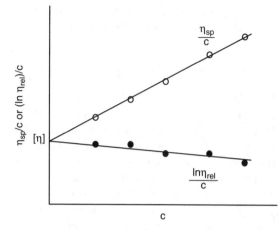

FIGURE 4.9
A schematic plot of η_{sp}/c (○) and $(\ln \eta_{rel})/c$ (●) vs. concentration c, and extrapolation to zero concentration to determine intrinsic viscosity $[\eta]$.

Figure 4.9. This intercept yields the intrinsic viscosity. In the cases where the two intercepts are not identical, $[\eta]$ is obtained as the average of the two intercepts.

4.3.3.2 Relationship between Intrinsic Viscosity and Molecular Weight

Intrinsic viscosity is a characteristic property of individual molecules in a given polymer–solvent pair. It is related to the conformation of the molecules in a way that higher value of intrinsic viscosity represents a more extended structure at a given chain length. The intrinsic viscosity measured in a specific solvent can be related to the viscosity average molecular weight, M_v, by the empirical Mark-Houwink equation:

$$[\eta] = kM_v^{\alpha} \tag{4.22}$$

where k and α are empirical constants which vary with type of polymer, solvent, and the temperature at which the viscosity is determined. For polysaccharides, parameters k and α are primarily dependent on the geometry of the interresidue linkages within the polymer chains. Most long chain polysaccharide molecules in solution take on a somewhat kinked or curled shape, which is described as random coils. Polysaccharides with expanded coil conformation give high values of k. Those with compact coil conformation give low values of k. The exponent α usually lies in the range of 0.5 to 0.8 for linear random coil polysaccharides and increases with increasing chain stiffness. The intrinsic viscosity and α value of a given polysaccharide vary in different solvents. In a good solvent, the chain is fairly loosely extended, and the intrinsic viscosity is high; the α value is close to 0.8 or higher. In a poor solvent, intramolecular interaction is stronger than the interaction

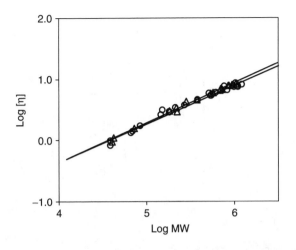

FIGURE 4.10
Mark-Houwink plots from oat (○) and barley (△) β-glucans measured in water solutions at 25°C. (Adapted from Wang et al., 2003.)[2]

between polymer and the surrounding solvent molecules. As a result, polysaccharide chains assume a more compact conformation, and the solution has a lower intrinsic viscosity. In this case, the α value is close to 0.5. For a rigid or rod like polymer, the chains are greatly extended in solution; the α value can be as high as 2.0. Therefore, α values can be used to distinguish if a polymer has a compact or extended conformation: low α values indicate a more compact conformation while high α values suggest that the polymer has a more extended conformation.

The Mark-Houwink equation is used for the estimation of molecular weight of polysaccharides. The constants k and α can be obtained from literature or determined experimentally by measuring the intrinsic viscosities of several polysaccharide samples of different molecular weight, preferably with narrow distribution in molecular weight. The molecular weights are determined by an independent method as described in previous sections. Since the viscosity average molecular weight is difficult to obtain directly, the weight or number average molecular weights are often used. A plot of log [η] vs log MW usually gives a straight line. The slope of this line is the α value and the intercept is equal to log k. Figure 4.10 shows the Mark-Houwink relationships for (1→3)(1→4)-β-glucan samples fractionated by precipitation with ammonium sulphate.[2] Values of k and α for many food polysaccharides can be found in the literature.[16]

4.3.3.3 Determination of Intrinsic Viscosity

4.3.3.3.1 Instrumentation

The most commonly used instrument for dilute solution is a glass capillary viscometer, usually of Ubbelohde type, which is a modification of the Oswald

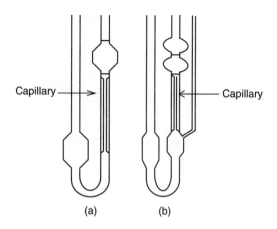

FIGURE 4.11
Ostwald (a) and Ubbelohde (b) glass viscometers.

viscometer (Figure 4.11). The advantage of using the Ubbelohde viscometer over the traditional Oswald type is that progressive dilutions can be made directly in the viscometer. The viscometer is suspended vertically in a circulated bath to maintain a constant temperature to within at least 0.01°C. The driving force is the hydrostatic head of the test liquid. The experiment measures the time needed for a constant volume of liquid to flow through the capillary. The relative viscosity is simply obtained as the ratio of the flow time for the solution (t) and the solvent (t_s):

$$\eta_{rel} = t/t_s \tag{4.23}$$

In order to obtain accurate results, preliminary test is carried out to find the range of polysaccharide concentrations which gives the quantity of η_{rel} in the range of 1.2 ~ 2.2.

The capillary viscometer can also be operated with application of external pressure. Sophisticated instruments, such as Viscotek Relative Viscometer, contain two stainless steel capillary tubes[17,18] that are connected in series and separated by a sample loading valve (Figure 4.12). Sample solution is loaded into a sample loop and then pushed into capillary 2 when loading valve is opened, while solvent remains in capillary 1 at all times. A steady state condition is reached when the sample solution completely fills capillary 2. The principle of operation is based on the measurement of pressure drops across the first solvent capillary (P_1) and across the second sample capillary (P_2) by differential pressure transducers (DPT). The relative viscosity of the sample solution is determined directly by the ratio of the pressure drops through the two capillaries:[17]

$$\eta_{rel} = k(P_2/P_1) \tag{4.24}$$

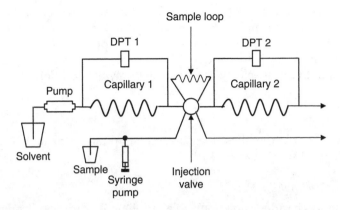

FIGURE 4.12
Schematic drawing of a dual-capillary viscometer. (Adapted from Wu et al., 2002.)[18]

where k is an instrument constant. The relative viscometer measures the solvent and sample solution simultaneously, therefore errors due to temperature fluctuation and solvent variations are avoided. It provides faster analysis and greater precision compared to the conventional glass capillary viscometers.

Based on a similar principle, a differential viscometer uses four identical stainless capillaries connected in a fluid analogue of a Wheatstone bridge (Figure 4.13).[19] A differential pressure transducer measures the pressure drop across the bridge (ΔP) and an inlet pressure transducer monitors the pressure drop through the bridge (P_i). When the effluent consists of pure solvent only, ΔP is zero. After the polymer solution is injected, the effluent containing

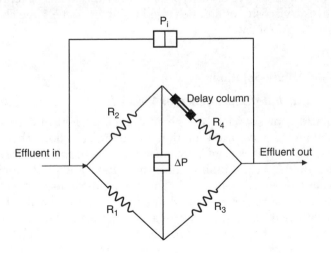

FIGURE 4.13
A simplified schematic drawing of a differential viscometer. (Adapted from Haney, 1985.)[19]

polymer fills capillaries R1, R2, and R3 while pure solvent from a delay reservoir remains in capillary R4. The slightly higher viscosity of the polymer containing effluent in R1 to R3 compared to the pure solvent in R4 causes a pressure imbalance in the bridge. The differential pressure ΔP is directly proportional to the specific viscosity of the polymer solution by:[19]

$$\eta_{sp} = 4\ \Delta P/(P_i - 2\Delta P) \qquad (4.25)$$

Knowing η_{rel} or η_{sp} from these measurements, $[\eta]$ can be calculated according to Equation 4.18 or Equation 4.19.

4.3.3.3.2   Electrolyte Effects

Some polysaccharides carry ionizable sulphate or carboxyl groups. These groups may dissociate in aqueous solution to form polyelectrolytes carrying charges of the same sign. The polymer chains tend to expand because of the electrostatic repulsion, leading to a higher intrinsic viscosity. The presence of counterions provides a shielding effect on electrostatic repulsion. This means that the intrinsic viscosity of such polysaccharides is dependant on overall ionic strength of the solution. The relationship between intrinsic viscosity and ionic strength (I) has been found to follow the empirical equation:[20]

$$[\eta] = [\eta]_\infty + SI^{-1/2} \qquad (4.26)$$

where S is a constant which could be used as a parameter of stiffness of the polyelectrolyte molecule. $[\eta]_\infty$ is the intrinsic viscosity at infinite ionic strength. It has been shown that the same values of $[\eta]_\infty$ is arrived at infinite ionic strength irrespective of the type of counterions, which is representative of the intrinsic viscosity of an uncharged polymer of the same chain length and backbone geometry.[21]

4.3.4   Semi-Dilute Solutions

4.3.4.1   Shear Rate Dependence of Viscosity

Polysaccharides of sufficiently high molecular weight and concentration form an entangled network in solution which impedes flow. These solutions may deviate substantially from Newtonian flow. There are three classes of non-Newtonian fluids: shear thinning, shear thickening, and plastic (or Bingham) fluids as illustrated in Figure 4.14. The apparent viscosity decreases with increasing shear rate for a shear thinning (or pseudoplastic) liquid, and increases with increasing shear rate for a shear thickening liquid. For a plastic fluid, a minimum shear stress known as the yield stress must be exceeded before flow begins. After the yield stress is exceeded, the liquid usually shows normal shear thinning behavior.

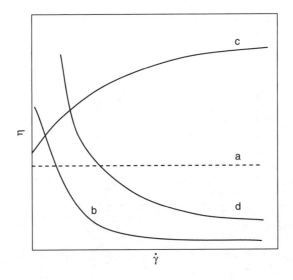

FIGURE 4.14
Viscosity η as a function of shear rate ($\dot{\gamma}$) for various viscous fluids. (a) Newtonian, (b) shear-thinning, (c) shear-thickening, and (d) plastic fluids.

In the vast majority of cases, polysaccharide solutions are pseudoplastic. The shear thinning behavior of a semi-dilute solution is caused by the decreasing number of chain entanglements as the chains orient themselves with the direction of the flow. At high shear rates, the newly formed entanglements cannot compensate for those being disentangled, hence decreased viscosity is observed. Figure 4.15 shows a typical steady shear flow curve in a bilogarithmic plane consisting of a limit of constant viscosity at very low shear rates (zero shear viscosity η_0), followed by a shear thinning region where viscosity decreases as shear rate increases, and a lower constant viscosity at the limit of very high shear rates (infinite shear viscosity η_∞). The viscosity plateau at high shear rates is rarely observed for most polysaccharide solutions except for highly stiff molecules such as xanthan gum.[22] The reason is, apart from the difficulty in obtaining experimental data at high shear rate region (>10^4), polysaccharides may be degraded at very high shear rate.

The degree of shear thinning of a polysaccharide solution depends on the intrinsic molecular characteristics including conformation, molecular weight, and charges for anionic polysaccharides. Extrinsic factors such as concentration, temperature, and pH may also influence the flow behavior. In general, solutions containing stiff polysaccharide chains are more pseudoplastic; the degree of pseudoplasticity increases with the polysaccharide concentration and molecular weight. For solutions of higher concentration or larger molecules, the start of shear thinning appears at lower shear rate and the viscosity reduction is more pronounced.

A most useful mathematical model for describing shear thinning behavior is the power law model:

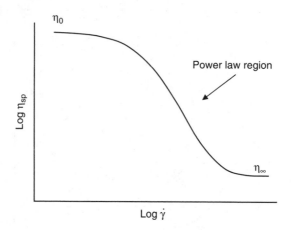

FIGURE 4.15
Idealized pseudoplastic flow curves for semi-dilute polysaccharide solutions.

$$\eta = k\dot{\gamma}^{n-1} \tag{4.27}$$

where n is called the flow behavior index or power law index and k is called the consistency index. Both n and k are concentration dependant. In general, when concentration or molecular weight increases n decreases, but k increases. A linear plot of n vs. log c has been obtained for a number of polysaccharide solutions.[22] When concentration approaches zero, n equals one, the power law relationship returns to the Newton's law (Equation 4.15). At high concentration, n falls in the range of 0.15 to 0.25.

The Cross equation can be used to describe both high and low shear rate regions of pseudoplastic polysaccharide solutions, which includes both the zero shear viscosity η_0 and the limiting viscosity at infinite shear rate η_∞:

$$\eta = \eta_\infty + \frac{\eta_0 - \eta_\infty}{(1+\alpha\dot{\gamma})^m} \tag{4.28}$$

where α is a time constant related to the relaxation time of a polysaccharide in solution and m is a dimensionless exponent. When $\eta_0 \gg \eta_\infty$ and η_∞ is small, the Cross equation represents a power law relationship (Equation 4.27) with exponent $m \sim (1 - n)$. The Cross equation generally describes well the shear rate dependence of aqueous polysaccharide solutions. Particularly, it has been widely used for predicting zero shear viscosity of solutions containing random coil polysaccharides.

Shear thickening of polysaccharide solution, although rare, is observed occasionally in concentrated suspensions of macromolecules or aggregates.

This phenomenon usually results from a shear-induced formation of ordered structures.[23]

Because of the shear dependency of viscosity, zero shear rate viscosity should be used for meaningful comparison of different systems. η_0 is obtained by measuring viscosity at a range of low shear rate and extrapolating to zero shear rate. In the literature, apparent viscosities are also compared at the same shear rate.

4.3.4.2 Time Dependence of Viscosity

For some non-Newtonian liquids, flow viscosity may depend on the length of flow time and flow history. In general, two types of flow can be distinguished, thixotropy and antithixotropy (Figure 4.16a and Figure 4.16b). In the former case, upon application of a constant shear rate, viscosity immediately reaches a peak value and then gradually decreases with time until it reaches a stationary state. In the later case, the viscosity gradually increases with time. While antithixotropy is less seen in food polysaccharide systems, thixotropic behavior is often exhibited by disperse systems in which the disperse phase is made up of molecular aggregates. The application of shearing conditions results a gradual reduction of aggregate dimensions, leading to a progressive decrease in viscosity with time. Both thixotropy and antithixotropy are reversible phenomena, i.e., the viscosity returns to its original value after cessation of shear with a delayed period rather than instantaneously. Thixotropy is usually associated with shear thinning behavior. For a typical thixotropic liquid, plot of viscosity vs. shear rate forms a hysteresis loop when the viscosity measurement is made first at increasing shear rate followed by a decrease in shear rate (Figure 4.17). In many polysaccharide solutions, thixotropy is a result of strong interchain association through hydrogen bonding. For example, freshly prepared cereal $(1{\rightarrow}3)(1{\rightarrow}4)$-$\beta$-D-glucan solutions usually exhibit little thixotropy. When these solutions are aged, during which large aggregates are formed, appreciable thixotropy is developed. High concentration solutions of methylcellulose also show apparently thixotropic behavior.[24]

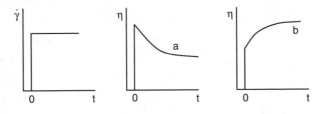

FIGURE 4.16
Viscosity as a function of flow time at a constant shear rate. (a) thixotropy and (b) antithixotropy.

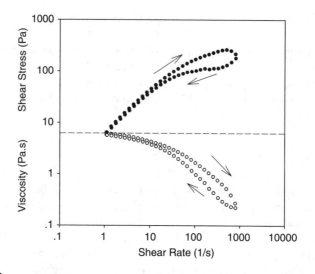

FIGURE 4.17

Hysteresis loop of shear stress and viscosity vs. shear rate for a thixotropic fluid (2.5% oat β-D-glucan aged for 12 h, 25°C).

4.3.4.3   Concentration and Molecular Weight Effects

Because of the shear rate dependence of viscosity, the effects of concentration and molecular weight on viscosity are mostly studied by using zero shear viscosity η_0 or more often zero shear specific viscosity $(\eta_{sp})_0$. At polysaccharide concentrations above the overlap concentration c^*, more pronounced increases in both zero shear viscosity and shear rate dependence of viscosity are observed. This is due to the formation of entangled networks by the macromolecules in semi-dilute solution. The viscosity of an entangled polysaccharide network is proportional to the extent of entanglements (expressed by the average number of entanglement points per molecule), which in turn is proportional to the chain length or molecular weight. Therefore, for a given polysaccharide, the viscosity of a semi-dilute solution also increases with molecular weight. The dependence of viscosity on concentration and molecular weight can be well represented by a power law relationship between zero shear specific viscosity $(\eta_{sp})_0$ vs. $c[\eta]$:

$$(\eta_{sp})_0 \propto (c[\eta])^n \qquad (4.29)$$

where the intrinsic viscosity $[\eta]$ is proportional to molecular weight for a given polymer solvent pair and the relationship can be described by, for example, the Mark-Houwink equation (Equation 4.22). The parameter $c[\eta]$ is called the coil overlap parameter. It characterizes the space occupancy of polysaccharide molecules in a solution. Equation 4.29 is valid for both dilute and semi-dilute solutions, but with different n values, the slopes of the plots

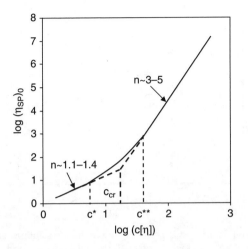

FIGURE 4.18
Schematic plot of zero-shear specific viscosity $(\eta_{sp})_0$ vs. dimensionless overlap parameter $c[\eta]$ in a bilogarithmic plane for polysaccharide solutions.

log $(\eta_{sp})_0$ vs. log $c[\eta]$, as shown in Figure 4.18. For almost all random coil polysaccharides reported in the literature, the magnitude of n is in the range 1.1~1.4 and 3.0~5.1 for dilute and semi-dilute solutions, respectively. There is a transition zone from dilute regime ($c < c^*$) to semi-dilute regime ($c > c^{**}$). The concentration c^{**} represents the start of semi-dilute regime. The concentration at which the two lines intersect is called critical concentration c_{cr}. For a wide range of polysaccharides, $(c[\eta])_{cr}$ are found to be in the range 2.5 to 4.[22] It has been shown that a master curve of log $(\eta_{sp})_0$ vs. log $(c[\eta])$ can be drawn for a number of random coil polysaccharides varying in structure and molecular weight. The slope of the plot is found to be 3.3 when $c > c^*$ and 1.4 when $c < c^*$. Higher n values observed for some polysaccharides, such as galactomannans, were attributed to specific intermolecular associations.[25]

4.3.4.4 Effects of Temperature

Another important factor that affects viscosity and flow behavior of polysaccharide solution is temperature. Under a given shear stress, the rate of flow is determined by polymer chain entanglement/disentanglement and slippage when chains pass each other. In most cases, increase of temperature promotes disentanglement of the chains, hence viscosity decreases. For an ideal Newtonian liquid, this is in accordance with the Arrhenius equation

$$\log \eta = \log A + \frac{E}{R}\frac{1}{T} \tag{4.30}$$

where E is the activation energy for viscous flow, R is the gas constant, and A is a constant. According to Equation 4.30, the viscosity of a Newtonian

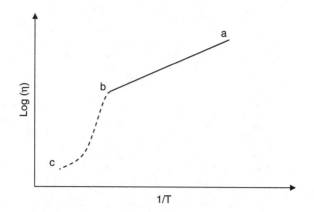

FIGURE 4.19
Idealized Arrhenius-type relationship between polysaccharide solution viscosity η and temperature. Note from points a to b, there is no conformational change; and from b to c, order to disorder conformational change causes abrupt decrease in viscosity.

solution decreases logarithmically with increasing temperature (Figure 4.19). Since polysaccharide solutions contain considerable secondary bonding, such as hydrogen bonding, a plot of log η vs. $1/T$ may not always be a straight line. The viscosity temperature relationship is usually reversible for the same gum solution but characteristic for each polysaccharide gum. This relationship is also dependant on the shear rate at which the viscosity is measured. Increasing temperature often leads to an increase of the onset shear rate at which the non-Newtonian region starts; in another word, shear thinning starts at higher shear rates. Change of temperature may cause conformational changes of polysaccharide molecules in solution. For example, for a number of polysaccharides such as gellan gum and xanthan gum, temperature increase may induce a reversible conformational change: from an ordered helix to a disordered coil. This conformational transition is accompanied by an abrupt drop in solution viscosity (dotted line in Figure 4.19).

4.3.4.5 Effects of pH and Ionic Strength

For polysaccharides with charged groups such as carboxylic acid group, shear viscosity is also sensitive to ionic strength and pH. Due to an expanded chain conformation caused by same charge repulsions, a higher viscosity is normally obtained at lower ionic strength. At higher ionic strength and in the presence of enough counterions, electrostatic repulsions are suppressed, leading to a less extended chain conformation and hence lower viscosity. The viscosity of neutral polysaccharide solution is not normally affected by pH unless degradation (thus decrease in viscosity) occurs at extreme pHs. For some polysaccharides containing anionic group (COO^-), lowering solution pH increases viscosity. In particular, there is usually a marked increase in viscosity when pH is below the pk_a value. This is because the repression

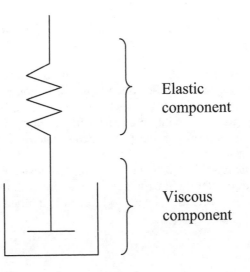

FIGURE 4.20
Maxwell model for a viscoelastic material under one-dimensional deformation.

of ionization of carboxylic acid groups lessens the electrostatic repulsion and promotes intermolecular association. Depending on the polysaccharide concentration and the way the pH is adjusted, the association may lead to precipitation or bring about gelation.

4.3.4.6   *Viscoelastic Properties*

Polysaccharide solutions are viscoelastic in nature, i.e., have both solid- and liquid-like characteristics. Among others the one-dimensional deformation Maxwell model is the simplest to describe the viscoelastic behavior. As shown in Figure 4.20, this model contains an elastic spring and a viscous damping element (dash pot) connected in series. It is assumed that the contributions of both the spring and dash pot to strain are additive upon application of a stress. The elastic property may be described by the Hooke's law, which states for an ideal elastic body the applied stress (σ) is proportional to the resultant strain (γ) but is independent of the rate of strain ($d\gamma/dt$).

$$\sigma = E\gamma \tag{4.31}$$

where E is the Young's modulus or elastic modulus. Combining Hooke's law with Newton's law (Equation 4.15), the motion of the Maxwell model can be expressed in the form of the following differential equation:

$$\frac{d\gamma}{dt} = \frac{1}{E}\frac{d\sigma}{dt} + \frac{\sigma}{\eta} \tag{4.32}$$

and

$$\eta = \tau E \tag{4.33}$$

where τ is called relaxation time. The Maxwell model basically demonstrates that viscoelasticity is a relaxation phenomenon. In comparison to the time scale of measurement, material with large τ (i.e., long relaxation time) is characterized as an elastic solid; material with small τ behaves as a viscous liquid. When the τ is comparable to the time scale of a measurement, the material is viscoelastic.

When a sample is subjected to a shear force (σ), instead of an one-dimensional deformation, there will be a three-dimensional deformation, that is a shear strain (γ). Similarly, the shear modulus G is defined by:

$$G = \frac{\sigma}{\gamma_0} \tag{4.34}$$

Dynamic oscillatory tests are used widely to characterize the viscoelasticity of a polymer solution. In this case, a sample is subjected to a small sinusoidally oscillating strain of amplitude γ_m and angular frequency ω. The resultant stress will be also sinusoidal with amplitude σ_0, but may have a phase difference of δ radians compared with the strain. When deformation is within the linear viscoelastic range, the following viscoelastic quantities are defined:[26]

$$G^* = G' + iG'' \tag{4.35}$$

$$G' = \frac{\sigma_0}{\gamma_0} \cos \delta \tag{4.36}$$

$$G'' = \frac{\sigma_0}{\gamma_0} \sin \delta \tag{4.37}$$

$$\tan \delta = \frac{G''}{G'} \tag{4.38}$$

$$\eta^* = \eta' - i\eta'' = \frac{G''}{\omega} - i\frac{G'}{\omega} \tag{4.39}$$

where G^* and η^* are called complex modulus and complex dynamic viscosity respectively. G' and G'' are the storage modulus and loss modulus representing the elastic and viscous contributions to the properties of a material respectively. The loss tangent, $\tan\delta$ is a measure of the ratio of viscous component vs. elastic component of the complex modulus. The elastic component is in phase with the strain; by contrast, the viscous component is 90° out of phase with the strain. Most of the polysaccharide systems fall between the two extremes. η' is the dynamic viscosity which is the in-phase component of the complex viscosity. It is related to the loss modulus G'' (Equation 4.39). For viscoelastic liquids, at low frequencies it approaches steady flow viscosity η_0. η'' is called the imaginary viscosity.

Evaluation of these viscoelastic quantities as a function of frequency within the linear viscoelastic range provides useful information about the rheological properties of polysaccharides. Based on the relative magnitudes of G' and G'' in a frequency sweep experiment, three types of polysaccharide systems may be distinguished: solution, weak gel, and gel.[27] As illustrated in Figure 4.21a, for dilute solutions of polysaccharides, $G'' > G'$ with $G'' \propto \omega$ and $G' \propto \omega^2$ at low frequencies. As oscillatory frequency or polymer concentration is increased, there is a cross over between G' and G'', implying that the system changes from being a viscous liquid to being a viscoelastic solid (Figure 4.21b). Both G' and G'' become less frequency dependent as the frequency is increased; a rubbery plateau of G' is observed at high frequency region. The mechanical spectrum of gels is shown in Figure 4.21c. $G' > G''$ at all frequencies and G' is independent of frequency while G'' only slightly increases with the increase of frequency. The tanδ of a weak gel is around 10^{-1}; for a true gel tanδ is much smaller (10^{-2}).

It has been frequently observed that for many polysaccharide solutions the complex dynamic viscosity $\eta^*(\omega)$ closely resembles the steady shear viscosity $\eta(\dot{\gamma})$ when the same numerical values of ω and $\dot{\gamma}$ are compared over several decades. This empirical correlation is called the Cox-Merz rule. For example, as illustrated by Figure 4.22, good superposition of $\eta^*(\omega)$ and $\eta(\dot{\gamma})$ was obtained for xyloglucan solutions of different concentrations at lower frequency range; small deviation was also noticed at higher frequency.[28] When ordered structures, such as aggregates of polymer chains or gel networks are present, considerable divergence between $\eta^*(\omega)$ and $\eta(\dot{\gamma})$ curves is observed as ω and $\dot{\gamma}$ increase. An example of this is the semi-dilute salt solutions of xanthan after heat treatment as shown in Figure 4.23.[29] Xanthan molecules normally adopt a rigid, ordered conformation in salt solution. This system has been recognized as having a weak gel structure that involves specific interchain couplings between the xanthan molecules. Therefore, the Cox-Merz rule can be used to distinguish a structured system from an isotropic polymeric solution.

4.3.4.7 Frequency Temperature/Concentration Superposition

For semi-dilute polysaccharide solutions, an increase in concentration results in an increase of both G' and G'', and also a shifting of the crossover point toward lower frequencies. When there are no conformational transitions or formations of supermolecular structure, the shape of the $G'(\omega)$ and $G''(\omega)$ curves may be alike for any given polysaccharide solutions of different concentration and measured at different temperature. A composite curve can

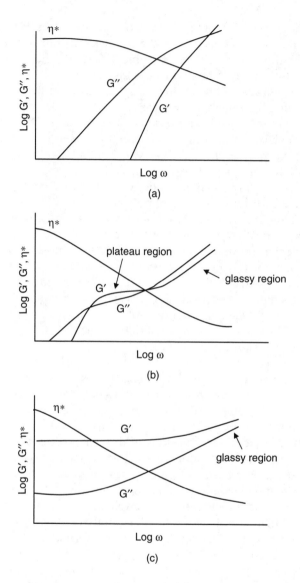

FIGURE 4.21
Mechanical spectra, storage modulus G', loss modulus G'' and dynamic viscosity η^* vs. frequency ω in a bilogarithmic plane. (a) a dilute solution, (b) a entangled semi-dilute solution and (c) a gel. (From Ross-Murphy, 1994. With permission.)[27]

be generated for the description of moduli frequency dependence by certain procedures. A general procedure is to reduce the measured moduli to a reference arbitrarily chosen temperature (T_0) and concentration (c_0) as following:

$$G'_r = G' \frac{T_0 c_0}{Tc} \qquad (4.40)$$

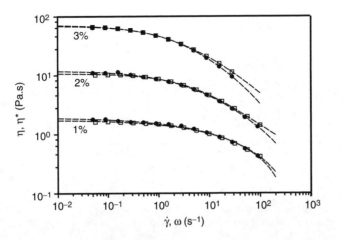

FIGURE 4.22
Steady shear viscosity η(●) and complex dynamic viscosity η*(□) plotted against shear rate ($\dot{\gamma}$) and frequency (ω) for 1%, 2% and 3% (w/w) detarium xyloglucan solutions. (Redrawn from Wang et al., 1997.)[28]

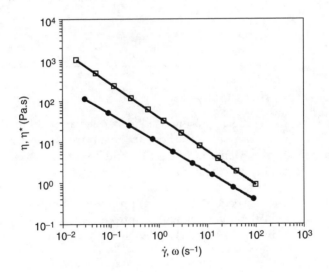

FIGURE 4.23
Steady shear viscosity η(●) and complex dynamic viscosity η*(□) plotted against shear rate ($\dot{\gamma}$) and frequency (ω) for 1% xanthan solutions. (Redrawn from Richardson and Ross-Murphy, 1987.)[29]

$$G_r'' = G'' \frac{T_0 c_0}{Tc} \tag{4.41}$$

The composite curves are then obtained by plotting the reduced moduli $G_r'(\omega)$ and $G_r''(\omega)$ against the reduced frequency $\omega_r = \omega\alpha$, where α is a temperature/concentration shift factor.[26] Figure 4.24 is an example of the

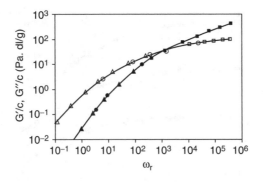

FIGURE 4.24
Frequency-concentration superposition for dynamic moduli of guar gum solutions: 1% (triangles), 2% (circles) and 3% w/w (squares), filled symbols: G', open symbols: G''. (Redrawn from Robinson et al., 1982.)[30]

superposition of viscoelastic data obtained from guar gum solutions at different concentrations.[30] The superposition procedure provides a device for enlarging the accessible frequency scale for experimental measurements of G' and G''.

4.4 Polysaccharides Gels

Polysaccharides as gelling agents are used in a wide variety of food products and the consumption of polysaccharides for this purpose is increasing rapidly with the increased demand of healthy and convenient foods. In order to improve the quality of existing products and processes, and to develop new ones, it is important to understand the mechanisms of gelation at a molecular level. In this section the general properties of polysaccharide gels and their characterisation methods will be introduced. The particular characteristics of each individual polysaccharide will be dealt with in Chapter 6.

4.4.1 General Characteristics

4.4.1.1 Gelation Mechanism

Food polysaccharide gels are prepared from aqueous solutions or dispersions of polysaccharides. The polymer chains are cross linked via covalent or noncovalent bonds so as to form a three-dimensional polymer network that fills the volume of the liquid medium. To induce gelation, polysaccharides have to be first dissolved or dispersed in a solution, in order to disrupt mostly the hydrogen bonds at the solid state. The subsequent transformation of the sols to gels is achieved by treatments such as change of temperature, addition of cations or cosolutes, and change of pH, etc. The purpose of these treatments is to decrease the intramolecular interactions and to increase the intermolecular interactions. Gelation of highly branched polysaccharide solutions such as gum arabic may simply be caused by effective molecular entanglements. However, adoption of regular three-dimensional structures, such as helices or flat ribbons, is a primary requisite for gelation of linear polysaccharides. Extended polysaccharide chains tend to tangle at higher concentrations. Similar molecules (or segments) can wrap around each other, forming multiple helices, without loss of hydrogen bonding but reducing conformational heterogeneity and minimizing hydrophobic surface contact with water. A minimum number of cross links need to be formed to overcome the entropy effect and to form a stable network. The cross linkages in polysaccharide gels usually involve extensive ordered segments from two or more polysaccharide chains, which form a well defined structure called junction zones. Several types of intermolecular interactions may contribute to the gelation of polysaccharides. These include hydrogen bonding, ionic or ion-dipole bonding, van der Waals attraction, and hydrophobic interactions. Figure 4.25 gives some examples of idealized junction zone models for polysaccharide gel networks. Figure 4.25a shows an egg box structure of cross linkages. The polysaccharide chains or their segments are associated into matched pairs in a two-fold ribbon-like conformation — the metal ions cooperatively bound, sitting inside the electronegative cavities like eggs in an egg box. Familiar examples of this type are gels from low methoxyl pectin and alginate. In Figure 4.25b the cross linkages involve the association of double helices through weak interactions such as van der Waals attractions or hydrogen bonding. In the presence of gel-promoting cations, these helices may also aggregate through ionic interactions as illustrated by Figure 4.25c. Figure 4.25d describes a junction zone model that is proposed for some neutral segments such as unsubstituted mannan segments in galactomannans and amidated pectin segments.

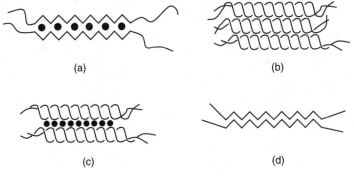

(a) (b)

(c) (d)

FIGURE 4.25
Examples of schematic models for junction zones of polysaccharide gels, (a) egg-box junction, (b) aggregated double helical junction, (c) cation promoted association of double helices (d) association of extended ribbon-like structure.

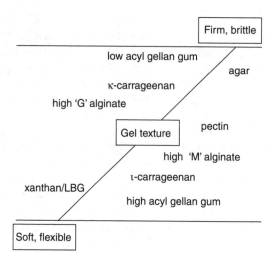

FIGURE 4.26
Qualitative comparison of the textures of polysaccharide gels. (Adapted from Williams and Phillips, 2000.)[31]

4.4.1.2   General Properties

Polysaccharides are able to form gels with a broad range of structures and properties. Gel textures characterized by mechanical properties could be from elastic to brittle, and from soft to hard as summarized by Williams and Phillips (Figure 4.26).[31] The texture of gels can be controlled by molecular characteristics of polysaccharides and/or by gelling conditions. Most polysaccharide gels are prepared by cooling a solution of a gelling polysaccharide, except for methylcellulose and its derivatives which can form gels upon heating. Gels formed by heating are called heat-setting gels; whereas gels formed by cooling are cold-setting gels. Gelation of some polysaccharides is thermoreversible, i.e., gel forms on cooling of hot solution and melts on heating, or vice versa. Under a given condition, each polysaccharide has a characteristic setting temperature (T_s) and a melting temperature (T_m) if thermoreversible gel is formed. The melting temperature of a gel can be equal or higher than the setting temperature. When $T_m > T_s$, a gel is said to exhibit thermal hysteresis. For example, while pronounced thermal hysteresis is seen for deacylated gellan gels (Figure 4.27a), high acyl gellan gels show no appreciable thermal hysteresis (Figure 4.27b).[32] Some other polysaccharides form thermally irreversible gels; examples are alginate gels formed on the addition of polyvalent cations and high methoxyl pectin gels formed at low pH.

Syneresis is not unusual for many polysaccharide gels, during which unbound water is eliminated from gel network. At early stages of gelation process, junction zones are relatively small in both number and extent. The

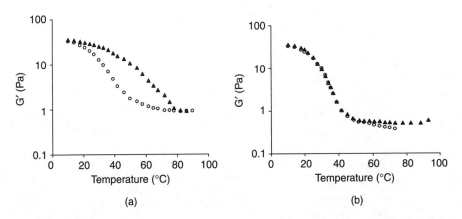

FIGURE 4.27
Variation of G' on cooling (●) and heating (▲) at 1°C/min for partially deacylated gellan preparations (0.5% wt% in deionised water) with acyl contents (% stoichiometric) of (a) 3% acetate and (b) 50% acetate. (From Morris et al., 1996. With permission.)[32]

size and number of junction zones grow with time and so does the gel strength. Further growth of junction zones may lead to formation of precipitates and elimination of water, i.e., syneresis of gel network. Therefore, some irregularities or kinks among the regular, linear chain segments help to limit the size of junction zones, and hence, reduce syneresis.

4.4.1.3 Factors Affecting Gelation

4.4.1.3.1 Structural Features

Chemical structure and molecular conformation of polysaccharides are the primary factors that determine their gel properties and gelation mechanisms. As mentioned earlier, in order to gel, polysaccharide chains or chain segments have to adopt ordered structures; these ordered structures crosslink with each other to form a stable three-dimensional network. For the same type of polysaccharides, the variation in fine structure of the polymer chains results in different gelation mechanisms and gels with different properties. Pectin is a typical example. As will be seen in the following chapters, the gelation mechanisms, thus gel properties, of high methyl pectins are very different from those of low methyl pectins. The degree of esterification of high methyl pectins determines the rate and temperature of gelation, as well as the amount of co-solute required for gelation. Another example is alginates. High D-mannuronic acid alginates form turbid gels with low elastic moduli; in contrast, high L-guluronic acid alginates form transparent, stiffer and more brittle gels.

4.4.1.3.2 Concentration and Molecular Weight Dependence

Gel formation only occurs at polymer concentrations exceeding a critical concentration (c_0). The gel strength, represented by G' (storage modulus) or

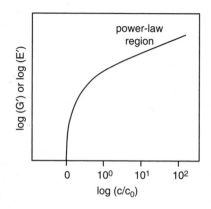

FIGURE 4.28
Idealized representation of the relationship between gel modulus and concentration for a polysaccharide gels.

E' (Young's modulus), typically increases with increasing concentration in a manner described in Figure 4.28. The value of log G' or log E' at first increases sharply when concentration reaches and is slightly above the critical concentration c_0, then an approximately power law relationship follows:

$$G' \text{ (or } E') \propto c^n \tag{4.42}$$

This is indicated by a linear region in the plot of log G' or log E' vs. log (c/c_0). The power index n is the slope of this plot which is in the range of 1.8 to 2.5 for most polysaccharide gels.

Molecular weight is also important for gelation. Intermolecular associations of polysaccharides are stable only above a minimum critical chain length necessary for the cooperative nature of the interaction, typically in the range 15 to 20 residues.[33] In general, the elastic modulus (G') increases with increasing molecular weight up to a certain point, then becomes independent of molecular weight, whereas the rupture strength continues to rise with increasing molecular weight.[33,34] However, it has been found that for some neutral polysaccharides, i.e., mixed linkage $(1\rightarrow3)(1\rightarrow4)$-β-D-glucans, gelation rate is inversely proportional to the molecular weight of the polysaccharides.[35-38] As a consequence, low molecular weight samples form stronger gels than high molecular weight samples within certain times. In fact, high molecular weight β-glucans ($> \sim5 \times 10^5$ g/mol) from oat do not gel within reasonable experimental time. It has been proposed that low molecular weight $(1\rightarrow3)(1\rightarrow4)$-β-D-glucans exhibit higher mobility in solution, which increases the probability of interacting with each other, hence, forming ordered structures (junction zones). These processes eventually may lead to the formation of gels.

4.4.1.3.3   Ionic Strength and pH Effects

For some polysaccharides, altering pH or the type and amount of counterions, can considerably change the properties of resultant gel. Gelation of anionic polysaccharides is particularly affected by the type and concentration of associated cations. In general, at low cation to polysaccharide ratios, increasing cation concentration improves gelling properties until an optimum is achieved. After the optimum ionic concentration is reached, further addition of cations will result in a decline of gel strength, syneresis or precipitation of polysaccharides. The association of charged tertiary structures may be promoted by specific counterions whose radius and charge are suitable for incorporation into the structure of the junction zones. For example, although both K^+ and Ca^{2+} induce gelation of either κ-carrageenan or ι-carrageenan, when both cations present, K^+ concentration is the factor determining the sol-gel transition of κ-carrageenan, whereas Ca^{2+} concentration determines the sol-gel transition of ι-carrageenan.[39]

4.4.1.3.4   Co-Solute Effects

The presence of some low molecular weight co-solutes such as sugars may enhance the gelation of polysaccharides. The presence of such co-solutes will compete for water with the polysaccharides, which promotes interactions between polysaccharide molecules. Gel properties usually increase with increasing concentration of co-solutes to reach maxima. Nickerson et al.[40] showed that the presence of co-solutes (0 to 50% sucrose and corn syrup) enhanced gelation of gellan gum during cooling, where the setting temperature and gel strength (G') both increased as co-solute levels increased (Figure 4.29). The amount of thermal hysteresis also increased with increasing levels of co-solutes in this range. The addition of excess levels of co-solutes (>70% w/w) caused a weakening of the gel structure, where gel strength (G') declined with increasing co-solute concentrations.

4.4.1.3.5   Mixed Gels

Mixed gels formed by two or three polysaccharides may impart novel and improved rheological characteristics to food products. Synergy is observed for a number of binary systems including pectin-alginate, galactomannans with polysaccharides of rigid conformations, such as xanthan, agarose, carrageenan, and yellow mustard gum. In these mixtures, synergism confers either enhanced gel properties at a given polysaccharide concentration, or gelation under conditions in which the individual components will not gel. By varying the ratios of the two or three polysaccharides, it is possible to create a series of gel structures. Figure 4.30 demonstrates the composition dependence of the gel strength of xanthan-locust bean gum mixed gels at a fixed total polymer concentration.[41] Although the gelation mechanisms for mixed polysaccharides are still controversial, there is evidence that some form of binding and structure compatibility has to be present between the two polysaccharides.[42]

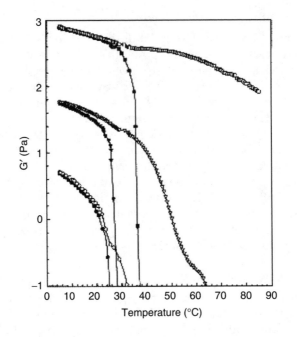

FIGURE 4.29
Variation of *G′* during cooling (solid) and heating (open) for 0.5% (w/w) gellan solutions in the presence of 0% (circle), 25% (triangle) and 50% (w/w) (square) co-solutes. (From Nickerson, et al., 2004. With permission.)[40]

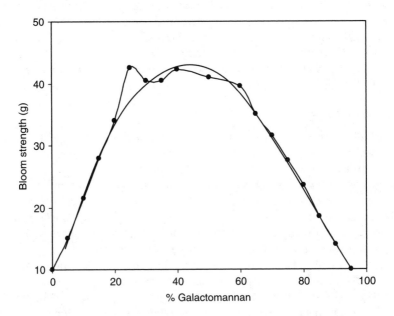

FIGURE 4.30
Composition dependence of gel strength in xanthan-locust bean gum gels at a fixted total polymer concentration of 1% w/v. (Adapted from Schuppner, 1967.)[41]

4.4.2   Characterization techniques

4.4.2.1   Rheological Methods

Fundamentally, rheological measurements of polysaccharide gels can be divided into two types: small or large deformation tests. These two types of measurements provide different information which is complementary but does not necessarily correlate. Small deformation tests characterize the viscoelastic properties of a gel; this method is more appropriate for elucidating the structural features and gelling mechanisms of polysaccharides. Large deformation tests, however, are usually associated with processing and usage of gels. In a large deformation test, the stress strain relationship obtained can be used not only to measure the elastic properties at small deformation, but also to characterize the inelastic properties, or failure properties of gels.

4.4.2.1.1   Small Deformation Test

Small deformation tests can be performed in both shear and compressive (or tensile) modes. Shear or Young's moduli are measured within the so called linear viscoelastic region in which the measured modulus is independent of strain. This limited strain region varies with the nature of the polysaccharide gels and has to be determined experimentally. Most polysaccharide based gels are essentially strain independent when deformation is controlled within 0.1 (10%).[43] Dynamic oscillatory rheometry has often been used to characterize the viscoelastic behavior of polysaccharide gels. As discussed previously (section 4.3.4.6), the storage modulus, loss modulus, and the phase angle obtained from a dynamic mechanical analysis give useful information about the rheological properties of a gel. In general, large values of storage modulus suggest that the gel is rigid and resistant to deformation (strong gel behavior). Because of its nondestructive manner, the dynamic mechanical analysis is often employed to monitor the gelation process such as those shown by Figure 4.27 and Figure 4.29. This technique has found extensive uses in conjunction with other techniques (such as differential scanning calorimetry (DSC) and nuclear magnetic resonance (NMR) on the study of gelation mechanisms of polysaccharides.

Alternative rheological approaches for characterization of gels are the creep and relaxation tests. In creep tests the time-dependent deformation induced by applying a constant stress and often the recovery of the compliance upon removal of the stress are monitored. Results of a creep test are generally expressed as a time-dependent function of creep compliance $J(t)$:

$$J(t) = strain(t)/stress \qquad (4.43)$$

For an elastic solid, after application of the stress, $J(t)$ rises to a constant instantly; on removal of the stress, $J(t)$ returns to zero. However, for a viscous liquid, $J(t)$ increases steadily as time proceeds; there is no recovery of the deformation after the stress is removed. For a viscoelastic material including

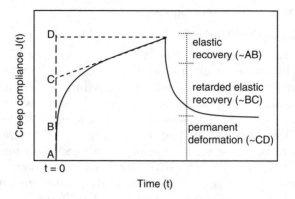

FIGURE 4.31
The creep compliance J(t) as a function of time t after applying a constant shear stress for a viscoelastic material.

biopolymer gels, a typical creep compliance curve consists of three principal regions.[44] As is shown in Figure 4.31, J(t) first increases instantaneously upon application of the stress (A to B), followed by a retarded elastic region (B to C), finally enters a linear region over long times (C to D). From the slope of the linear region the viscosity of the tested sample can be calculated. On sudden removal of the stress, the curve contains an elastic recovery region, followed by a retarded elastic recovery region, then a permanent deformation region. Analysis of the compliance in each region provides important information on the viscoelastic properties of tested samples.

In a relaxation test, the stress decay with time is recorded following the application of a constant deformation to the sample (Figure 4.32). The stress

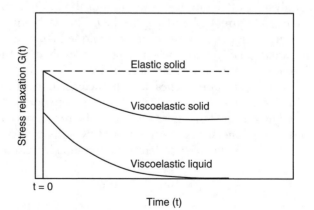

FIGURE 4.32
Stress relaxation experiment showing response for an elastic solid, a viscoelastic solid and a viscoelastic liquid. (Adapted from Mitchell, 1980.)[45]

required to maintain the deformation is measured and results are expressed in terms of the stress relaxation G(t):

$$G(t) = \text{stress}(t)/\text{strain} \tag{4.44}$$

For an ideal elastic solid, G(t) is equal to the elastic modulus, which stays constant with time. For a viscoelastic material, G(t) will decay rapidly at the beginning and then gradually slow down. Ultimately, G(t) may become zero (viscoelastic liquid) or approach a finite value (viscoelastic solid).[45] The stress relaxation results can be analyzed and quantified by calculating the relaxation spectrum[26] or more often by using different mechanical models, for example, a generalized Maxwell model.[46]

4.4.2.1.2 Large Deformation Tests

Uniaxial compression of a gel sample between two flat parallel plates is the most simple and popular technique for large deformation tests. Another type of popular empirical approach uses a plunger that may be driven into a gel sample. In both methods, sample is either compressed between two plates (the compression test) or penetrated by a plunger (penetrometer test) at a constant rate and the required force is measured (Figure 4.33). In some instruments, simply the force required to press a plunger of specified distance into the gel is measured (Bloom Gelometer) and used as an indication of the hardness of a gel. The Young's modulus may be calculated from the initial slope of the stress strain curve. The gel's brittle character is described by the maximum force (fracture stress) and the accompanying strain (fracture strain) at fracture. These parameters are correlated with what is commonly referred to as gel strength. The area below the stress strain curve corresponds to the total energy of fracture. A variety of instruments that work in a similar

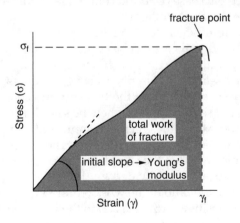

FIGURE 4.33
Schematic plot of stress σ vs. strain γ for a gel material under nonlinear deformation up to rupture. σ_f and γ_f are fracture stress and fracture strain respectively.

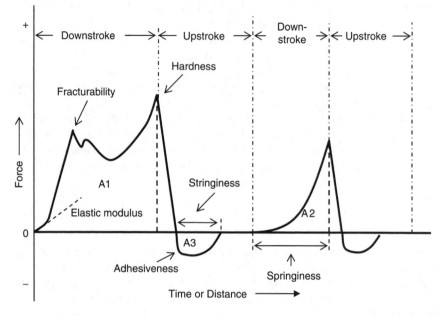

FIGURE 4.34
Typical profile for a texture profile analysis for a gel. (Adapted from Bourne, 2002.)[50]

principle were developed and found extensive uses in the characterization of food gels.[47]

In an attempt to relate the mechanical tests to sensory assessments of food texture, an instrumental technique called texture profile analysis (TPA) was developed.[48,49] It is possible to derive a whole series of apparent measures of a gel texture from a TPA test. In such a test, a sample of standard size and shape is placed on a base plate and compressed and decompressed twice by an upper plate attached to the drive system. The force vs. deformation curve is recorded from which seven textural parameters may be extracted: hardness, elasticity, adhesiveness, cohesiveness, brittleness, chewiness, and gumminess.[50] A typical TPA curve is shown in Figure 4.34. Similar to the single point compression test described above, hardness is obtained from the maximum force exerted during the first compression cycle and fracturability is the amount of strain applied when the sample fractures. Springiness is a measure of the extent to which a compressed sample returns to its original size after the force of the first compression cycle is removed. It is calculated as the distance that the sample is compressed during the second compression to the peak force divided by the initial sample height. Cohesiveness is defined as the ratio of the work required to compress the sample on the second bite (positive force area A2) to the work required to compress the sample on the first compression (positive force area A1). It measures how much the sample structure is broken down by the first compression, thus represents the degree of difficulty in breaking down the gel's internal structure. The negative force area of the first compression (A3) corresponds to the work required to pull the upper plate away from the sample and is defined as adhesiveness. The gumminess is calculated as hardness × cohesiveness and related to the energy required to disintegrate a sample so that it is ready for swallowing. The chewiness is defined as the product of gumminess × springiness.

4.4.2.2 Thermal Analysis

4.4.2.2.1 Principles of Thermal Analysis

As described earlier, some polysaccharides form gels upon cooling. Thermoreversible gels will melt if the gels are heated to exceed a certain temperature (melting temperature). There are a number of physical events occurring during the heating and cooling processes. For example, endothermic transition occurs when heat flows into a sample caused by disruption of ordered structures, such as melting of a gel or crystalline. Exothermic transition occurs when heat is released from a sample system during the formation of ordered structures, such as gels and crystallites. Method used to monitor thermal events as a function of temperature/time is called thermal analysis (TA). Of the various TA techniques available (differential thermal analysis, DTA; differential scanning calorimetry, DSC; dynamic mechanical analysis, DMA; thermogravimetric analysis, TGA; thermomechanical analysis, TMA), DSC is the most popular technique used for characterization of polysaccharide gels. Information provided by DSC measurements usually includes temperatures for glass transition (T_g), melting (T_m) and gelation (T_s); enthalpies (ΔH) during melting and gelation are also obtained. An example of typical DSC thermograms and detected transitions for polysaccharide gels is illustrated in Figure 4.35. Heating polysaccharide gels at very high temperature may result in oxidation and/or decomposition of the polymer chain.

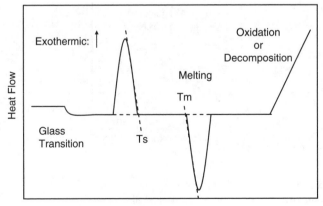

FIGURE 4.35
Typical DSC thermogram and detected transitions for polysaccharide gels.

TA techniques can provide qualitative and quantitative data about any physical and chemical changes in polysaccharide gels that involve endothermic or exothermic processes. These methods are fast, easy to operate, and simple in sample preparation. In addition, TA methods can be applied to both liquid and solid samples.

4.4.2.2.2 Application of DSC Analysis in Polysaccharide Gels

Most of polysaccharide-based gels are cold-setting gels, except methylcellulose and its derivatives, which form gels upon heating. These gels could be thermo-reversible or irreversible depending on gelation mechanisms. A cold-setting thermo-reversible gel has a gelling and a melting temperature while a thermo-irreversible gel only has a setting temperature. An example of using DSC to monitor the development of a polysaccharide gel by cooling the polysaccharide solution and consequently melting of the gel by heating is presented in Figure 4.36.[51] The cooling curve of 3.2% gellan gum exhibited a sharp exothermic peak at ~40°C (gel setting temperature, T_s) while the heating curve has a sharp endothermic peak at temperature (T_m) slightly

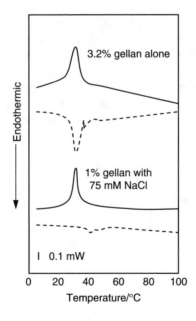

FIGURE 4.36
The DSC cooling (solid line) and heating (dashed line) curves of 3.2% and 1% gellan gums. (Redrawn from Miyoshi et al., 1996.)[51]

higher than the T_s. In addition, a small endothermic peak is also observed at higher temperature indicating a possible second ordered structure present in the gel. DSC can also be used to monitor changes in thermal properties of polysaccharide gels caused by other factors. For example, the addition of salt to 1% gellan gum aqueous system resulted in multiple endothermic peaks and the shift of T_m to a higher temperature (Figure 4.36).[51]

Many polysaccharide gels do not give a sharp melting temperature in a DSC thermogram; instead, multiple endothermic peaks are frequently observed with a broad gel-sol transition period. For example, a series of cryogels prepared from cereal β-D-glucans exhibited a broad range of gel-sol transition temperature (55 to 75°C) (Figure 4.37) suggesting different degrees or types of ordered structures or junction zones were formed during the gelation process. The endothermic peaks of cereal β-D-glucans are affected by molecular weight as well as structural features: the broadness of the endothermic peaks decreased with decreasing in molecular weight of β-D-glucans. Although samples with higher molecular weight seem to form less densely ordered junction zones (lower ΔH) due to diffusional constraints, the observed elevated melting temperature (T_m) indicates that longer segments of interchain associations were formed in higher molecular weight β-D-glucans, which are resistant to disruptions at lower temperatures.[37]

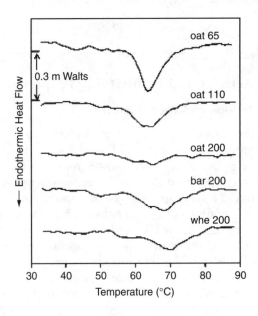

FIGURE 4.37
DSC thermograms of a series cryogels prepared from cereal β-D-glucans exhibited broad range of gel-sol transition temperature (55–75 °C). (From Lazaridou and Biliaderis, 2004. With permission.)[37]

The apparent melting enthalpy values (ΔH) reflect energy required for disrupting the ordered junction zones. For cereal β-D-glucans, the ΔH values of similar molecular weight correspond well with the relative amount of cellotriosyl unit (one of the major oligosaccharide units in cereal β-D-glucans; see Chapter 3 for details) in the polymer chain. For example, ΔH values of wheat, barley, and oat β-glucans are 3.6 mJ/mg, 3 mJ/mg and 2 mJ/mg, respectively, which are coincident with the order of the ratio of cellotriosyl (DP3) to cellotetrosyl (DP4) units in cereal β-D-glucans (which is 4.4, 3 and 2.3 for wheat, barley, and oat β-D-glucans, respectively). The sum of DP3 and DP4 is about 92 to 93% of the total polymer weight.[37,52] Apparently, the higher proportion of cellotriosyl unit in wheat β-D-glucans furnishes a higher density and/or extension of the ordered junction zones, which has been translated into more cooperative three-dimensional networks in the cryogel as reflected by a higher ΔH and T_m values from DSC analysis (Figure 4.37).[37]

Unlike other gelling systems, e.g., proteins and starch gels, nonstarch polysaccharide gels are often prepared at low polymer concentrations. Therefore, high sensitivity of instrumentation and large sample size will help to give reproducible data. DSC can also be used to determine other physicochemical properties of polysaccharide gels, such as thermal stability, reaction kinetics, glass transitions (in some cases), crystallisation time and temperature, percent of crystallinity, and specific heat capacity. These thermal properties are important parameters in food processing.

4.5 Surface Activity and Emulsifying Properties

4.5.1 Surface Tension and Food Emulsions

A molecule in the interior of a liquid is attracted by other surrounding molecules to all directions, as shown in Figure 4.38. However, a molecule at the surface will be pulled inward because the number of molecules per unit volume is greater in the liquid than in the vapor (Figure 4.38). As a result, the surface of a liquid always tends to contract to the smallest possible area. This is the reason that drops of a liquid always become spherical because it gives the minimum surface area. In order to increase the area of a liquid, energy is required to do the work. The work needed to increase the surface area by 1 cm^2 is defined as the free surface energy. The contracting nature of surface molecules enhanced the intermolecular attractive forces at the surface, i.e., surface tension. Surface tension is defined as the force in dynes required to break a surface film of length 1 cm. It is also described as the energy required for the increase of surface area of a liquid by a unit amount. Although surface energy is the fundamental property of a surface, in practice surface tension is more often used for convenience; surface energy in ergs/cm^2

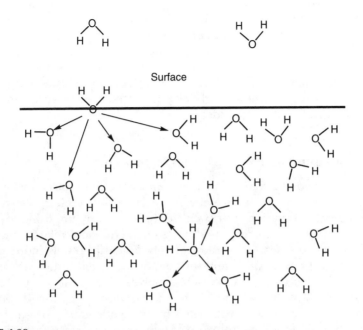

FIGURE 4.38
Surface attraction tension of water molecules inside a liquid and at the surface.

is numerically equal to surface tension in dynes/cm. The main molecular force in an aqueous system is the hydrogen bond; this gives water a very high surface tension of 72.8 dynes/cm at 20°C (compared to 22.3 dynes/cm for ethanol).

The existence of a surface implies a separation between two media, e.g., liquid and gas, oil and water in emulsions. Both cases are commonly seen in many food applications. For example, foam in beer has a liquid and gas interface while salad dressings and creams have an oil and water interface. The surface tension between two liquids is also called interfacial tension. Some polysaccharides exhibit surface activity by reducing the surface tension of water and stabilizing emulsions, and thereby find broad applications in the food industry as emulsion stabilizers.

Oil-in-water or water-in-oil emulsions are widely used in the food industry. Nevertheless, the thermodynamically unstable nature of oil-in-water emulsions causes quality problems during production and storage of such food products. The oil phase tends to separate from the water phase due to immiscibility and density differences between the two liquids. In industrial practice, emulsifiers are used to reduce the interfacial tension, thereby, to slow the creaming process and to extend shelf life of the food products. Frequently, polysaccharide gums are added to such emulsions to stabilize the system. However, polysaccharides are mostly hydrophilic and not conventionally perceived as surface active compounds. The universal role of gums in emulsion systems is to thicken the continuous phase, thereby inhibiting or slowing droplet flocculation and/or creaming of the oils. There is evidence that a number of polysaccharide gums actually exhibit surface activity. These gums have a double role in an emulsion system: they act as both an emulsifier and a thickener. In this section, polysaccharides are divided into two groups: nonadsorbtion polysaccharides (no surface activity) and surface active polysaccharides.

4.5.2   Nonadsorption Polysaccharides in Emulsions

Nonadsorption polysaccharides in oil-in-water or water-in-oil emulsions may have both positive and negative effects. When added to oil-in-water emulsion, some polysaccharides can stabilize the system by imposing viscosity to the aqueous phase, thereby modifying the texture and retarding droplet creaming. In other cases, the addition of polysaccharides could induce destabilisation of dispersions and emulsions by the mechanism of depletion flocculation.[53] The flocculation caused by the polysaccharides is possibly driven by the exclusion of the polymer from the intervening gap associated with an attractive interparticle force created by the tendency of solvent to flow out from the gap under the influence of local osmotic pressure gradient.[53,54] Whether a polysaccharide is actually stabilizing the emulsions or promoting depletion flocculation depends on concentrations and the structural features of the polysaccharides, e.g., effective volume ratio (R_v, defined as the ratio of effective volume in solution over geometric volume of the polysaccharide molecules).[54] The effective volume ratio for globular molecules is ~1. The effective volume ratio of highly extended structures is much larger than that of compact structures. An example showing the dependence of depletion attraction between oil droplets on polysaccharide concentration and structural features (R_v) is depicted in Figure 4.39. At sufficiently low concentrations, no significant flocculation (depletion attraction < 4 kT) is observed. This is because the entropy loss linked to polymer aggregation outweighs the depletion effect; therefore, the system remains stable. However, when polysaccharide concentration exceeds a critical level, i.e., the critical flocculation concentration (CFC), flocculation occurs (defined as depletion attraction = 4 kT).[54] It is also clearly indicated in Figure 4.39 that the higher the effective volume ratio, the lower the critical flocculation concentration for the polymer. When polysaccharide concentrations are in an intermediate range, the oil droplets flocculate and cause creaming instability by a depletion mechanism. This effect predominates over the stabilizing effect contributed by the increase in viscosity (which retards creaming).[53,54] Xanthan gum is an example which has a high R_v; its presence at a very low concentration (0.025 or 0.05 wt%) is enough to induce depletion flocculation, leading to rapid phase separation (complete within 1–2 h) of a low viscosity liquid-like emulsion system.[53]

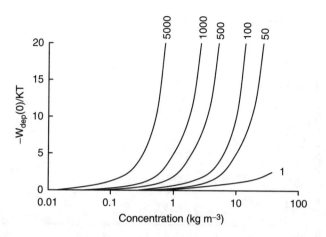

FIGURE 4.39
Dependence of depletion attraction (-Wdep(0)/kT) between oil droplets on polysaccharide concentration and structural features (expressed as effective volume ratio R_v, marked above the curves). (Adapted from McClements, 2000.)[54]

When polysaccharide concentration is sufficiently high, the viscosity of the continuous phase becomes so large that the droplets cannot move. In such situation polysaccharides form a three-dimensional network through intermolecular entanglements that entraps the oil droplets and effectively inhibits their movement, therefore prolonging the stability of the emulsion system.

4.5.3 Polysaccharides with Surface Activities

There are a number of naturally produced or chemically modified polysaccharides that are surface active. They may act as emulsifiers and emulsion stabilizers through adsorption at oil–water interfaces forming protective layers.[53] The ability of such polysaccharides to reduce the surface tension of water follows the general trend: the initial addition of polysaccharides will dramatically reduce the surface tension; further increase of polysaccharide concentration will result in continued decrease of the surface tension of water until a saturated concentration is reached; above which, no further reduction of surface tension should be observed (Figure 4.40). However, when determining the surface tension of a polysaccharide solution using the Du Nouy (ring) method, the surface tension of the solution could be overestimated at high polymer concentrations due to excessive adhesive or viscous nature of the solution.

According to Dickinson, a biopolymer should have the following characteristics to become effective in stabilizing dispersed particles or emulsion droplets:[53]

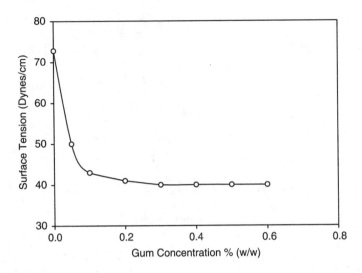

FIGURE 4.40
Reduction of surface activity of water as a function of polysaccharide concentration.

- Strong adsorption at the oil-water interface. The polysaccharides should be amphiphilic; that means in addition to the hydroxyl group (hydrophilic), the polymer should have a substantial degree of hydrophobic character (e.g., nonpolar side chains or a peptide/protein moiety) to keep it permanently anchored to the interface.

- Complete surface coverage. This requires that the polysaccharides are soluble in water so that there is sufficient polymer chain present in the continue phase to fully cover the surface of the oil droplets.

In addition, the formation of a thick steric and/or charged protecting layer will help to stabilize the emulsion systems.

A number of polysaccharides have been converted to amphiphilic polymers by introducing hydrophobical groups to the sugar units; these include methylcellulose and hydroxypropyl methylcellulose, polypropylene glycol alginate, and hydrophobically modified starches (see Chapter 8). There are also naturally occurring polysaccharides containing hydrophobic groups, for example, acetylated pectin from sugar beet. Such polysaccharides can be used as both emulsifier and stabilizer in the food industry.[55]

There is another group of polysaccharide gums which do not possess the desired amphiphilic properties, but still have substantial surface activity and emulsification properties. For example, gum arabic is the most commonly recognized hydrocolloid emulsifier which is widely used in the soft drinks industry for emulsifying flavor oils (e.g., orange oil) under acidic conditions. Gum arabic is a genuine emulsifier which can form a macromolecular stabilizing layer around oil droplets due to its well known film forming ability.[53] The protein-rich high molecular weight species are preferentially adsorbed onto the surface of oil droplets while the carbohydrate portion

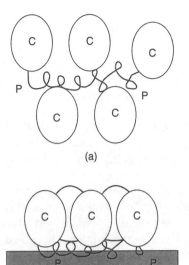

FIGURE 4.41
Models of gum arabic as an emulsifier in oil-in-water emulsions: (a) Sketch representation of gum arabic structure: c and p represent the carbohydrate and protein moieties respectively; (b) Behavior of gum arabic at the oil/water interface. (Redrawn from Connolly et al., 1988.)[56]

inhibits flocculation and coalescence by electrostatic repulsions and steric forces. This active fraction of gum arabic contains arabinogalactan–protein complexes, in which several polysaccharide units are linked to a common protein core. A wattle blossom model of the arabinogalactan–protein complexes and the stabilizing mechanism in oil-in-water emulsions is depicted in Figure 4.41.[56]

Galactomannans are a group of polysaccharides with rigid hydrophilic backbone (mannan) and grafted galactose units. Since the carbohydrate structure does not suggest the presence of any hydrophobic group, it is generally assumed that this type of hydrocolloid stabilizes emulsions by modifying the rheological properties of the aqueous phase between the dispersed particles or droplets.[53] However, numerous studies have reported that galactomannans exhibit surface activity and are able to stabilize oil-in-water emulsions. Some researchers have attributed the emulsion stabilizing properties of galactomannans to the presence of small amounts of hydrophobic proteins associated with the gums. Nevertheless, purified guar or locust bean gum still have the capacity to emulsify oils and stabilize fairly coarse emulsions at a moderately low gum/oil ratio.[55] Fenugreek gum, another member of the galactomannan family, exhibited unusual strong surface activity and emulsification properties.[57,58] Further purification of the gum by physical or enzymatic (pronase hydrolysis) treatments reduced nitrogen content to less than 0.1% (converted to protein of about 0.6%). The enzymatic purification process reduced the surface activity, however, the purified fenugreek gum still exhibited substantial surface activity and was able to make stable emulsions of moderately low average droplet size (~3 μm).[58] This raises some questions about how much of the emulsification properties of galactomannans were from the protein moieties associated with the gums? Is there any other hydrophobic character in the galactomannan molecule which contributed to the emulsification properties of the gum? It is worth noting that the structure of fenugreek gum is significantly different from those of other galactomannans in that almost all the mannosyl residues of the backbone chain are substituted by a single galactosyl residue. Does this play a role in the unique emulsifying property of this polymer? Further research is required to provide answers to these questions.

Other polysaccharide gums have also been used to stabilize emulsions in the food industry, such as pectins, partially hydrolyzed pectins, and soluble soybean polysaccharides. The mechanisms involved in the stabilizing processes of these polysaccharides remain unclear.

In real food systems, emulsions usually contain mixtures of hydrocolloids, proteins, and small molecule surfactants. The nature of the hydrocolloid–surfactant interactions and the interactions between polysaccharides and proteins will influence both rheology and stability of the system. In such a complex system, fully understanding these interactions will be useful in making long lasting, stable, emulsion-based food products.

References

1. Carraher, C. E., *Seymour/Carraher's Polymer Chemistry*. New York: Marcel Dekker, 52–95. 2000.
2. Wang, Q., Wood, P. J., Huang, X., and Cui, W. Preparation and characterization of molecular weight standards of low polydispersity from oat and barley (1,3)(1,4)-β-D-glucan. *Food Hydrocolloids*, 17, 845–853. 2003.
3. Tanford, C. *Physical Chemistry of Macromolecules*. New York: John Wiley & Sons Inc. 210–221. 1961.
4. Harding, S. E., Vårum, K. M., Stokke, B. T., and Smidod, O. Molecular weight determination of polysaccharides. *Advances in Carbohydrate Analysis*, 1, 63–144. 1991.
5. Zimm, B. H. The scattering of light and the radial distribution function of high polymer solutions. *Journal of Chemical Physics*, 16, 1093–1099. 1948.
6. Burchard, W. Light scattering. In: S. B. Ross-Murphy. ed. *Physical Techniques for the Study of Food Biopolymers*. New York: Blackie Academic & Professional, 151–214. 1994.
7. Tabor, B. E. Preparation and clarification of solutions. In: M. B. Huglin, ed. *Light Scattering from Polymer Solutions*. London: Academic Press, 1–25. 1972.
8. Benoit, H., Grubisic, Z., Rempp, P., Decker, D., and Zilliox, J. G. Liquid-phase chromatographic study of branched and linear polystyrenes of known structure. *Journal of Chemical Physics*, 63, 1507–1514. 1966.

9. Haney, M. A., Mooney, C. E., and Yau, W. W. A new right angle laser light scattering detector for size exclusion chromatography in the triple detector configuration. *International Labmate*, 18, 1–3. 1992.

10. Wang, Q., Ellis, P. R., and Ross-Murphy, S. B. Dissolution kinetics of guar powders-II. Effects of concentration and molecular weight. *Carbohydrate Polymers*, 53, 75–83. 2003.

11. Wang, Q., Wood, P. J., and Cui, W. Microwave assisted dissolution of beta-glucan in water — implications for the characterization of this polymer. *Carbohydrate Polymers*, 47, 35–38. 2002.

12. Devotta, I., Badiger, M. V., Rajamohanan, P. R., Ganapathy, S., and Mashelkar, R. A. Unusual retardation and enhancement in polymer dissolution: Role of disengagement dynamics. *Chemical Engineering Science* 50, 2557–2569. 1995.

13. Ouano, A. C. Dissolution kinetics of polymers: Effect of residual solvent content In: R. B. Seymour and G. A. Stahl, eds. *Macromolecular Solutions. Solvent-Property Relationships in Polymers*. New York: Pergamon Press, 208–217. 1992.

14. de Gennes P.G. *Scaling Concepts in Polymer Physics*. New York: Cornell University Press. 1979.

15. Steiner, E., Divjak, H., Steiner, W., Lafferty, R. M., and Esterbauer, H. Rheological properties of solutions of a colloid-disperse homoglucan from *Schizohyicum commune*. *Progress in Polymer Science*, 217–220. 1988.

16. Harding, S. E. Dilute solution viscometry of food biopolymers. In: S. E. Hill, D. A. Ledward, and J. R. Mitchell, eds. *Functional Properties of Food Macromolecules*. Gaithersburg, MD: Aspen Publisher, 1–49. 1998.

17. Hitchcock, C. D., Hammons, H. K., and Yau, W. W. The dual-capillary method for modern-day viscometry. *American Laboratory*, 26, 26–31. 1994.

18. Wu, C., Senak, L., Bonilla, J., and Cullen, J. Comparison of relative viscosity measurement of polyvinylpyrrolidone in water by glass capillary viscometer and differential dual capillary viscometer. *Journal of Applied Polymer Science*, 86, 1312–1315. 2002.

19. Haney, M. A. A new differential viscometer. *American Laboratory*, 17, 41–56, 116–126. 1985.

20. Pals, D. T. and Hermans, J. J. Sodium salts of pectin and carboxy methyl cellulose in aqueous sodium chloride. *Recueil des Travaux Chimiques des Pays-Bas*, 71, 433–437. 1952.

21. Smidsrød, O. Solution properties of alginate. *Carbohydrate Research*, 13, 359–372. 1970.

22. Launay, B., Doublier, J. L., and Cuvelier, G. Flow properties of aqueous solutions and dispersions of polysaccharides. In: J. R. Mitchell and D. A. Ledward, eds. *Functional Properties of Food Macromolecules*. New York: Elsevier Applied Science Publishers. 1–78. 1986.

23. Lapasin, R. and Pricl., S., *Rheology of Industrial Polysaccharides*. London: Blackie Academic and Professional, 250–494. 1995.

24. Young, S. A. and Smart, J. D. An investigation of the rheology of methylcellulose solutions using dynamic oscillatory, flow and creep experiments. *European Journal of Pharmaceutical Sciences*, 4 (S1), S65. 1996.

25. Morris, E. R., Cutler, A. N., Ross-Murphy, S. B., Rees, D. A., and Price, J. Concentration and shear rate dependence of viscosity in random coil polysaccharide solutions. *Carbohydrate Polymers* 1, 5–21. 1981.

26. Ferry, J. D. *Viscoelastic Properties of Polymers* (3rd edition). New York: John Wiley. 1980.

27. Ross-Murphy, S. B. Rheological methods. In: S. B. Ross-Murphy, ed. *Physical Techniques for the Study of Food Biopolymers*, New York: Blackie Academic Professional, 343–392, 1994.
28. Wang, Q., Ellis, P. R., Ross-Murphy, S. B., and Burchard, W. Solution characteristics of the xyloglucan extracted from detarium senegalense Gmelin. *Carbohydrate Polymers*, 33, 115–124. 1997.
29. Richardson, R. K. and Ross-Murphy, S. B. Non-liner viscoelasticity of polysaccharide solutions. 2: Xanthan polysaccharide solutions. *International Journal of Biological Macromolecules*, 9, 257–264. 1987.
30. Robinson, G. Ross-Murphy, S. B., and Morris, E. R. Viscosity-molecular weight relationships, intrinsic chain flexibility, and dynamic solution properties of guar galactomannan. *Carbohydrate Research* 107[1], 17–32. 1982.
31. Williams, P. A. and Phillips, G. O. Introduction to food hydrocolloids. In: P. A. Williams, G. O. Phillips. eds. *Handbook of Hydrocolloids*. Boca Raton: CRC Press, 1–20. 2000.
32. Morris, E. R., Gothard, M. G. E., Hember, M. W. N., Manning, C. E., and Robinson, G. Conformational and rheological transitions of welan, rhamsan and acylated gellan. *Carbohydrate Polymers*, 30, 165–175. 1996.
33. Whistler, R. L. Solubility of polysaccharides and their behavior in solution. *Advances in Chemistry Series*, 117, 242–255. 1973.
34. Mitchell, J. R. and Blanshard, J. M. V. On the nature of the relationship between the structure and rheology of food gels. In: P. Sherman, ed. *Food Texture and Rheology*. London: Academic Press, 425–435. 1979.
35. Böhm, N. and Kulicke, W. M. Rheological studies of barley $(1\rightarrow3)(1\rightarrow4)$-$\beta$-glucan in concentrated solution: mechanistic and kinetic investigation of the gel formation. *Carbohydrate Research*, 315, 302–311. 1999.
36. Vaikousi, H. , Biliaderis, C. G., and Izydorczyk, M. S. Solution flow behavior and gelling properties of water-soluble barley $(1\rightarrow3)(1\rightarrow4)$-$\beta$-glucans varying in molecular size. *Journal of Cereal Science*, 39, 119–137. 2004.
37. Lazaridou, A. and Biliaderis, C. G. Cryogelation of cereal β-glucans: structure and molecular size effects. *Food Hydrocolloids*, 18, 933–947. 2004.
38. Tosh, S. M., Wood, P., Wang, Q., and Weisz, J. Structural characteristics and rheological properties of partially hydrolyzed oat β-glucan: the effects of molecular weight and hydrolysis method. *Carbohydrate Polymers*, 55, 425–436. 2004.
39. Stanley, N. F. Carrageenans. In: P. Harris, ed. *Food Gels*. New York: Elsevier Applied Science, 79–119. 1990.
40. Nickerson, M. T., Paulson, A. T., and Speers, R. A. Time-temperature studies of gellan polysaccharide gelation in the presence of low, intermediate and high levels of co-solutes. *Food Hydrocolloids*, 18, 783–794. 2004.
41. Schuppner, H. R. Jr. Heat reversible gel and method. Australian Patent 401 434. 1967.
42. Morris, V. J. Gelation of polysaccharides. In: S. E. Hill, D. A. Ledward, and J. R. Mitchell. eds. *Functional Properties of Food Macromolecules*. Gaithersburg, MD: Aspen Publisher, 143–226. 1998.
43. Mitchell, J. R. Rheology of gels. *Journal of Texture Studies*, 7, 313–339. 1976.
44. Sherman, P. *Industrial Rheology*. New York: Academic Press, 12–20. 1970.
45. Mitchell, J. R. The rheology of gels. *Journal of Texture Studies*, 11, 315–337. 1980.
46. Gross, M. O., Rao, V. N. M., and Smit, C. J. B. Rheological characterisation of low methoxyl pectin gel by normal creep and relaxation. *Journal of Texture Studies*, 11, 271–290. 1980.

47. Rolin, C. and de Vries, J. *Food Gels*. London: Elsevier Applied Science Publishers Ltd, 401–434. 1990.

48. Szczesniak, A. Classification of textural characteristics. *Journal of Food Science*, 28, 385–389. 1963.

49. Friedman, H. H., Whitney, J. E., and Szczesniak, A. S. The texturometer: A new instrument for objective texture measurement. *Journal of Food Science*, 28, 390–396. 1963.

50. Bourne, M. C. *Food Texture and Viscosity: Concept and Measurement*. New York: Academic Press. 182–186. 2002.

51. Miyoshi, E., Takaya, T., and Nishinari, K. Rheological and thermal studies of gel-sol transition in gellan gum aqueous solutions. *Carbohydrate Polymers*, 30, 109–119. 1996.

52. Cui, W. *Polysaccharide Gums from Agriculture Products: Processing, Structure and Functional Properties*. Lancaster, PA: Technomic Publishing Company Inc. 103–166. 2001.

53. Dickinson, E. Hydrocolloids at interfaces and the influence on the properties of dispersed systems. *Food Hydrocolloids*, 17, 25–39. 2003.

54. McClements, D. J. Comments on viscosity enhancement and depletion flocculation by polysaccharides. *Food Hydrocolloids*, 14, 173–177. 2000.

55. Gaonkar, A. G. Surface and interfacial activities and emulsion characteristics of some food hydrocolloids. *Food Hydrocolloids*, 5, 329–337. 1991.

56. Connolly, S. , Fenyo, J.-C., and Vandevelde, M.-C. Effect of a proteinase on the macromolecular distribution of acacia senegal gum. *Carbohydrate Polymers*, 8, 23–32. 1988.

57. Garti, N., Madar, Z., Aserin, A., and Sternheim, B. Fenugreek galactomannans as food emulsifiers. *Food Science and Technology*, 30, 305–311. 1997.

58. Brummer, Y., Cui, W., and Wang, Q. Extraction , purification and physicochemical characterization of fenugreek gum. *Food Hydrocolloids*, 17, 229–236. 2003.

5

Understanding the Conformation of Polysaccharides

Qi Wang and Steve W. Cui

CONTENTS

5.1   Basics of Polysaccharide Conformation

5.1.1   Introduction

Previous chapters covered the basic chemistry and physical properties of food polysaccharides. In order to understand the molecular basis of polysaccharides in relation to their functional properties, it is important to appreciate higher levels of structures of polysaccharides based on the shape or conformation of these polymers in nature. This chapter will introduce some basic concepts, terminologies, and characterization methodologies used in studying the conformation of polysaccharides. For advanced studies of polymer conformation analysis, readers are referred to literatures at the end of this chapter.

5.1.2   Concepts and Terminologies

Polysaccharide chains are generally formed by a repeated sequence of monomers or oligomers. The monosaccharide units in a polysaccharide are restricted to rotate about the glycosidic bonds. The relative orientations between any two participating monosaccharide units are defined by two or three torsion angles around the glycosidic bonds, same as defined for disaccharides (Chapter 1, Figure 1.24). The conformation of any individual monosaccharide is relatively fixed in the polysaccharide chain, however, the sugar residues linked through glycosidic linkage will rotate around the glycosidic bond and often tend to adopt an orientation of lower or lowest energies. As a result, a polysaccharide chain consisting of specific primary structure will assume a characteristic geometrical shape in space, such as ribbon or helix. These shapes or conformations polysaccharides assume are described as the secondary structures.

There are two general classes of conformation for polysaccharides: ordered conformation and disordered conformation. In the ordered conformation, the values of the torsion angles are fixed by cooperative interactions between residues, such as in solid or gel states. In the disordered conformation there exists continued fluctuations of local conformation (between each pair of neighboring units and individual sugar rings) and overall chain conformation. Random coil conformation is a typical property of linear polysaccharides in solutions.

Polysaccharide chains with well-defined secondary structures may interact with each other to form tertiary structures, which are ordered organizations

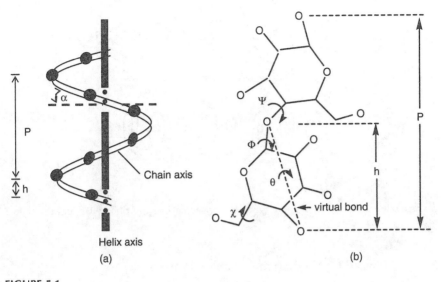

FIGURE 5.1
Schematic illustrations of helices: (a) five fold helix; (b) two fold helix.

involving a group of polysaccharide chains. For example, some polysaccharides exist in solution as bundles or micelles containing from a few to several tens of polysaccharide chains. Further association of these ordered entities leads to large three-dimensional networks — i.e., the quaternary structures.

Conventionally, the regular conformation of polysaccharides is described in terms of a helix as schematically explicated in Figure 5.1. Helices or helix characteristics can be described using the following parameters:

- n = number of repeating units per turn
- h = unit height (translation per repeating unit along the helix axis)
- p = pitch height of helix = $n \cdot h$

Since the ring geometry is usually fixed, it is convenient to treat the entire sugar ring as a single virtual bond. The chirality of a helix is expressed by assigning a positive or negative sign either to n or h: a positive value for a right handed screw and a negative value for a left handed screw.

Another important parameter in characterizing polysaccharide conformation is the characteristic ratio C_∞, which is defined as the ratio of the mean-square end-to-end distance of coil molecules at unperturbed condition to that of freely jointed chains. It describes the degree of chain expansion of the polymer in a given solvent. A high value of C_∞ indicates the polymer having an extended conformation; by contrast, low values of C_∞ suggest the polymer taking a compact conformation.

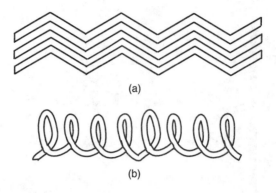

(a)

(b)

FIGURE 5.2
Linkage patterns and conformation types of polysaccharides: (a) ribbon-like and (b) hollow helix.

5.1.3   Ordered Conformation of Polysaccharides

5.1.3.1   *Conformation of Homopolysaccharide Chains*

The best characterized polysaccharide group in terms of molecular conformation is linear homoglycans which contain a single type of sugar residue in a single type of linkage. The ordered conformation of polysaccharides consisting of N repeating units can be represented by (N-1) sets of dihedral angles (Φ_i, ψ_i) between the repeating units. According to Rees,[1] the secondary structure of homoglycans are classified into four conformational types based on the dihedral angle formed between the bonds to and from each residue across the sugar ring: ribbon-like, hollow helix, crumpled, and loosely jointed types. The most common types, ribbon-like and hollow helix types of conformation, are illustrated in Figure 5.2. The overall chain shape of polysaccharides is primarily determined by the geometrical relationships within each sugar unit rather than the interaction energies between them. Representations of some geometrical relationships for various sugar residues are shown in Figure 5.3. The following generalized interrelationships exist: polymer chains with zig-zag arrangements (Figure 5.3a,e) almost always fall into the ribbon family; whereas polymers with U-turn arrangements (Figure 5.3b,f) usually belong to the helix family.

5.1.3.1.1   *Ribbon-like Type Conformation*

Polysaccharide chains in which the dihedral angle is close to 180° are designated as type A.[1] For a (1→4) linked β-D-glucosyl residue and a (1→4) linked α-D-galactosyl residue, the bonds from one sugar residue to its two bridging oxygens are parallel to each other and form a zig-zag configuration (Figure 5.3a,e). Polysaccharide chains with zig-zag type of linkages usually have a ribbon-like shape (Figure 5.2a). Examples of these types of homoglycans are found in plant cell walls, such as cellulose, xylan, and mannan.

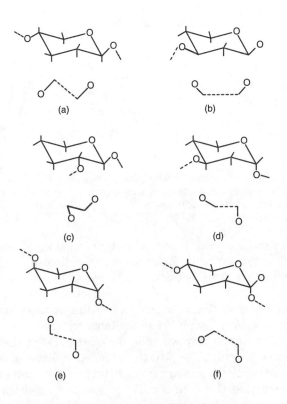

FIGURE 5.3
Geometrical relationships across sugar units and conformation types: (a) zig-zag relationship across 1→4-linked β-D-glucopyranose; (b) U-turn across 1→3-linked β-D-glucopyranose; (c) twisted form across 1→2-linked β-D-glucopyranose; (d) zig-zag relationship across 1→3-linked α-D-glucopyranose; (e) 1→4-linked α-D-galactopyranose; and (f) U-turn across 1→4-linked α-D-glucopyranose. (Adapted from Rees, 1977.)[1]

Ribbon-like chains are most easily aligned and closely packed through numerous hydrogen bonds and van der Waals forces. The resultant compact structures essentially prevent solvent penetration and retain insolubility in water. Cellulose, which is a (1→4)-linked linear polymer of β-D-glucose, has a flat and extended ribbon conformation characterized by a two-fold helix, which can not be solubilized even in alkaline solution (for structure details, see Chapter 6). Another important factor contributing to the strength of binding in cellulose is the stiffness of the chain. The extended ribbon-like secondary structure of cellulose has a very high value of characteristic ratio C_∞ (close to 100) and the unit height (h) is close to the actual length of the sugar residue.

Conformational analysis based on x-ray diffraction data indicates (1→4)-β-D-xylan chains adopt a twisted ribbon-like conformation which can be described by a three-fold, left-handed helix with a pitch of 15 Å.[2] The three-fold helix with 120 degree rotation between residues converts the ribbon

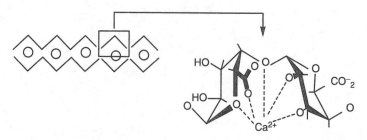

FIGURE 5.4

Proposed packing of polyguluronate chains with interstices filled with Ca^{2+} ions. Left, schematic representation of two buckled chains (each by a zig-zag shape) with Ca^{2+} ions represented by circles. Right, detailed conformational representation of a possible mode of cation coordination by one chain. (Adapted from Rees, 1977.)[1]

chain into a cylinder and distributes the two remaining hydroxyl groups more evenly around the helix than they are in cellulose. This renders the xylan chain being less densely packed, thereby, more soluble in water than cellulose. In cereal arabinoxylans, the xylan backbone carries many single or short arabinosyl side chains which abolish the structural characteristics of homoglycan and bring solubility to the heteroglycan.

Another type of ribbon conformation is called buckled ribbon, which is adopted by some anionic (1→4)-diaxially linked polysaccharides, such as poly-α-D-galacturonic acid sequences in pectin and poly-α-L-guluronic acid sequences in alginates. This type of chain arrangement may leave interstices when they pack together that are usually stabilized by accommodating metal cations (e.g., Ca^{2+}). This form of association is described as egg-box model as shown in Figure 5.4. Several gelling polysaccharides, including pectins and alginate, form ordered structures via the egg-box model.

5.1.3.1.2   Hollow Helix Type Conformation

If the bonds to and from each unit are no longer parallel to one another, but adopt a U-turn form, as in (1→3)-linked β-D-glucopyranose and (1→4)-linked α-D-glucopyranose (Figure 5.3b,f, respectively), the resultant structure is likely to be a hollow helix type (Figure 5.2b). Since this type of arrangement has empty voids along the helix axis, the polymer chains usually are stabilized by twisting around each other to form double or triple helix, or by closely packing together to fill the voids (Figure 5.5). The hollow helix structure may also be stabilized by forming complexes with small molecules of suitable size. (1→4)-linked α-D-glucopyranose (i.e., amylose) is a good example of this type. In the natural form of starch granules, both A- and B-type amyloses adopt a six-fold, left-handed parallel double helix conformation. The double helix structure is stabilized by interchain hydrogen bonds within the double helix and between the double helices. When amylose fiber is regenerated in the presence of solvents (i.e., DMSO), or complexed with small molecules, such as iodine and alcohols, it adopts a left-handed, six-fold single helix conformation known as V-type amylose. This helix is stabilized

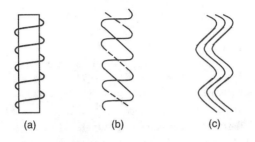

FIGURE 5.5
Schematic representation of hollow helices: (a) inclusion complex; (b) double helix; (c) nesting. (Adapted from Rees, 1977.)[1]

by both intra- and interchain hydrogen bonds with a shallow shape (low pitch value) but a large inner diameter so that the complexing molecules can be easily accommodated. For more details on conformation and structure of amylose, please refer to Chapter 7 (Section 7.3).

Helical structure can also arise from (1→3) linkages. For example, the bacterial polysaccharide curdlan is a (1→3)-linked linear polymer of β-D-glucose, which can form coaxial triple helices. In the hollow helical type polysaccharides, the number of residues per turn of helix (n) may vary from 2-10, whereas the unit height (h) is very small (approaches zero). Consequently, the polymer will give a small value of characteristic ratio (C_∞ is close to 10) indicating a rather compact conformation (compare to the extremely extended conformation of cellulose which has a C_∞ value near 100).

5.1.3.1.3 Other Types of Conformation

Theoretically, (1→2)-linked polysaccharides will have the dihedral angle around 60° and will assume crumpled ribbon conformations with restricted flexibility.[1] Such linkages are rare, although they do occur in the rhamno-galacturonic acid sequences in pectin, and in some bacterial polysaccharides. The (1→6)-linked polysaccharides have an additional element of flexibility because the residues are linked through three rather than two covalent linkages (Figure 1.24). This additional bond tends to favor extended flexible coil conformations, which are less amenable to ordered packing.

5.1.3.2 Conformation of Complex Polysaccharide Chains

As discussed in previous chapters, periodic type polysaccharides may consist of large repeating units which have more than one type of monosaccharides and linkages. They may also contain short branches. The repeating sequences could be separated by irregular sequences (i.e., kinks) to give interrupted sequences of polysaccharide chains. The absolute conformations of these complex polysaccharides are more difficult to elucidate because of the complexity associated with multi-dimensional space. The conformations of two regular sequences are discussed in this section.

5.1.3.2.1 *Periodic Type Conformation*

The simplest regular periodic polysaccharides are alternating sequences, such as ι-carrageenan, agarose, and hyaluronic acid, as illustrated in Figure 5.6. All the three polymers discussed here contain a 3-linked and a 4-linked sugar unit in the alternating polymer chains and each of the two sugar units belongs to different conformation families. The (1→4)-linked unit, if present alone in a homopolysaccharide, would generate a ribbon conformation; whereas the (1→3)-linked unit would generate a hollow helix in a homoglycan. The conformation of the two types of sugar units in an alternating sequence cannot be easily predicted. X-ray diffraction analysis suggested that hyaluronate has an extended hollow helix conformation. Both agarose and ι-carrageenan give less extended conformation in which two chains are packed in double helices.[1] Nevertheless, the two double helices are different in detailed structures: agarose double helix is more compressed and the chain is left-handed due to the presence of L-configuration of the (1→4)-linked sugar unit; by contrast, the double helix formed by carrageenan is right-handed since the 4-linked sugar unit is in D-form.

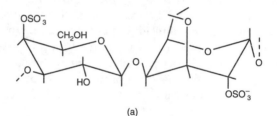

(a)

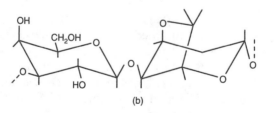

(b)

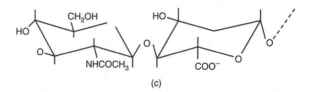

(c)

FIGURE 5.6

Structural features of alternating polysaccharides: (a) ι–carrageenan, polymer of 3-linked β-D-galactopyranose 4-sulphate and 4-linked 3,6-anhydro-α-D-galactopyranose 2-sulphate; (b) agarose, polymer of 3-linked β-D-galactopyranose and 4-linked 3,6-anhydro α-L-galactopyranose; (c) hyaluronate, polymer of 3-linked 2-acetamido-2-deoxy-β-D-glucopyranose and 4-linked β-D-glucopyranosyluronate. (Adapted from Rees, 1977.)[1]

5.1.3.2.2 Interrupted Sequence Type

There is another group of polysaccharides which exhibit high regularity in some regions of the polymer chains, and the repeating units are separated by sequences of different structural features. κ-carrageenan is a good example of interrupted sequences. It has regular alternating structure regions same as in ι-carrageenan, as shown in Figure 5.6a. The regular structure is interrupted from time to time by a 2,6-sulphate galactosyl residue in replacing the regular 3,6-anhydrogalactose 2-sulphate. This interrupted structure has profound implications on the overall shape and conformation of this polysaccharide. For example, carrageenan chains with alternating sequences are able to form double helices under favorable conditions. The presences of different sequences or structural features interrupt the ordered conformation and act as kinks. These kinks will force the polymer chains into cross-linked networks, which essentially lead to the gelation of κ-carrageenan from aqueous solutions. The formation of double helices and the three-dimensional networks of carrageenan is schematically demonstrated in Figure 5.7.

The mechanism for gelation outlined above for carrageenan can also be extrapolated to other interrupted sequence structures. There are several polysaccharides that form gels because of the presence of interrupted sequence structures. In alginates, for example, the interruptions are made by insertion of sugar units of inverted chair conformation. In galactomannans, the interruptions are provided by branched β-D-mannosyl units.

The formation of a higher level of ordered structures from different types of polymer chains could involve the associations of helix to helix, ribbon to ribbon, or helix to ribbon, which can be generalized as the formation of junction zones, as illustrated in Figure 5.8.

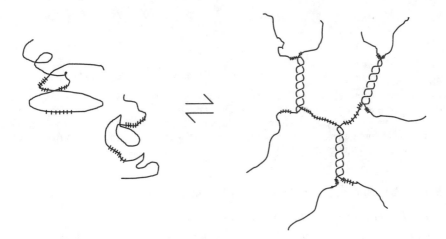

FIGURE 5.7
Schematic illustration of formation of double helix in aqueous system. The double helices formed are represented by the smooth part of each line and the kinking sequences by the crossed lines. (Adapted from Rees, 1977.)[1]

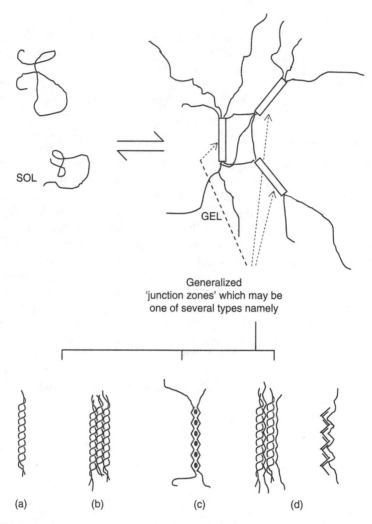

FIGURE 5.8

Generalized scheme for network formation by chains of interrupted type: (a) cross-linkage by double helices as in ι-carrageenan; (b) by bundles of double helices as in agarose; (c) by ribbon-ribbon associations of the egg-box type as in alginate; (d) helix-ribbon associations as in mixed systems containing agarose and certain galactomannans or xanthan gum and certain gluco-mannans. (Adapted from Rees, 1977.)[1]

5.1.4   Disordered Conformation of Polysaccharides

5.1.4.1   The Concept of Random Coil

There are many polysaccharides with much less sequence regularities compared to the polysaccharides described in previous sections. There could be sections in the polymer chain where repeating units can be found, however, in most of the cases, these regular chain segments are not long enough to form ordered structures, such as junction zones. As a result, this group of polysaccharides takes the shape of a random coil, i.e., statistical coil, in solution, where the polymer chain could rotate or change directions about the glycosidic bond fairly freely, and there exists continued fluctuations of local conformation (between each pair of neighboring units and individual sugar rings) and overall chain conformation. This disordered state is stabilized by conformational entropy.

The random coil conformation is a typical property of linear polysaccharides in solutions. In fact, most of the polysaccharides, even with certain structural regularities, will assume a random coil conformation in dilute solutions (provided they are soluble in the solvent). Examples of such polysaccharides include κ-carrageenan, alginates, and locust bean gum, which are soluble in water and give random coil conformation in dilute solutions. The difference is that polymer chains with certain structural regularity could form ordered conformations (Figure 5.8) under favored conditions such as high enough concentration, presence of binding agents, or change of temperatures and pHs. However, for irregular sequenced linear polysaccharides, the only conformation that can be observed is the random coil type, because the sequence of the polysaccharide chain does not allow the formation of ordered structure or junction zones. For highly branched polymers, the polysaccharides usually assume a compact globular shape (e.g., gum arabic), the solution of which usually exhibits very low viscosities and Newtonian flow behavior.

Most linear polysaccharides in solutions may take on numerous conformations of similar energy levels. Constant changes of conformations are also observed due to Brownian motions. The quantities used for describing the conformation and geometrical size of polysaccharides are thus time average values calculated by statistical mechanics. In order to characterize the dimensions and shapes of random coil type polysaccharides, several concepts are introduced in the following sections.

5.1.4.1.1 *Average End-to-End Distance*

The average dimensions of random coil polysaccharide chains are commonly described by two parameters: average end-to-end distance (R) and average radius of gyration (R_g). According to Flory,[3] the distance between one end of a polymer chain and the other for a coiled polymer is called end-to-end distance or displacement length; the distance between one end of the polymer chain and the other when the polymer is stretched out is called contour length (L). The relationship between end-to-end distance and contour length is demonstrated in Figure 5.9. The contour length is the maximum possible end-to-end distance.

The approximate average of R is simply given by the root-mean-square distance between the two ends of the backbone chain in a given conformation:

$$R = \left\langle R^2 \right\rangle^{1/2} = \left\langle nl^2 \right\rangle^{1/2} = n^{1/2}l \quad \text{or} \quad R^2 = \left\langle R^2 \right\rangle = nl^2 \qquad (5.1)$$

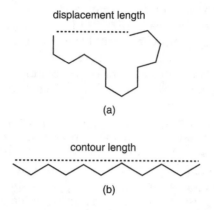

FIGURE 5.9
Schematic presentation of end-to-end distance (displacement length) and the contour length (L) of a polymer chain. (Adapted from Tanford, 1967.)[4]

where n is the number of monomers in the polymer chain and *l* is the length of the monomer.

5.1.4.1.2 Radius of Gyration

For a given configuration, the radius of gyration (R_g) of a random coil polysaccharide is defined as the square root of the weight average of r_i^2 for all the mass elements:

$$R_g = \sqrt{\left(\frac{1}{n} \sum_i r_i^2 \right)} \qquad (5.2)$$

where n is the number of entities in the chain and r_i is the distance of the ith mass vector from the center of the mass, as depicted in Figure 5.10.

R and R_g are interrelated by the following equation in an ideal situation in which there is no physical interference:

$$R_g^2/R^2 = 1/6 \qquad (5.3)$$

5.1.4.1.3 Excluded Volume (V_e)

Excluded volume (V_e) is another important parameter for characterizing the dimension of polysaccharide chains in solutions. Excluded volume is defined as the volume from which a macromolecule or a segment of it in a dilute solution effectively excludes all other macromolecules. For polymer chains with long-range interactions, the space occupied by one segment is excluded for all other segments. This is described as intramolecular excluded volume, which expands the polymer chain. V_e is different from the dimension of individual molecules, but a reflection of sterical/geometrical characteristics of dynamic interactions in solutions: it is the sum of contributions from

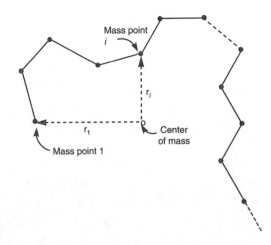

FIGURE 5.10
Representation of the vectors used in calculating the radius of gyration. (Adapted from Tanford, 1967.)[4]

molecular dimensions, molecular conformation, and supermolecular phenomena due to interactions. Therefore, investigation of V_e will reflect the conformation, dimensions and interactive properties about polysaccharides in solutions. The effect of excluded volume usually increases with increasing molecular weight.

5.1.4.2 Conformation of Polysaccharides in Unperturbed Conditions[4,5]

5.1.4.2.1 Flexible Polysaccharide Chains

Free Rotating Linear Polymer Chain — For easy understanding, we first consider a completely unrestricted linear polymer chain under unperturbed conditions. Take an example of a polymer chain which has n+1 equivalent elements joined by n bonds; all the bonds have fixed but different lengths (this is true for nonhomo-polysaccharides), and assume all the bonds can freely rotate and all bond angles are equally probable (this is not true for polysaccharides). Assign θ as the angle between the positive directions of any two successive bonds. If the length and direction of each bond is represented by a vector l_i (Figure 5.11), the square of the average end-to-end distance $\overline{R^2}$ can be expressed as:

$$\overline{R^2} = \overline{R \cdot R} = \overline{\left(\sum_{i=1}^{n} l_i\right) \cdot \left(\sum_{j=1}^{n} l_j\right)} = \sum_{i=1}^{n} \sum_{j=1}^{n} \overline{(l_i \cdot l_j)} \qquad (5.4)$$

For a completely unrestricted polymer chain, two successive bonds may take any angle (θ) directions at equal probabilities. As a result, the average product of two successive bonds at all configurations must equal to zero:

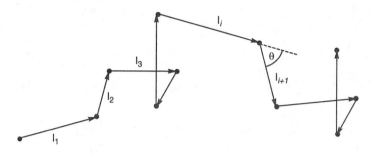

FIGURE 5.11
Schematic presentation of a completely unrestricted polymer chain: the bond lengths l_i are arbitrary, and the angles θ may take on all possible values with equal probability. (Adapted from Tanford, 1967.)[4]

$$\overline{l_i \cdot l_{i+1}} = l_i \cdot l_{i+1}\overline{\cos\theta} = 0 \tag{5.5}$$

From Equation 5.5 it can be derived that all terms with i different from j in Equation 5.4 will equal to zero. The only terms remaining in the equation is $l_i l_i$. Therefore, Equation 5.4 can be modified to:

$$\overline{R^2} = \sum_{i=1}^{n} \overline{l_i^2} = n\overline{l_i^2} = nl_{av}^2 \tag{5.6}$$

where $(l_i^2)^{1/2} = l_{av}$ is the average (root-mean-square) bond length in the polymer chain. If all the bond lengths are identical, l_{av}^2 can be replaced by l^2.

Effect of Bond Angle — For a polymer as shown in Figure 5.11, let all the bonds have the same length l (constant bond length); any restriction of the bond angle will invalidate equations described earlier (Equation 5.4 and Equation 5.6). The three successive bonds in such a polymer are depicted in Figure 5.12. The average end-to-end distance is corrected by an angle related term to Equation 5.6:[4]

$$\overline{R^2} = nl^2 \cdot \frac{1+\cos\theta}{1-\cos\theta} \tag{5.7}$$

For a polymer chain with a fixed bond angle, e.g., $\theta = 60$ degree ($\cos\theta = 0.5$), $\overline{R^2} = 3nl^2$.

This example demonstrates that the presence of fixed bond angles will increase the average end-to-end distance. Equation 5.7 can be extended to flexible polysaccharides joined by bonds of fixed length and fixed bond angle. If the bond angle $\theta = 90$ degree, Equation 5.7 is the same as Equation 5.6. A general influence of bond angle on the average end-to-end distance can be summarized as following:

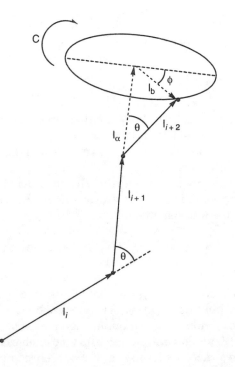

FIGURE 5.12
Three successive bonds in a polymer chain: the first two are in the plane of the figure, the terminus of l_{i+2} may lie anywhere on the circle C. (Adapted from Tanford, 1967.)[4]

- If the angle between successive bonds (i.e., 180-θ) is obtuse, $\overline{R^2}$ is always greater than Equation 5.6 would predict
- If the angle is acute, $\overline{R^2}$ is always smaller than Equation 5.6 would predict
- Equation is invalid for the angle close to zero as it would represent more rigid/rod-like conformation with little flexibility

Effect of Restricted Rotation — The equations described in previous sections were derived based on the assumption that the polymer has a freedom of rotation about the glycosidic bonds. In practice, particularly for polysaccharides, the rotation about a glycosidic bond is restricted sterically. As illustrated in Figure 5.12, some of the rotation angles (φ) become more probable than the others. This leads to the introduction of another factor to Equation 5.7:[6]

$$\overline{R^2} = nl^2 \frac{1+\cos\theta}{1-\cos\theta}\frac{1+\overline{\cos\phi}}{1-\overline{\cos\phi}} \tag{5.8}$$

where $\overline{\cos\phi}$ is the average value of cosφ. For completely free rotation, $\overline{\cos\phi} = 0$, Equation 5.8 is the same as Equation 5.7.

However, Equation 5.8 is considered not practical due to difficulties encountered for the calculation of $\overline{\cos\phi}$. Therefore, empirical treatments have been used to evaluate the effect of restricted rotation on the conformation of polysaccharides.

Chain Stiffness and Effect of Chain Length on End-to-End Distance ($\overline{R^2}$) — As described in previous sections, there are two types of factors affecting polymer dimensions. The first type is based on the primary structure of the polymer, including bond length l, bond angle θ, and degree of restriction to rotation (ϕ); the second type of factor affecting polymer dimension is their chain length, i.e., the molecular weight. The chain length can be expressed as the number of monomer units n. The above generalized relationship can be expressed by the following equation:

$$\overline{R^2} = \beta^2 \cdot n \tag{5.9}$$

where β^2 is a constant reflecting the characteristics of primary structure of the polymer. For polysaccharides, β^2 is affected by the primary structures such as monosaccharide composition, linkage patterns, and sequences.

Comparison of Equation 5.9 and Equation 5.8 enables one to treat flexible polymer chains restricted by rotation as if they were completely unrestricted. The apparent bond length β will be larger than the true average bond length l_{av}. Therefore, the ratio of β/l_{av} can be regarded as a measure of rotational restriction or chain stiffness. The apparent bond length β can be expressed by:

$$\beta^2 = l_e^2/m \tag{5.10}$$

where l_e is the root mean square of the statistical segment length, which is frequently referred to as Kuhn length, and m is the number of units in the statistical segment.

The chain stiffness of semi-flexible or wormlike chains can be described by a parameter called persistence length (l). The persistence length is the average projection of end-to-end vector of a macromolecule on the tangent to the chain contour at a chain end in the limit of infinite chain length. The persistence length (l) is related to the Kuhn statistical segment length l_e by:

$$l_e = 2l \tag{5.11}$$

For a polymer with molecular weight M, the number of units n can be substituted by M/M_0, where M_0 is the mass of repeating units. Equation 5.9 can be rewritten as:

$$\overline{R^2} = \beta^2 \cdot \frac{M}{M_0} = \frac{\beta^2}{M_0} \cdot M \tag{5.12}$$

In polysaccharides, M_0 could be a single sugar unit or an oligosaccharide repeating unit (may contain several monosacchairdes). Therefore, once the primary structure or nature of the polymer is determined, and the repeating segment is identified, the relationship between end-to-end distance and the molecular weight of the polymer is governed by Equation 5.12.

5.1.4.2.2 Conformation of Rigid and Stiff Polysaccharide Chains

In an extended rigid polymer chain, the contribution of each successive segment to the end-to-end distance is equal. Therefore, a direct simple relationship between molecular weight and end-to-end distance is observed:[4,5]

$$R = k M \tag{5.13}$$

where k is a constant. This relationship is applicable both to fully extended rigid polymer chains and to perfectly helical chains.

For stiff polymer chains, which may be considered as in between a completely rigid and flexible chains, the following relationship exists:

$$R = k M^{0.60-0.99} \tag{5.14}$$

The exponent of M may decrease slowly with the increase of molecular weight, because longer chains will introduce more flexibility or freedom to the polymer chains.

5.1.4.3 Effect of Solvent Quality, Concentration, and Temperature

Previous discussions on polymer conformation were in ideal conditions in which secondary interactions were not considered, such as interactions between different polymer chains, between segments of a single chain, and between polymer chains and solvent molecules. However, in real polymer solutions, the conformation of a polymer chain is significantly influenced by polymer concentrations (chain interactions), solvent quality, and temperature. This section will discuss these concerns.

5.1.4.3.1 Effect of Solvent Quality

In a dilute polymer solution, in which intermolecular interaction is not significant, a polymer may have a preferred configuration depending on solvent quality. In a good solvent, the attractions between solvent and polymer molecules are greater than that between polymer segments themselves. By contrast, in a poor solvent, the attraction between segments of the same polymer chain is greater than that between the solvent and the polymer chain. There exists an ideal solvent condition in which there is no preferred interactions. In other words, the interactions between solvent and polymer molecules has little or no contribution to the conformation of polymer chains. This solvent is called an indifferent solvent or θ solvent in which the second virial coefficient A_2 is 0. For a polymer in θ solvent, the equations described for unperturbed conditions can be used without modification. However, for good or poor solvents, further modification of the equations described in previous sections is required.

If the ideal value of the mean square end-to-end distance is $\overline{R_0^2}$, the true mean square end-to-end distance $\overline{R^2}$ will be greater or smaller than R_0^2. According to Flory and Fox,[5,7] the following relationship exists:

$$\overline{R^2} = \alpha^2 \overline{R_0^2} = \alpha^2\beta^2 n = \alpha^2\beta^2 \frac{M}{M_0} \qquad (5.15)$$

where α is an empirical constant related to intramolecular excluded volume. In a good solvent, solvent molecules are attracted to the polymer chains, which make the presence of segments adjacent to one another less likely. In a poor solvent, the presence of adjacent segments to one another is preferred. Therefore, α indicates the relationship between effective exclusion volume and physical excluded volume:

- In a good solvent, $\alpha > 1$, the effective exclusion volume is greater than the physical excluded volume.
- In a θ solvent, $\alpha = 1$, the effective exclusion volume equals to the physical excluded volume.
- In a poor solvent, $\alpha < 1$, the effective exclusion volume is smaller than the physical excluded volume.

In a poor solvent ($\alpha < 1$), the polymer solution is unstable: aggregation or precipitation may occur and the system will be separated into two or more phases. Therefore, in practical applications, $\alpha = 1$ is essentially the minimum value and α is an increasing function of molecular weight:

$$\alpha = kM^{0.10} \qquad (5.16)$$

5.1.4.3.2 Effect of Concentration and Temperature

Concentration Effect — Previous descriptions of random coil chains were basically in dilute solutions where each polymer chain is individually separated from other chains. As the concentration increases, the distance between two polymer chains becomes smaller; when the polymer concentration is increased to a critical point (critical concentration, C^*), polymer chains start to penetrate with each other and form the so-called chain crossover or chain entanglement phenomena, thus interchain interactions occur. A parameter ξ is introduced to describe the interactions between polymer chains. ξ refers to the distance between two entanglement points. A power law relationship is found between ξ and polymer concentration c:

$$\xi = k\, c^{-3/4} \qquad (5.17)$$

In a semi-dilute solution, polymer solution dynamics is predominated by the formation of entanglements. As a result, ξ plays a more important role in describing the conformation of polysaccharides.

As the concentration is increased to concentrated solutions, the length of ξ is usually smaller than the coil size.

Temperature Effect — Temperature has a similar effect on polymer chain conformation as solvent does. Based on Flory's theory of mean field approximation, de Gennes[8] described the effect of temperature on chain conformation by the following equation:

$$\alpha^5 - \alpha^3 - \frac{y}{\alpha^3} = kn^{1/2}\frac{w_1}{l^3} \tag{5.18}$$

where $y = k^2w_2l^{-6}$, w_1 and w_2 are successive virial coefficients, n is the number of monomers along the polymer chain and l is the length of the monomer. Similar to θ solvent, there is theta temperature (θ) at which the conformation of the polymer chain is in ideal condition.

The term α^5-α^3 represents the elasticity of the polymer chain; the negative term -α^3 controls the polymer from large swelling. Term α^5-α^3 is important at $T > \theta$ in which the polymer chain swells (expands). Term y/α^3 represents a hard core repulsion that slows down the collapse of the polymer chain. When $T < \theta$, chain collapsing is likely to occur, therefore, term y/α^3 becomes important. Term $kn^{1/2} w_1/l^3$ is related to the reduced temperature (τ), which is defined as:

$$\tau = \frac{T - \theta}{T} \quad \text{if } T > \theta;$$

$$\tau = \frac{\theta - T}{\theta} \quad \text{if } T < \theta$$

Term $kn^{1/2} w_1/l^3$ is important both for $T > \theta$ (swelling of the chain) and $T < \theta$ (contraction of the chain).

The expansion parameter α is a function of a reduced variable $\tau n^{1/2}$:

$$\alpha = f(\tau n^{1/2}) \tag{5.19}$$

At $T = \theta$, $\tau = 0$, the average end-to-end distance is near the ideal condition, i.e.:

$$R \sim R_\theta \sim n^{1/2}$$

and polymer chain conformation is in ideal condition.

At $T < \theta$,

$$R \sim n^{-1/3}\,\tau^{-1/3},$$

indicating a globular and compact chain conformation.
    At $T > \theta$,

$$R \sim n^{3/5}$$

in which the polymer chain takes an extended conformation.[5]

5.1.4.3.3 Effect of Polydispersity

The equations described in preceding sections are applicable only to monodisperse molecular weight polymers. However, polysaccharides are polydisperse in nature, therefore, the applications of those equations to polysaccharides will need additional averaging processes. As described in Chapter 4, the number average, weight average, and zeta average were used for descriptions of molecular weights. These averages can be applied to express the average mean square end-to-end distance $\langle \overline{R^2} \rangle$ of different molecular weights (N_i molecules with molecular weight M_i):

$$\left\langle R^2 \right\rangle_n = \sum_i N_i \overline{R^2} \Big/ \sum_i N_i \tag{5.20}$$

$$\left\langle R^2 \right\rangle_w = \sum_i N_i M_i \overline{R^2} \Big/ \sum_i N_i M_i \tag{5.21}$$

$$\left\langle R^2 \right\rangle_z = \sum_i N_i M_i^2 \overline{R^2} \Big/ \sum_i N_i M_i^2 \tag{5.22}$$

where the subscript n, w and z represent number, weight and z averages respectively.

Whether R, R_g or $\overline{R^2}$ is used to express the dimensions of polymers depends largely on the procedures used for obtaining the parameters. Usually high quality data can be obtained from well fractionated polymers; in this case, all polymer chains can be considered identical in degree of polymerization and assuming single characteristic values of R and R_g.

The dimensions of branched polymers are best expressed as the average radius of gyration since the concept of end-to-end distance may not be applicable for highly branched polymers. It is expected that the R_g values for branched polymers is smaller than their linear counter part when compared at the same molecular weight. For example, one point of random branching in a polymer chain would reduce the R_g^2 values from $1\times (R^2/6)$ to about $0.9 \times (R^2/6)$; two points of branching would reduce the value to to $0.83 \times (R^2/6)$.[9]

5.2 Methods for Conformation Analysis

5.2.1 X-Ray Diffraction

5.2.1.1 Introduction

X-ray diffraction analysis is the most direct and powerful experimental means to determine molecular structure at atomic resolution. Much information about molecular geometry has come from the application of this technique to single crystals. Diffraction is essentially a scattering phenomenon. When a monochromatic x-ray beam travels through a testing specimen, a small proportion of the radiation is scattered by individual electrons in the sample. These scattering waves interfere with each other resulting in mutual reinforcement of some scattered rays. Depending on the arrangement of the atoms within the sample, the resultant x-rays may form a regular pattern if recorded by means such as flat film camera. Information on molecular structure may be derived from the analysis of diffraction patterns.

For almost all monosaccharides and some small oligosaccharides, it is not a problem to prepare single crystals for x-ray measurement. X-ray characterized structures are available for most of these molecules.[10,11] However, high molecular weight polymers including polysaccharides rarely form single crystals that are large enough for x-ray analysis. Instead, x-ray studies on polysaccharides are usually carried out using oriented fibers or films prepared from concentrated solutions. During the fiber preparation, the polysaccharide molecules are forced to align approximately parallel along the helix axes, but the orientation of the molecules about their helix axes is random. Although, this organization is artificial, the information obtained may help to understand the ordered structures of polysaccharides that occur in solutions and gels.

5.2.1.2 Basics of Fiber Diffraction Analysis of Polysaccharides

Analog to single crystals, a fiber diffraction pattern can be linked to a lattice arrangement which is built up by periodic unit cells. The lattice constants are defined by the lengths a, b, and c of the unit cell edges and the angles α, β, and γ between the axes (Figure 5.13). The interpretation of diffraction patterns contains two aspects: the analysis of diffraction directions yields information about the lattice structure and screw symmetry; the analysis of the scattering intensity as a function of diffraction angle may provide information about relative placement of the atoms in the polymer chain.

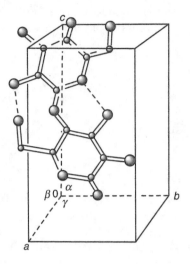

FIGURE 5.13
Schematic drawing of a unit cell and its contents. (Adapted from Rao et al., 1998.)[12]

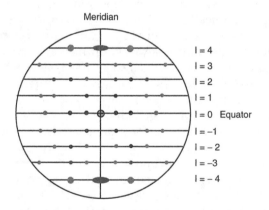

FIGURE 5.14
An idealized diffraction pattern from a polycrystalline polysaccharide fiber showing Bragg reflections on layer lines. (Adapted from Rao et al., 1998.)[12]

When a tested sample is crystalline or sufficiently regular, the scattered beams are concentrated around a number of sharply defined scattering directions. As illustrated in Figure 5.14, an idealized diffraction pattern recorded by a flat film camera from a polysaccharide fiber consists of arrays of discrete spots (Bragg reflections) varying in intensity. The horizontal lines are called layer lines, and the vertical line passing through the center of the pattern is called the meridian and is parallel to the fiber axis. The position of every reflection is directly related to the unit cell parameters. The spacing distance (d-spacing) between adjacent reflection planes d_{hkl} and diffraction angle 2θ are related by the Bragg Law:

$$\frac{2\sin\theta}{\lambda} = \frac{n}{d_{hkl}}, \quad n = 1,2,3,\ldots \tag{5.23}$$

where λ is the wavelength of the x-ray radiation. The observed reflections are indexed in terms of Miller indices (h, k, l) for each crystallographic plane that gives rise to the reflections. These are then used to determine lattice constants a, b, c and angles among them, as well as the space group symmetry. It is clear that the molecule in Figure 5.14 has $n = 4$ repeating units per turn with a unit height of $h = c/4$.

To obtain information about the positions of all atoms in the unit cell, the structure amplitudes F(s) are determined from the intensities of observed Bragg reflections. The intensity can be measured with a microdensitometer or digitized with a scanner and processed. The intensity of scattering from a crystal is nonzero only when the scattering vector s coincides with the reciprocal lattice. The square root of the intensity I_{hkl} provides the absolute magnitude of the structure amplitudes F_{hkl}:

$$\left|F_{hkl}\right| = \sqrt{I_{hkl}} \tag{5.24}$$

Unlike in single crystal structure analysis, a fiber diffraction pattern does not contain enough numbers of reflections for resolving the exact positions of individual atomic coordinates in the unit cell. Usually, structural data of relevant sugar residue that has been determined by the single crystal analysis together with other stereochemical constraints are used to construct all possible types of molecular and packing models. Once the monosaccharide ring shape is selected, the geometry of the corresponding polysaccharide helix can be defined by the two or three conformational angles ϕ, ψ and χ. This helix is then used to construct fiber structures. There are two representative programs developed for the generation and refinement of fiber structures for biopolymers: linked-atom least-squares (LALS) analysis[13] and the variable virtual bond (PS79) method.[14] Using these programs, possible crystal model(s) are constructed and evaluated against the stereochemical restraints and the x-ray diffraction data; refinement of the models are made by allowing changes in glycosidic conformation angles and parameters describing the crystal packing. Finally, water molecules and other ions are located by standard methods (i.e., difference Fourier or R factor maps).[15]

5.2.1.3 Examples of Fiber Diffraction Structure of Polysaccharides

Plant polysaccharide galactomannans have a $(1{\rightarrow}4)$-β-D-mannan backbone which is substituted with $(1{\rightarrow}6)$-linked α-D-galactosyl units. The galactose to manose ratio (G/M ratio) varies with the source of the plants. Despite the difference in their species and G/M ratios, the x-ray fiber diffraction patterns of galactomannans were found very similar to each other.[16–18] Here we use

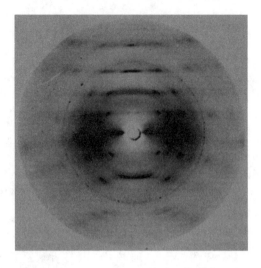

FIGURE 5.15
A typical x-ray diffraction pattern from a polycrystalline and well-oriented fiber of guar galactomannan. (From Chandrasekaran et al., 1998. With permission.)[19]

the work of Chandrasekaran[19,20] on guar galactomannan as an example. The fiber was prepared by slowly drying of galactomannan solution (1 mg/ml) under controlled relative humidity (70 to 95%), followed by gradually stretching. Figure 5.15 is a typical diffraction pattern from a polycrystalline and well-oriented fiber of guar galactomannan. There are a total of 20 non-meridional reflections up to 2.5 Å resolution. All these spots are indexable with an orthorhombic unit cell of dimensions a = 9.26, b = 30.84 and c = 10.38 Å. The two meridional reflections on the 2nd and 4th layer lines suggest that the polysaccharides have a twofold helix symmetry and a pitch of 10.38 Å. The space group symmetry is consistent with *I*222. The unit cell has room for four one-turn galactomannan helices and four water molecules.

The x-ray data, geometry of monosaccharides, and other stereochemical constraints were used to construct and refine a twofold helix of pitch 10.38 Å using the LALS program. There are six major conformation angles: (ϕ_1, ψ_1, χ_1) in the backbone and (ϕ_2, ψ_2, and χ_2) in the side chain (Table 5.1). Two packing parameter were used to generate the crystal structure having the space group *I*222. An initial helical model characterized by χ_1 in the trans domain was identified. Finally, the crystal structure was refined using the difference Fourier maps to locate ordered water molecules and flexible sugar rings. The major conformation angles of the helix in the final model are given in Table 5.1 together with those of mannans[21] for comparison. The main chain of the galactomannan helix has a fully extended mannan-like conformation. As shown in Figure 5.16, adjacent mannose residues within a backbone are linked by O-3H ⋯ O-5 hydrogen bonds which gives structure stability. The atom O-5 of the galactosyl residue forms a hydrogen bond with atom O-3 of the adjoining mannosyl residue in the reducing end. The galactomannan

TABLE 5.1

Major Conformation and Bridge Bond Angles (and e.s.d.) in Degrees in Guar Galactomannan and Mannan Helices[a]

Parameter	Guaran	Mannan I	Mannan II	Remarks
ϕ_1 (O-5M-C-1M-O-4'M-C-4'M)	−98(2)	−90	−88	$\beta(1\rightarrow4)$
ψ_1 (C-1M-O-4'M-C-4'M-C-5'M)	−143(2)	−149	−153	$\beta(1\rightarrow4)$
X_1 (C-4M-C-5M-C-6M-O-6M)	158(3)	180	−23	hydroxymethyl
T_1 (C-1M-O-4'M-C-4'M)	117(1)	117	117	bridge
ϕ_2 (O-5G-C-1G-O-6M-C-6M)	149(2)			$\alpha(1\rightarrow6)$
ψ_2 (C-1G-O-6M-C-6M-C-5M)	−111(3)			$\alpha(1\rightarrow6)$
X_2 (C-4G-C-5G-C-6G-O-6G)	− 63(6)			hydroxymethyl
T_2 (C-1G-O-6M-C-6M)	116(2)			bridge

[a] The letters M and G, and symbol ' refer to mannose, galactose, and the reducing end, respectively.

Source: From Chandrasekaran et al., 1998.[20]

chain appears to have a sheet-like structure of width ~15.4 Å. The four helices in a unit cell form two parallel sheets as shown in Figure 5.17. There are four intersheet hydrogen bonds, three of which are among the galactosyl side chains and the other is between the main chains. Each chemical repeat (1.0 mannosyl and 0.6 galactosyl units) is associated with 2.75 structured water molecules. The water bridges further strengthen the attractive interactions within and between the galactomannan helix sheets. This network of hydrogen bonds is critical for the stability of the crystal structure of galactomannan.

5.2.2 Scattering Methods

5.2.2.1 *Introduction*

Scattering techniques, especially small angle x-ray scattering (SAXS) and light scattering (LS) are commonly used to study the size and shape of macromolecules in dilute solutions. The principles on which these scattering techniques depend are basically similar: the angular dependence of the normalized scattering intensity provides information on the size and shape of macromolecules. The resolving power of scattering techniques is related to the wavelength of radiation source, i.e., the shorter the wavelength, the higher the resolution. The wavelengths are 0.1–0.3 nm for small angle x-ray scattering, 0.2–1.0 nm for small angle neutron scattering (SANS) and 436~633 nm for light scattering respectively. Conventional light scattering typically reveals only the global dimensions of a macromolecule, which may be tens to hundreds of nanometers for a typical polysaccharide. SAXS and SANS can probe molecular structure at closer range of about 2–25 nm.[22] Light scattering is effective in measuring the angular dependence of intensity typically in the range 30° to 135°. It is often used to measure the radius of

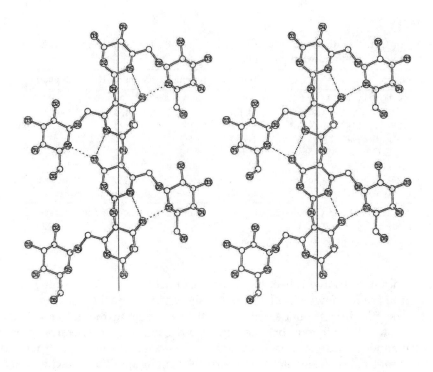

FIGURE 5.16

Two mutually perpendicular stereo views of the guar galactomannan helix. Two turns are shown at the top. The helix is stabilized by a series of hydrogen bonds (dashed lines) involving both main and side chain atoms. The vertical line, helix axis, measures 2c. An axial projection of one turn at the bottom shows the flat nature of the galactomannan chain whose thickness is only 4 Å. (From Chandrasekaran et al., 1998. With permission.)[20]

gyration, molecular weight, and second virial coefficient. SAXS can be carried out at very small angles, typically less than 1°, and is thus superior for the determination of size and shape of macromolecules, but is less convenient for the determination of molecular weight and second virial coefficient.

5.2.2.2 Basics of Scattering Techniques

5.2.2.2.1 Light Scattering

Light scattering measurements can be carried out in two modes, static and dynamic. The former measures the average scattering intensity within a selected time period, whereas the latter measures the fluctuation of the intensity over time. It has been discussed in Chapter 4, (Section 4.2.2.2), that from a dilute solution of macromolecules the weight average molecular weight

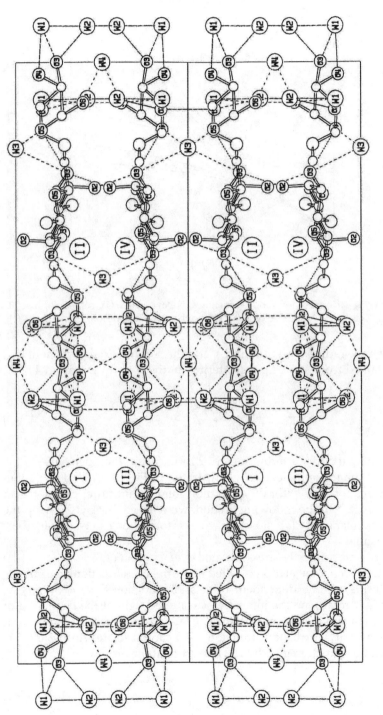

FIGURE 5.17

Packing arrangement, as viewed down the c-axis, of the galactomannan helices in two unit cells (2a is down and b across the page). Both direct and water-mediated hydrogen bonds (dashed lines) involving the hydroxyl groups (labeled) of mannose and galactose are responsible for association of the sheets of helices in pairs within and between unit cells. (From Chandrasekaran et al., 1998. With permission.)[20]

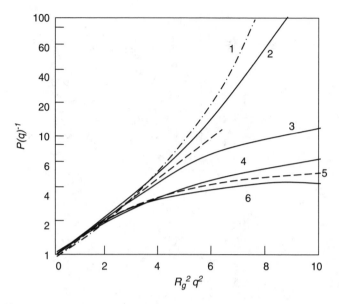

FIGURE 5.18
Particle scattering factor P(θ) for various structures: hollow sphere (1), sphere (2), disc (3), monodisperse coils (4), polydisperse coils (5), rigid rods (6). (Adapted from Burchard, 1978.)[23]

(M_w), z-average radius of gyration (R_g), and the second virial coefficient (A_2) can be obtained from static light scattering on the basis of Equation 4.8:

$$\frac{Kc}{R(q)} = \frac{1}{M_w P(q)} + 2A_2 c + 3A_3 c^2 + \ldots \qquad (5.25)$$

in which, the particle scattering factor or form factor, $P(q)$, depends on the geometry of the macromolecules as illustrated in Figure 5.18.[23] The mathematical expression of $P(q)$ for different molecular architectures can be found in the reference.[24] In the following section, we will see that by studying the angular dependence of $P(q)$, more structural information can be obtained for macromolecules.

If two dynamic properties are correlated over a period of time, the function that describes such a correlation is called the correlation function. In dynamic light scattering, a time dependent correlation function $G_2(t) = \langle i(0)i(t)\rangle$ is measured, where $i(0)$ are the intensity of scattered light at the beginning of the experiment and $i(t)$ that at certain time later, respectively. In dilute solutions, the intensity correlation function $G_2(t)$ can be related to the electric field correlation function $g_1(t)$ by the Siegert relationship:

$$G_2(t) = B \cdot \left(1 + f^2 \left| g_1(t) \right|^2 \right) \qquad (5.26)$$

where B and f are constants. For dilute solutions of monodisperse small particles, $g_1(t)$ decays as a single exponential function:

$$\left|g_1(t)\right| = \exp(-\Gamma \cdot t) \tag{5.27}$$

where Γ is called the decay or relaxation rate, which is the parameter of interest that needs to be extracted by fitting the correlation function. A number of models and approaches have been developed to fit the correlation functions. For more detail, the reader is referred to the reference.[25] In the situation that the decay is entirely due to translational diffusive motion of the center of mass of macromolecules, it can be shown that:

$$\Gamma = D \cdot q^2 \tag{5.28}$$

where D is the translational diffusion coefficient and q is the magnitude of the scattering wave vector. Knowing the translational diffusion coefficient, the hydrodynamic radius (R_h) can be calculated through the Stokes-Einstein equation:

$$D = \frac{kT}{6\pi\eta R_h} \tag{5.29}$$

where k is the Boltzmann's constant, T is the absolute temperature and η is the viscosity of the solvent.

Since the static and hydrodynamic dimensions vary characteristically with the structure of the macromolecules, the dimensionless ratio, $\rho = R_g / R_h$, may provide information on the architecture of the macromolecules and their aggregates.[26] Typical ρ values for some molecular architectures were listed in Table 5.2. Generally, the values of ρ decrease with increasing branching density; but the increasing polydispersity compensates the effect of increasing branching.

5.2.2.2.2 Small Angle X-Ray Scattering (SAXS)

The fundamental function for SAXS data analysis follows the Guinier law. For an isolated scattering particle in the limit of small q, this is given by:

$$I(q) = \rho_0^2 v^2 \exp\left(-\frac{1}{3}q^2 R_g^2\right) \tag{5.30}$$

$$q = \frac{4\pi}{\lambda}\sin\theta$$

where $I(q)$ is the intensity of independent scattering by a particle. q is the scattering vector and 2θ is the scattering angle. ρ_0 is the average scattering

TABLE 5.2

The ρ Parameter for Selected Polymer Structures

	ρ
Homogeneous sphere	0.788
Random coil, monodisperse	
θ conditions	1.50
Good solvent	1.78
Random coil, polydisperse (Mw/Mn=2)	
θ conditions	1.73
Good solvent	2.05
Regular stars	
θ conditions, f = 4	1.33
θ conditions, f $\gg$ 1	1.079
Randomly branched structures, polydisperse	
θ -non swollen	1.73
Good solvent, swollen	2.05
Rigid rod	
Monodisperse	>2.0
Polydisperse (Mw/Mn = 2)	$\gg$2

Source: Adapted from Burchard, 1994.[26]

length density and υ is the volume of the particle. The Guinier law indicates that when the logarithm of $I(q)$ is plotted against q^2, the initial slope gives $R_g^2/3$. Depending on the way it is plotted and the molecular model used, the Guinier law can provide information about the size of macromolecules in various shapes. For instance, the cross-sectional radius of gyration R_c for a rodlike molecule can be obtained from a linear plot of $\ln qI(q)$ vs. q^2 according to Equation 5.31.

$$qI(q) \sim \exp\left(-R_c^2 q^2/2\right) \qquad (5.31)$$

The slope of this plot is proportional to R_c^2.

5.2.2.3 Application to the Polysaccharide Structure Study

Interpretation of scattering data from polymer solutions is usually carried out by employing various molecular models, such as Gaussian chain model and wormlike chain model, and by plotting the data in appropriate forms.[26] By examining the dependence of particle scattering factor $P(q)$ on the scattering vector q, structural parameters other than those that are discussed above may be derived. The description of the shape of a polymer chain is in fact relative to the size scale with which we examine it. According to Kratky and Porod,[27] the scattering intensity of a macromolecule chain of infinite length to radius ratio presents a different behavior in each of three accessible angular ranges as shown schematically by the diagrams in Figure 5.19a. At very small q, the whole molecule is seen as a particle characterized

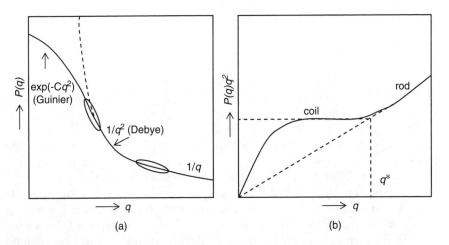

FIGURE 5.19
Schematic presentation of the characteristics of the single particle scattering function in different q regions. (Adapted from Roy, 2000.)[27]

by R_g, the scattering intensity falls exponentially according to the Guinier law. As q is increased, the molecule is seen as a flexible random coil, and $P(q)$ decreases as q^{-2} (Debye function). At still larger q, the molecule can be considered as a more or less rigid rod consisting of short sequence of monomers, the intensity is proportional to q^{-1} as a behavior of a thin rod. The regions of transition between the three sections of the angular variation are emphasized in a Kratky plot (Figure 5.19b) where $P(q)q^2$ is plotted against q. The value of q^* at which the transition from the coil to rod regions occurs is inversely related to the persistence length l_k by:

$$q^* = \frac{12\pi}{l_k} \tag{5.32}$$

Another useful approach is to plot $q \cdot P(q)$ against q or $q \cdot R_g$, the so-called Holtzer plot. For a rigid rod the plot reaches a plateau at large q; for a semiflexible chain, a plateau is reached after a maximum at intermediate q, and for a flexible coil the plot only yields a maximum. The height of the asymptotic plateau (divided by π) in these plots yields the mass per unit length, i.e., the linear mass density (M_l). It has been shown that the number of Kuhn segments per chain (N_k) can be calculated from the ratio of the maximum height divided by the height of the plateau.[28] The Kuhn length (l_k) can then be calculated together with the measured molecular weight (M_w) as:

$$l_k = \frac{M_w}{M_l \cdot N_k} \tag{5.33}$$

Moreover, the position of the maximum coincides with the polydispersity index.

The studies carried out on native and modified xanthan gum in dilute solutions are good examples of application of static and dynamic light scattering to polysaccharides.[29] In this study, static and dynamic light scattering experiments were carried out on a native xanthan (NX), a pyruvate-free xanthan (PFX), and an acetyl-free xanthan (AFX). The molecular weight, radius of gyration, and second virial coefficient were extracted from the static scattering data through the Zimm plot approach. The Holtzer plots for the three forms of xanthans exhibited a shape characteristic of semiflexible molecules (Figure 5.20). From these plots the number of Kuhn segments, the linear mass density, and the Kuhn length were obtained by the methods discussed above. From dynamic measurement the translational diffusion coefficient D and the hydrodynamic radius R_h were determined, hence the structural parameter ρ (R_g/R_h). Some of the structural parameters obtained from the study are summarized in Table 5.3.

As mentioned above, small angle x-ray scattering may also provide information on the cross-sectional radius of polymer chains. This provides a direct experimental tool for measuring the strandness of the helix. For example, Figure 5.21 is a cross-sectional Guinier plot from gellan gum solutions at temperature of 60°C and 10°C respectively.[30] The cross-sectional radius of gyration was found to increase from 3.0 Å at 60°C to 4.0 Å at 10°C, indication of a transition from single helix to double helix as the solution was cooled. Scattering techniques are often used to probe the conformational transition of polysaccharides in solution. For example, the thermal transition evident in low ionic strength xanthan solutions was followed by dynamic light scattering.[31] It was observed that the apparent hydrodynamic radii significantly decreased with increasing temperature in the vicinity of the helix-coil

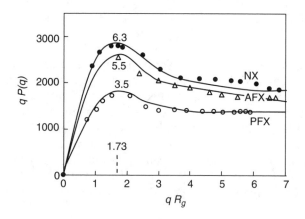

FIGURE 5.20

Holtzer plots of the particle scattering factor for a native xanthan (NX), a pyruvate-free xanthan (PFX) and an acetyl-free xanthan (AFX). The solid lines represent the calculated curves. (Adapted from Coviello et al., 1986.)[29]

TABLE 5.3

Structural Parameters Obtained from Static
and Dynamic Light Scattering

Parameter	NX	PFX	AFX
$M_w \times 10^{-6}$, Da	2.94	1.37	1.77
R_g, nm	289.5	240.8	210.0
$A_2 \times 10^4$, cm³·mol·g⁻²	4.94	5.22	6.00
M_l, Da·nm⁻¹	1830	1240	1623
l_k, nm	255	310	198
N_k	6.3	3.5	5.5
$D \times 10^8$, cm²·s⁻¹	1.55	2.60	2.60
R_h, nm	137.2	81.5	81.5
ρ (R_g/R_h)	2.17	2.95	2.58

Note: For a native xanthan (NX), a pyruvate-free xanthan (PFX) and an acetyl-free xanthan (AFX)

Source: Data is from Coviello et al., 1986.[29]

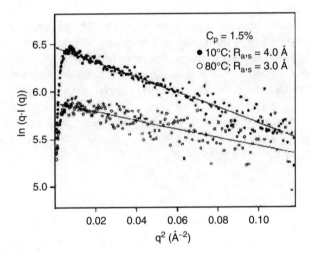

FIGURE 5.21
Cross-sectional Guinier plots from gellan gum aqueous solutions at 10°C and 60°C. The concentration is 1.5% wt% and the estimated cross-sectional radius of gyration is indicated in the figure. (From Yuguchi et al., 1996. With permission.)[30]

transition temperature. Light scattering is also useful in investigating aggregation properties of polysaccharides.

5.2.3 Applications of NMR Spectroscopy (Relaxation Times and NOE) in the Conformation Analysis of Carbohydrates

Nuclear magnetic resonance (NMR) spectroscopy has been widely used in structural analysis in carbohydrate research: it is used to identify monosaccharide composition, elucidate linkage patterns and glycosidic configurations, and establish the sequence of individual sugar units (for details, please see Chapter 3). Various NMR techniques have also been developed to study the molecular motions and conformations of polysaccharides, such as relaxations (T_1, T_2 and NOE, see definitions below), coupling constant and diffusion coefficients, etc. These techniques, especially NOE and relaxation times, have been successfully applied in analyzing polysaccharide conformations and molecular dynamics in solutions. In this section, basic principles and examples using ^{13}C relaxations (T_1, T_2 and NOE) in the conformation analysis of polysaccharides will be discussed.

5.2.3.1   NMR Relaxations: Basic Concepts and Principles

5.2.3.1.1   Relaxation Times

T_1 relaxation, or spin-lattice relaxation, is a measure of the time taken for the molecules to reach thermal equilibrium by transferring energy from the spin system to other degrees of freedom, such as molecular motions. Reversely, the measurement of T_1 could provide structural information of the polysaccharides caused by molecular motions. The detection of T_1 relaxation requires that the dipole to dipole interactions, which are derived from local changing magnetic fields developed in the molecules due to thermal dynamic motions, must: 1) be time dependent; 2) cause transitions between different spin statuses. The detection of T_1 can be achieved by an inversion recovery method using a repeated radio frequency (RF) pulse sequence.[32]

T_1, as a function of correlation time, indicates the degree of molecular motion of polysaccharide in solutions, and it is influenced by viscosity and temperature of the sample system:

$$\frac{1}{T_1} = k \cdot \tau_c \qquad (5.34)$$

where k is a constant and τ_c is the correlation time. τ_c is approximated by:

$$\tau_c = \frac{4\pi^3 a^3 \eta}{3kT} \qquad (5.35)$$

where a^3 is the effective volume, η is viscosity.

T_2 relaxation, or spin to spin relaxation, is characterized by exchange of energy of spins at different energy levels. It is different from T_1 in that it does not lose energy to the surrounding lattice. In other words, T_2 is a process where the random forces modulate the spin energy levels at low frequencies without causing transitions.

The time scale in NMR experiments is slow; this makes molecular motions such as overall rotatory diffusion, segmental motion (oscillation about the glyconsidic bonds), and internal motions (librations of C-H bonds) detectable, which are important factors for understanding the conformation and molecular dynamics of polysaccharides.

5.2.3.1.2 Nuclear Overhauser Effect (NOE) Phenomenon and Conformation Analysis

Nuclear Overhauser effect (NOE) is an incoherent process in which two nuclear spins cross-relax through dipole to dipole interactions and other mechanisms. There is a strong polarization for a saturated spin which will create a dipolar field or an extra magnetic field that can be felt by all the other spins in the vicinity. In other words, the cross relaxations could cause changes in one spin through perturbations of the others. There are two major factors affecting NOE intensity:

- Molecular tumbling frequency
- Inter-nuclear distance

Therefore, the analysis of NOE intensity could provide information to the three dimensional structure and molecular dynamics of polysaccharides chains.

Proton to proton NOE techniques are widely used in elucidating the structure of polysaccharides, whereas ^{13}C relaxation measurements (including NOE) have been extensively used for conformation analysis and for studying the molecular dynamics of polysaccharides in solutions.[32-34]

Distance Dependence of the NOE — NOE is highly dependent on distance: only protons close in space (within about 4–5 Å) will give NOE signals. It gives rise to changes in the intensities of NMR resonances of spin I when the spin population differences of neighboring spin S are altered from their equilibrium values (by saturation or population inversion). The intensity of the NOE is proportional to r^{-6}, where r is the distance between the two spins.

Motion Dependence of the NOE — NOE phenomenon does not only depend on space, but is also influenced by molecular motions, including molecular tumbling or rotatory diffusions, segmental motions and internal rotations. For small molecules in water at room temperature, molecules tumble very quickly and the correlation time τ_c is very small ($\sim 10^{-11}$ sec). For large molecules, including oligosaccharides and polysaccharides (i.e., MW > 3000 Da), tumbling rates of molecules are much slower in solutions and produce large and negative NOE responses and larger correlation times τ_c ($\sim 10^{-5}$ to 10^{-9} sec).[32] For mid-sized molecules (e.g., 1000–3000 Da) tumbling rates are intermediate, in which the NOE crosses from the positive to the negative regime, thus give rather small NOEs (close to zero). In this case, other techniques have to be used to overcome the weak signals, such as rotating-frame NOE (ROE)

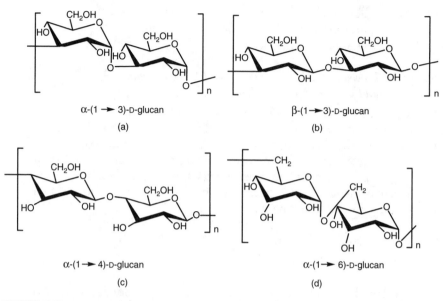

FIGURE 5.22
Structure features of four linear polysaccharides: (a) α-(1→3)-D-glucan; (b) β-(1→3)-D-glucan; (c) α- and β-(1→4)-D-glucan; (d) α-(1→6)-D-glucan.

techniques or alternating solution conditions to change the motional properties of the molecules (temperature or solvent properties). The analysis of motion behavior of molecules using NOE techniques provides the basis for studying the molecular dynamics and conformation of biopolymers.

5.2.3.2 Conformation, Molecular Dynamics, and ¹³C NMR Relaxations

There are only limited number of polysaccharides that have been studied by NMR relaxation experiments aimed at understanding the conformations and molecular dynamics. Among the polysaccharides examined, four linear polysaccharides were studied by Dais's group.[32-34] These four polysaccharides have the same monosaccharide repeating unit (glucose) but different linkage patterns, as shown in Figure 5.22.

5.2.3.2.1 Frequency and Temperature Dependence of T_1, T_2, and NOE

In ¹³C NMR relaxations, T_1, T_2, and NOE values of each ring carbon in polysaccharides varies with temperature and frequencies. The field dependence of ¹³C spin to lattice and spin to spin relaxation times (ms) and NOE values for the various carbon nuclei of amylose in DMSO solution at 80°C is presented in Table 5.4. Further studies of relaxation parameters as functions of temperature and frequency over four linear polysaccharides (Figure 5.22) allowed observations of the following common motional characteristics:[34]

TABLE 5.4

Field Dependence of the ^{13}C Spin-Lattice T, and Spin-Spin Relaxation Times T_2 (ms) and NOE Values for the Various Carbon Nuclei of Amylose in $(CD_3)_2SO$ Solution at 800°C

Carbon Atom	Beta (Degrees)	100 MHz			75 MHz			50 MHz			20 MHz		
		T_1	T_2	NOE	T_1	T_2	NOE	T_1	T_2	NOE	T_1	T_2	NOE
C-1	82.0[a] 37.0[b]	224	105	1.58	182	91	1.7	127	75	1.9	79	58	2.24
C-2	99.0 75.6	255	119	1.63	197	107	1.69	150	84	1.86	86	65	2.2
C-3	72.3 75	253	119	1.68	203	105	1.72	154	87	2.02	90	65	2.1
C-4	75.3 67	251	115	1.59	194	108	1.71	144	87	1.83	88	64	2.08
C-5	79.9 57.3	249	117	1.65	204	108	1.75	151	89	1.89	88	61	2.12
C-6	— —	152	82	1.64	130	68	1.81	90	54	2.02	62	38	2.25

[a] These values represent the angle β between the C-H vector and the z-axis.
[b] Angles formed between the C-H vector and the virtual bond.

Source: Adapted from Dais, 1987.[33]

- As the temperature decreases, the T_1 values decrease monotonically at all fields (frequencies). After reaching a minimum, T_1 increase with further decrease in temperature.

- At a given temperature and solvent, T_1 values increase with increasing magnetic field; the difference in T_1 values among the magnetic fields become more pronounced as the temperature decreases (slow motion regime).

- T_2 values decrease continually with decreasing temperature, and T_2 values are always much smaller than the corresponding T_1 values, especially at low temperatures.

- NOE values decrease with increasing magnetic field, although they tend to converge as temperature decreases to reach the slow motion regime.

- At high temperatures, where motions become faster, NOE values are higher, but still much less than the theoretical maximum of 2.988.

Based on these observations, it is not possible to describe the dynamics of these polysaccharides by a single exponential time correlation function (TCF).

5.2.3.2.2 ^{13}C Relaxations and Molecular Dynamics

There are three types of motions that are important for modeling the molecular dynamics of polysaccharides:[32]

- Overall rotatory diffusion or tumbling of the polysaccharide chains
- Segmental motion, or local chain motions, such as oscillatory motions about the glycosidic bond
- Internal motions of substituents (pendent groups), such as hydroxymethyl groups exocyclic to the polymer backbone or other substitutes

Each of these motions contributes to the conformation of carbohydrates independently and the composite time correlation function (TCF) is the product of the correlation functions associated with each motion.

The overall rotatory diffusion (molecular tumbling) is MW dependent. For example, for a relatively low MW polysaccharides, the overall rotatory diffusion is the dominate relaxation source. As MW increases, the tumbling rate will gradually decrease; at sufficiently high MW, this motion becomes so slow that its contribution to the relaxation parameter could be neglected. For amylose chains of MW = 3.3×10^5 in DMSO, the corresponding correlation time is fairly long (7.0×10^{-6} s).[34] Generally speaking, the TCF of the overall motion for a polymer chain of high molecular weight is a slowly decaying function (correlation time about 10^{-5} to 10^{-8} s), whereas the TCF for the local motions is a much faster decaying function (correlation time 10^{-9} to 10^{-11} s).[32]

The second independent motion (segmental motion) has been described by several models to reflect the geometric constraint that controls chain flexibility of polysaccharides. Among those models proposed, a modified bimodal $\log(\chi^2)$ distribution was successful in reproducing relaxation data for α-(1→3)-, and β-(1→3)-glucans, but not suitable for dextrans in DMSO and water solutions.[32] This is because the modified $\log(\chi^2)$ distribution describes the segmental motion only but is ineffective for counting other type of motions, such as local libration of the C-H vectors; these motions are apparently significant for a flexible polymers like dextrans.

To overcome the shortcomings of the modified $\log(\chi^2)$ distribution model, a model was developed by Dejean, Laupretre, and Monnerie, which is known as the DLM model.[35] The DLM model covers both the local anisotropic motion such as fast libration of the C-H vectors and the segmental motion. The DLM time-correlation function governs two motional processes: (1) a diffusion process along the chain that occurs via conformational transitions characterized by two correlation time τ_0, τ_1 for isolated single-bond conformational transitions and for cooperative transitions, respectively; (2) bond librations that are the wobbling in a cone motion of the backbone internuclear C-H vector as illustrated in Figure 5.23. The librational motion is associated with a correlation time τ_2. The extent of the libration about the rest position of the C-H bond, which coincides with the axis of the cone, is determined by the cone half-angle, θ (Figure 5.23).[32,33] The experimental data agrees well with the calculated values based on DLM model throughout the entire temperature range, and the fitting parameters for the DLM model of the four linear polysaccharides are summarized in Table 5.5.

Comparing the τ_1 values of the four polymers it can be concluded that the segmental motion of dextran is about 1.5 times faster than that of the α-(1→3)-D-glucan, and 2 to 2.5 times faster than that of the β-(1→3)-D-glucan at all temperatures (except for at 25°C), confirming the greater flexibility of dextran polymer chains compared to the other two glucans. The glycosidic linkage through C1-O-C6-C5 in dextran (three bonds) has more motional freedoms than the simple glycosidic linkage through C1-O-C3 (two bonds)

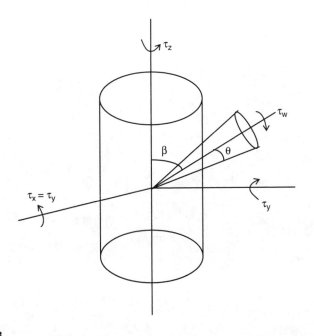

FIGURE 5.23
Internal motion as wobbling in a cone — diffusion in a cone model. (Adapted from Dais, 1987.)[32]

in the α- and β-(1→3)-D-glucan (Figure 5.22). The τ_1 values of amylose (α-1→4-linkage) are comparable to that of the β-(1→3)-D-glucan at all temperatures, indicating the two polysaccharide chains have similar flexibility, although the τ_1 values of amylose are slightly higher than that of the β-(1→3)-D-glucan. The apparent τ_1 values for dextran in water are about 2 times of that in DMSO. However, after considering the viscosity effect of solvent, the corrected τ_1 values are indistinguishable, suggesting that dextran does not have any appreciable difference in mobility in the two solvents. For more detailed information in this area, the readers are referred to the references.[32–34]

For sufficiently high molecular weight random coil polymers (above a critical value of 1000–10,000 Da, depending on chemical structure), the overall motion is much slower than the chain local motions, and thus makes a negligible contribution to the T_1 values of the backbone and side-chain carbons. The molecular weights of most polysaccharides of interest to food scientists are much greater than this limit. The correlation time (τ_R) of the overall motion cannot be determined from relaxation data alone because all carbons in the carbohydrate chain relax via both the overall and local motions. Therefore, the overall motion correlation time τ_R is derived from the following hydrodynamic equation:

$$\ln\left(\tau_R/\tau_R'\right) = k'[\eta]c \tag{5.36}$$

TABLE 5.5

Correlation Times (10^{-10} s), Librational Angles ($\theta°$), Diffusional Angle ($\chi°$) Obtained from the DLM and Restricted Diffusion TCFs for Linear Polysaccharides

	Amylose			Dextran		α-(13)-D-glucan			β-(13)-D-glucan		
t (°C)	τ_1	τ_t	$\chi°$	$\tau_1{}^a$	$\tau_1{}^b$	τ_i	τ_t	$\chi°$	τ_t	τ_t	$\chi°$
25	45.73	0.67	37	36.88	14.96	42.74	0.52	57	42.34	0.64	39
35					11.15						
40	39.28	0.54	39	17.91		22.09	0.37	55	32.31	0.64	48
45					7.45						
55					5.35						
60	24.38	0.45	46	9.64		13.61	0.22	59	21.35	0.34	59
65					3.46						
75					2.49						
80	14.13	0.32	52	6.26		7.45	0.2	69	14.19	0.26	68
85					2.03						
95						4.82	0.17	82	9.35	0.18	72
100	9.28	0.2	55	3.91							
110											
120	5.85	0.18	57	2.26		3.37	0.12	77	4.63	0.1	74
τ_0/τ_1	7			3	8	3			5		
τ_1/τ_2	100			200	140	30			100		
θ (C-1)	21.5										
θ (C-3)											
θ (C-4)					24						
θ (C-5)											
θ (C-6)											
θ (C-ave)	27			28.5	28	28.5			29.5		

a In Me$_2$SO-d_6
b In D$_2$O

Source: Adapted from Tylianakis et al., 1999.[33]

in which, τ'_R is a single correlation time of an isotropic motion in infinite dilution:

$$\tau'_R = \frac{2Mw[\eta]\eta_0}{3RT} \tag{5.37}$$

where Mw is the molecular weight, $[\eta]$ is the intrinsic viscosity of the polymer and η_0 is the solvent viscosity; k' is the Huggins constant (see Chapter 4 for definition). The calculated τ_R values for the four polysaccharides are summarized in Table 5.6. Since these parameters represent polysaccharide behavior in the high molecular weight limit, they are dominated by local polymer dynamics.[33]

5.2.4   Summary of Conformation Analysis

The characterization of secondary, tertiary, and quaternary structural levels of organizations of polysaccharides is essential for understanding the molecular

TABLE 5.6

Molecular Weight (Mw), Intrinsic Viscosities (η), Huggins Constant κ' and Molecular Correlation Times (τ_R) of Overall Motion of Four Linear Polysaccharides at 30°C

Polysaccharide	Solvent	$Mw \times 10^3$	$[\eta]$, mLg^{-1}	κ'	τ_R ($\times 10^{-7}$)
Amylose	Me^{2}SO-d$_6$	330	104.7	0.414	7
Dextran	Me^{2}SO-d$_6$	35	68.6	2.512	143.6
	D$_2$O		46.1	2.525	3.6
α-(1→3)-D-glucan	Me^{2}SO-d$_6$	7.5	50.0	2.128	2
β-(1→3)-D-glucan	Me^{2}SO-d$_6$	400	502.8	3.11	105

Source: Adapted from Tylianakis et al., 1999.[33]

basis of their functional properties. Recent development in computer technology and software significantly simplified the data analysis of x-ray diffraction, scattering, and NMR methods. However, most of the methods used were based on empirical calculations and models, therefore, adjustment is required to obtain accurate data and reflect the real molecular conformation. The most significant development in polysaccharide conformation analysis is the rapid development of computational methods. There are several review articles and numerous research papers covering the progress and development of molecular modeling.[36–39] The trend of conformation analysis of polysaccharides is the combination of experimental approaches and computer modelings, which appears to yield more valid and accurate information that can be used to establish the relationship of polysaccharide structures and their physical and functional properties.

References

1. Rees, D. A. *Polysaccharide Shapes.* John Wiley & Sons, New York. 1977.
2. Settineri, W. J. and Marchessault, R. H. Derivation of possible chain conformations for poly Cβ-1→4-anyhydroxylose, *Journal of Polymer Science*, Part C, 11, 253–264. 1965.
3. Flory, P. J. *Principles of Polymer Chemistry.* Cornell University Press, Ithaca, N.Y. 1971.
4. Tanford, C. *Physical Chemistry of Macromolecules*, John Wiley & Sons, Inc. New York,. 138–179. 1967.
5. Sun, S. F. *Physical Chemistry of Macromolecules, Basic Principles and Issues*, A Wiley-Interscience Publication, John Wiley & Sons, Inc. New York, 96–120 and 371–398. 2004.
6. Benoit, H. and Sadron, C. Compléments à l'étude de la statistique des chaînes moléculaires en solution diluée. *Journal of Polymer Science*, 4, 473–482. 1949.
7. Flory, P. J. and Fox, T. G. Treatment of intrinsic viscosities. *Journal of the American Chemical Society*, 73, 1904–1908. 1951.
8. de Gennes P.G. *Scaling Concepts in Polymer Physics.* Cornell University Press, New York, 1979.

9. Zimm, B. H. and Stockmayer, W. H. The dimensions of chain molecules containing branches and rings, *Journal of Chemical Physics*, 17, 1301. 1949.

10. Jeffrey, G. A. and Sundaralingam, M. Bibliography of crystal structures of carbohydrates, nucleosides, and nucleotides. *Advances in Carbohydrate Chemistry and Biochemistry*, 43, 203–421. 1985.

11. Brady, J. Oligosaccharides geometry and dynamics. In: P. Finch, ed. *Carbohydrates: Structures, Syntheses and Dynamics*. Kluwer Academic Publishers, London, 228–257. 1999.

12. Rao, V. S. R., Qasba, P. K., Balaji, P. V., and Chantrasekaran, R. *Conformation of Carbohydrates*. Harwood Academic Publishers, Amsterdam, 223–254. 199.

13. Smith, P. J. C. and Arnott, S. A linked-atom least-squares reciprocal-space refinement system incorporating stereochemical restraints to supplement sparse diffraction data. *Acta Crystallogr.*, Sect. A 34, 3–11. 1978.

14. Zugenmaier, P. and Sarko, A. The variable virtual bond: Modeling technique for solving polymer crystal structures. *American Chemical Society Symposium Society Symoposium Series*, 141, 225–237. 1980.

15. Ogawa, K. and Yui, T. X-ray diffraction study of polysaccharides. In: S Dumitriu. ed. *Polysaccharides: Structural Diversity and Functional Versatility*. Marcel Dekker, New York, 101–129. 1998.

16. Chien, Y. Y. and Winter, W. T. Accurate lattice constants for tara gum. *Macromolecules*, 18, 1357–1359. 1985.

17. Song, B. K., Winter, W. T., and Taravel, F. R. Crystallography of highly substituted galactomannans: fenugreek and lucerne gums. *Macromolecules*, 22, 2641–2644. 1989.

18. Kapoor, V. J., Chanzy, H., and Taravel, F. R. X-ray diffraction studies on some seed galactomannans from India. *Carbohydrate Polymers*, 27, 229–233. 1995.

19. Chandrasekaran, R., Radha, A., and Okuyama, K. Morphology of galactomannans: an x-ray structure analysis of guaran. *Carbohydrate Research*, 306, 243–255. 1998.

20. Chandrasekaran, R., Bian, W., and Okuyama, K. Three-dimensional structure of guaran. *Carbohydrate Research*, 312, 219–224. 1998.

21. Chandrasekaran, R. Molecular architecture of polysaccharide helices in oriented fibers. *Advances in Carbohydrate Chemistry and Biochemistry*, 52, 311–439. 1997.

22. Brant, D. A. Novel approaches to the analysis of polysaccharide structures. *Current Opinion in Structural Biology*, 9, 556–562. 1999.

23. Burchard, W. Light scattering techniques. In: F. Happey, ed. *Applied Fiber Science*. Academic Press, New York. 381–420, 1978.

24. Burchard, W. Static and dynamic light scattering from branched polymers and biopolymers. *Advances in Polymer Science*, 48, 1–124. 1983.

25. Stepanek, P. Data analysis in dynamic light scattering. In W. Brown, ed. *Dynamic Light Scattering: The Method and Some Applications*. Oxford Science Publications, Oxford, 177–241. 1993.

26. Burchard, W. Light scattering . In: S. B. Ross-Murphy, ed. *Physical Techniques for the Study of Food Biopolymers*. Blackie Academic & Professional, New York, 151–214. 1994.

27. Roy, R. J. *Methods of X-Ray and Neutron Scattering in Polymer Science*. Oxford University Press, Oxford, 155–209, 2000.

28. Schmidt, M., Paradossi, G., and Burchard, W. Remarks on the determination of chain stiffness from static scattering experiments. *Makromolekulare Chemie. Rapid Communications*, 6, 767–772. 1985.

29. Coviello, T., Kajiwara, K., Burchard, W., Dentin, M., and Crescenzi, V. Solution properties of xanthan gum. I. Dynamic and static light scattering from native and modified xanthan in dilute solutions. *Macromolecules*, 19, 2826–2831, 1986.

30. Yuguchi, Y., Mimura, M., Urakawa, H., Kitamura, S., Ohno, S., and Kajiwara, K. Small angle x-ray characterization of gellan gum containing a high content of sodium in aqueous solution. *Carbohydrate Polymers*, 30, 83–93. 1996.

31. Southwick, J. G., Jamieson, A. M., and Blackwell, J. Quasielastic light scattering studies of xanthan in solution. In: D. A. Brant, ed. *Solution Properties of Polysaccharides*. ACS, Washington, D.C. 1–13. 1981.

32. Dais, P. ^{13}C Nuclear magnetic relaxation and motional behaviour of carbohydrate molecules in solution. *Advances in Carbohydrate Chemistry and Biochemistry*, 51, 63–131. 1995.

33. Tylianakis, M., Spyros, A., Dais, P., Taravel, F. R., and Perico, A. NMR study of the rotational dynamics of linear homopolysaccharides in dilute solutions as a function of linkage position and stereochemistry. *Carbohydrate Research*, 315, 16–34. 1999.

34. Dais, P. Carbon-13 magnetic relaxation and local chain motion of amylose in dimethyl sulfoxide. *Carbohydrate Research*, 160, 73–93. 1987.

35. Dejean de la batie, R., Laupretre, F., and Monnerie, L. Carbon-13 NMR investigation of local dynamics in bulk polymers at temperatures well above the glass transition temperature. 1. Poly(vinyl methyl ether). *Macromolecules* 21, 2045–2052. 1988.

36. Perez, S., Kouwijzer, M., Mazeau, K., and Engelsen, S. B. Modelling polysaccharides: present status and challenges. *Journal of Molecular Graphics*, 14, 307–321. 1996.

37. Mazeau, K. and Rinaudo, M. The prediction of the characteristics of some polysaccharides from molecular modeling. Comparison with effective behaviour. A review article. *Food Hydrocolloids*, 18, 885–898. 2004.

38. Van Gunsteren, W. F. and Berendsen, H. J. C. Molecular dynamics computer simulation. Method, application, and perspectives in chemistry. *Angewandte Chemie*, 102, 1020–1055. 1990.

39. Boutherin, B., Mazeau, K., and Tvaroska, I. Conformational statistics of pectin substances in solution by a Metropolis Monte Carlo study. *Carbohydrate Polymers*, 31, 255–266. 1997.

6

Polysaccharide Gums: Structures, Functional Properties, and Applications

Marta Izydorczyk, Steve W. Cui, and Qi Wang

CONTENTS

6.1   Introduction

Polysaccharides are composed of many monosaccharide residues that are
joined one to the other by O-glycosidic linkages. The great diversity of
structural features of polysaccharides, which originates from differences in
the monosaccharide composition, linkage types and patterns, chain shapes,
and degree of polymerization, dictates their physical properties including
solubility, flow behavior, gelling potential, and/or surface and interfacial
properties. The structural diversity also dictates the unique functional properties exhibited by each polysaccharide. Polysaccharides, which are commercially available for use in food and nonfood industries as stabilizers,
thickening and gelling agents, crystalization inhibitors, and encapsulating
agents, etc., are also called hydrocolloids or gums. Polysaccharide gums
occur in nature as storage materials, cell wall components, exudates, and
extracellular substances from plants or microorganisms. Chemical modification of cellulose and chitin provides additional sources of hydrocolloids or
gums with improved functionality. This chapter will introduce some polysaccharides that are significant in nature and have been widely used in food
and nonfood applications.

6.2 Plant Polysaccharides

6.2.1 Cellulose and Derivatives

6.2.1.1 Sources and Structures

Cellulose is the most abundant naturally occurring polysaccharide on earth. It is the major structural polysaccharide in the cell walls of higher plants. It is also the major component of cotton boll (100%), flax (80%), jute (60 to 70%), and wood (40 to 50%). Cellulose can be found in the cell walls of green algae and membranes of fungi. *Acetobacter xylinum* and related species can synthesize cellulose. Cellulose can also be obtained from many agricultural by-products such as rye, barley, wheat, oat straw, corn stalks, and sugarcane.

Cellulose is a high molecular weight polymer of (1→4)-linked β-D-gluco-pyranose residues (Figure 6.1). The β-(1→4) linkages give this polymer an extended ribbon-like conformation. Every second residue is rotated 180°, which enables formation of intermolecular H-bonds between parallel chains. The tertiary structure of cellulose, stabilized by numerous intermolecular H-bonds and van der Waals forces, produces three-dimensional fibrous crystalline bundles. Cellulose is highly insoluble and impermeable to water. Only physically and chemically modified cellulose finds applications in various foodstuffs.

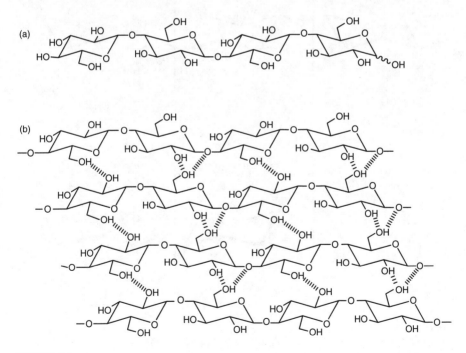

FIGURE 6.1
Basic structures of cellulose: (a) repeating units; (b) illustration of intermolecular hydrogen bonds.

6.2.1.1.1   Microcrystalline Cellulose

Microcrystalline cellulose (MCC) is prepared by treating natural cellulose with hydrochloric acid to partially dissolve and remove the less organized amorphous regions of this polysaccharide. The end product consists primarily of crystallite aggregates. MCC is available in powder form after drying the acid hydrolysates. Dispersible MCC is produced by mixing a hydrophilic carrier (e.g., guar or xanthan gum) with microcrystals obtained through wet mechanical disintegration of the crystallite aggregates.

6.2.1.1.2   Carboxymethylcellulose

Carboxymethylcellulose (CMC) is an anionic, water-soluble polymer capable of forming very viscous solutions. CMC is prepared by first treating cellulose with alkali (alkali cellulose), and then by reacting with monochloroacetic acid. The structure of CMC is depicted in Figure 6.2a. The degree of substitution (DS) with the carboxyl groups is generally between 0.6 to 0.95 per monomeric

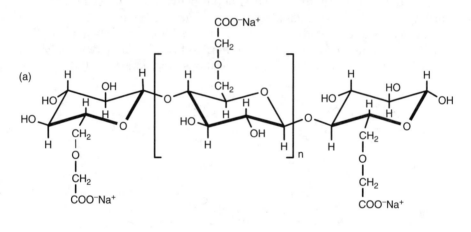

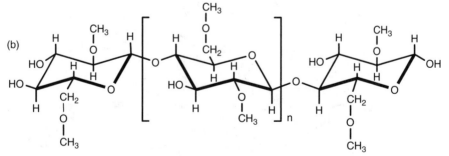

FIGURE 6.2
Structural features of chemically modified cellulose: (a) Carboxymethylcellulose (CMC); (b) Methylcellulose (MC).

unit (maximum DS is 3), and occurs at O-2 and O-6, and occasionally at O-3 positions.

6.2.1.1.3 *Methylcellulose and Its Derivatives*

Methylcellulose (MC) is a nonionic cellulose ether (Figure 6.2b) with thickening, surface activity (due to hydrophobic groups), and film forming properties. MC is prepared by treating alkali cellulose with methyl chloride. Other MC derivatives are also available, of which hydroxypropylmethylcellulose (HPMC) has been widely used. Hydroxypropylcellulose (HPC) and hydroxyethylcellulose (HEC) are used primarily in industrial applications.

6.2.1.2 **Functional Properties and Applications**

6.2.1.2.1 *Microcrystalline Cellulose (MCC)*

Colloidal MCC dispersions exhibit thixotropic flow behavior because of powdered aggregates. The aqueous dispersions are also thermally stable. MCC finds applications in pharmaceutical, food, and paper industries. For example, MCC is used as a bulking agent for modifying food textures and as a fat replacer in emulsion based food products. It is also a good suspending agent for particles and solids and also adds creaminess to various products (e.g., chocolate drinks). Adding MCC in batters and breadings can improve cling, reduce drying time and fat absorption during frying.

6.2.1.2.2 *Carboxymethylcellulose*

Carboxymethylcellulose (CMC) is soluble in cold water and the solubility is pH dependent: CMC is more soluble in alkali conditions and insoluble in acidic conditions. CMC solutions could be pseudoplastic or thixotropic, depending on molecular weight degree of polymerization (DP) and degree and pattern of substitution. For example, medium to high viscosity CMC solutions exhibit pseudoplastic flow behavior; whereas the same type of solution can be in thixotropic if the DS is between 0.4 to 0.7 and the substituents are unevenly distributed. CMC is able to form gels in the presence of trivalent metal ions such as aluminum. The gelling properties of CMC are influenced by polymer concentration, DP, pH, type, and level of metal cations.

CMC is used in the food industry as a thickener, stabilizer, and suspending agent. For example, CMC can improve the loaf volume during baking by stabilizing gas bubbles. It is also used as a stabilizer and ice crystal growth inhibitor in frozen desserts and soft-serve ice creams. In pet food industry, CMC is used to absorb water, thicken gravy, aid extrusion, and bind fines. CMC is a good coating for powders and tablets in the pharmaceutical industry because of its insoluble nature in the acidic stomach fluids and its soluble nature in alkaline intestinal fluids. CMC also finds applications in the cosmetic industry to stabilize hand lotions and vitamin-oil emulsions.

6.2.1.2.3   Methylcellulose and Its Derivatives (MC and HPMC)

Two unique functional characteristics of MC and HPMC are their inverse temperature solubility and thermal gelation properties. Gels are formed at elevated temperatures due to hydrophobic interactions between highly-substituted regions, which consequently stabilize intermolecular hydrogen bonding. The firmness of the heat setting gels is dependent on the degree of methylation. MC also has good film forming properties and it behaves as a true emulsifier because it contains both hydrophilic and hydrophobic groups.

Because of their multifunctional properties, MC and HPMC are used as emulsifiers and stabilizers in low-oil and/or no-oil salad dressings. The film forming and thermal gelation characteristics of MC are applied in fried foods to reduce oil absorption. MC and HPMC are used in baked goods to prevent boil-over of pastry fillings and to aid gas retention during baking. MC is also used in pharmaceutical industry for coating tablets and controlling drug release. MC and its derivatives are widely used as adhesives for wallpapers, release control agents for pesticides and fertilizers, and emulsifiers and stabilizers in shampoo and hair conditioners.

6.2.2   Hemicelluloses

6.2.2.1   Sources and Structures

Hemicelluloses are a heterogeneous group of polysaccharides constituting the cell walls of higher plants; these polysaccharides are often physically entangled, covalently and/or noncavalently bonded to cellulose and lignins. The structure of hemicelluloses may vary depending on their origin, but they can be divided into four groups based on composition of their main backbone chain: D-xylans with (1→4)-linked β-D-xylose; D-mannans, with (1→4)-linked β-D-mannose; D-xyloglucans with D-xylopyranose residues attached to the cellulose chain; and D-galactans with (1→3)-linked β-D-galactose. The first three groups are very similar to cellulose in having the main chain backbone linked via (1→4)-diequatorial linkages and capable of adopting extended ribbon conformations. Most of the hemicelluloses, however, are substituted with various other carbohydrate and noncarbohydrate residues, and unlike cellulose, they are heteropolysaccharides. This departure from uniformity because of various side branches renders them at least partially soluble in water.

6.2.2.1.1   Mannans and Galactomannans

The cell walls of seeds are especially rich in mannans and galactomannans (Figure 6.3). D-Mannans, found in tagua palm seeds, have a backbone composed of linear (1→4)-linked β-D-mannose chains. The best known D-galactomannans, locust bean, guar, and tara gums have the same linear mannan backbone but they are substituted with α-D-Gal*p* side units linked to O-6. The degree of substitution in galactomannans, which profoundly affects their

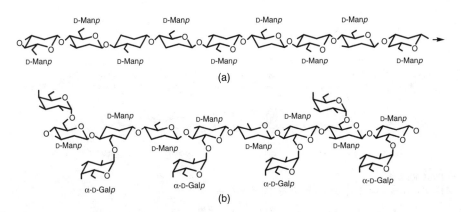

FIGURE 6.3
Structures of mannans (a) and galactomannans (b) from plants.

TABLE 6.1

Botanical Origin and Main Structural Features of Galactomannans

Galactomannan	Species of Origin	Man:Gal Ratio
Locust bean gum	*Ceratonia siliqua*	3.5
Senna gum	*Senna occidentalis*	3.5
Guar gum	*Cyamopsis tetragonolobus*	1.6
Tara gum	*Caesalpinia spinosa*	1.3
Fenugreek gum	*Trigonella foenum graecum*	1

solution properties, differs in galactomannans extracted from various plants (Table 6.1).

6.2.2.1.2 Xyloglucans

Xyloglucans, like cellulose, have linear backbones of (1→4)-linked β-D-gluco-pyranoses. Numerous xylopyranosyl units are attached along the main backbone. In many plant xyloglucans, the repeating unit is a heptasaccharide, consisting of a cellotetraose with three subtending xylose residues.[1] Some xylose residues may carry additional galactosyl and fucosyl units (Figure 6.4). A few plants may have arabino- instead of fucogalactosyl-groups attached to the xylose residues. One of the best characterized is the xyloglucan from the cotyledons of the tamarind seed (*Tamarindus indica*).[2]

6.2.2.1.3 Glucomannans

Glucomannans are linear polymers of both (1→4)-linked β-D-mannose and (1→4)-linked β-D-glucose residues. Glucomannans are obtained from dried and pulverized root of the perennial herb *Amorphophallus konjac*. Acetyl groups scattered randomly along the glucomannan backbone promote water solubility. Konjac glucomannan is a high molecular weight polymer (>300 kDa) which can form viscous pseudoplastic solutions. It can form a

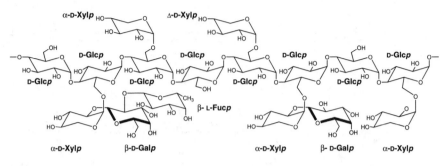

FIGURE 6.4
Partial structure of xyloglucans from tamarind seed.

gel in the presence of alkali. Konjac glucomannans, like locust bean and guar galactomannans, have synergism with xanthan gum and κ-carrageenans and can form thermally reversible gels.

6.2.2.1.4 Arabinoxylans

D-Xylans are composed of (1→4)-linked β-D-xylopyranoses with various kinds of side branches, the most common being 4-*O*-methyl-D-glucopyranosyl uronic acid linked mostly to O-2 of β-Xyl*p* units and α-L-Araf linked to O-3 of β-Xyl*p* units (Figure 6.5). The amount of arabinose and glucuronic acid in glucuronoarabinoxylans may vary substantially, ranging from substitution at almost all Xyl*p* to polymers having more than 90% of unsubstituted β-Xyl*p* units. Many cereal (wheat, barley, rye, oats) arabinoxylans do not carry glucuronic acid units. Arabinose side branches, however, may have ferulic acid groups esterified to O-5 of Ara*f* residues. α-L-Arabinose residues are attached at O-2 and O-3 to the xylose units.

6.2.2.1.5 β-D-Glucans

Mixed linkage (1→3), (1→4) β-D-glucans are present in the grass species, cereals, and in some lichens (e.g., *Cetraria islandica*). Cereal β-D-glucans contain predominantly (1→4)-linked β-D-Glc*p* units (~70%) interrupted by single (1→3)-linked β-D-Glc*p* units (~30%). The distribution of β-(1→4) and β-(1→3) linkages is not random; this leads to a structure of predominantly β-(1→3)-linked cellotriosyl and cellotetraosyl units. There are also longer fragments of contiguously β-(1→4)-linked glucose units (cellulose fragments) in the polymer chain (Figure 6.6). β-D-Glucans are high molecular weight, viscous polysaccharides. The main source of food β-D-glucans are the kernels of oats, barley, wheat, and rye. β-D-Glucans have been ascribed cholesterol and blood glucose lowering properties.

6.2.2.1.6 Arabinogalactan

Arabinogalactan is a major D-galactan obtained from soft-woods such as pine, larch, cedar, and spruce. This polymer has a main backbone of (1→3)-linked β-D-galactopyranosyl residues with β-(1→6)-linked disaccharides of

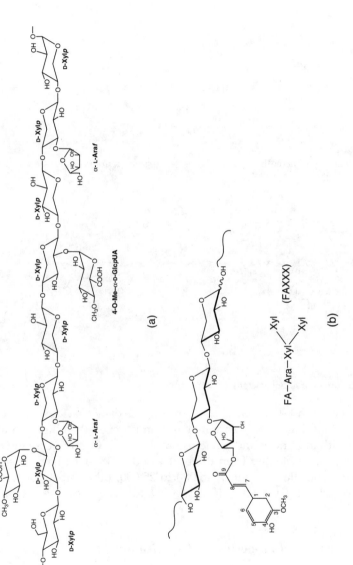

FIGURE 6.5
Illustration of partial structures of arabinoxylans from cereals: (a) arabinoxylans; (b) attachment of ferulic acid to the arabinoxylan chain.

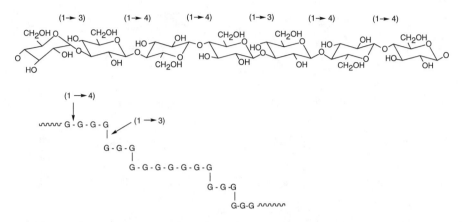

FIGURE 6.6
Structural features of mixed-linkage (1→3)(1→4)- β-D-glucans from cereals.

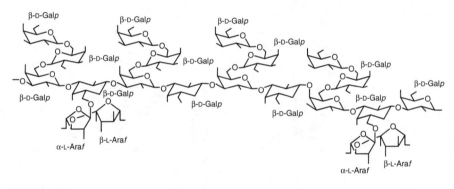

FIGURE 6.7
Partial structure of arabinogalactans.

β-D-Gal*p*-(1→6)-β-D-Gal*p* and α-(1→6)-linked disaccharides of β-L-Ara*f*-(1→3)-α-L-Ara*f* (Figure 6.7). Arabinogalactan is generally a highly branched polymer with arabinose and galactose ratio of 1:6. Commercially available arabinogalactan, obtained from the butt wood of Western larch, has a relatively low molecular weight of 15,000 to 25,000, little impact on viscosity, color, and taste. It is used as a low-calorie additive in beverages to increase the fiber content.

6.2.2.2   Functional Properties and Applications

6.2.2.2.1   Galactomannans

In aqueous solutions galactomannans assume random coil conformations and their flow behavior is concentration and shear-rate dependent. At low concentrations and/or sufficiently low shear rate, all galactomannan gums exhibit Newtonian flow behavior where the viscosity is independent of shear rate. At higher concentrations and higher shear rate, the solutions are pseudoplastic in that the viscosity decreases with the increase of shear rate, a typical non-Newtonian flow behavior (see Chapter 4 for details). The solubility of galactomannan gums is significantly affected by the degree and pattern of substitution of the galactosyl units. The linear mannan is similar to cellulose which is virtually insoluble in water. The presence of galactosyl residues in locust bean, guar, tara, and fenugreek gums significantly improves their solubility. The solubility of these gums increases with the increasing degree with substitution with galactosyl units, i.e., fenugreek gum > guar gum > locust bean gum (Table 6.1). The solubility and hydration rate are also influenced by particle size, pH, ionic strength, temperature, co-solutes, and agitation methods. Coincidentally, the thickening power decreases in the order of increasing galactose content, i.e., locust bean > tara > guar > fenugreek when the comparison is made at the same molecular weight and concentration of polymers.

Galactomannan gums are compatible with most hydrocolloids. There is a useful synergistic increase in viscosity and/or gel strength by blending galactomannan gums with certain linear polysaccharides including xanthan, yellow mustard mucilage, κ-carrageenan, and agarose. The extent of the synergistic interactions between galactomannans and other polysaccharides also follows the order of locust bean > tara > guar > fenugreek, which coincides with the increasing degree of substitution of the backbone chain. Unique surface activity has been observed for fenugreek gum that makes it an effective stabilizer for oil/water emulsions.

Guar gum and locust bean gums are extensively used in the food industry as thickening and stabilizing agents, usually in amounts of <1% of the food weight. For example, they have been used to improve shelf-life and prevent creaming or settling in salad dressings, soft drinks, and fruit juices. Locust bean gum is also used to improve freeze-thaw behavior of frozen products including ice cream and frozen desserts. Guar gum, used in baked products and pastry, reduces degree of starch retrogradation and improves texture and shelf-life. Low grades guar and locust bean gums and their derivatives also found applications in other industries including mining, paper, textile, and oil drilling.

6.2.2.2.2 Xyloglucans

Xyloglucan is soluble in cold-water and can form a gel only in the presence of alcohol or a substantial amount of sugar. The flow behavior of xyloglucan solutions is similar to galactomannans: they exhibit Newtonian flow behavior at low concentrations and non-Newtonian flow behavior at higher concentrations (ca. >0.5% w/w).[3] Dynamic rheological properties of xyloglucan solutions are similar to that of random coil polysaccharides in which the mechanical spectra changed from a spectrum typical of dilute polymer solutions ($G' < G''$ at all frequencies) at low concentrations to that of semi-dilute polymer solutions ($G' < G''$ at low frequencies and $G' > G''$ at higher frequencies) at higher concentrations.[3,4] Gelation is observed for tamarind seed xyloglucan in the presence of alcohol[5] or high levels of sugar (40 to 70% w/w).[6]

Sugar-induced gels are elastic and have good water holding properties and freeze-thaw processes can make the gels harder and more elastic. Gels can also be formed when tamarind xyloglucan is mixed with catechin, a polyphenolic compound.[1]

Only tamarind seed xyloglucan is commercially available. It has been used as a starch replacer in food products with reduced calories. It also can be used in the pharmaceutical industry for controlling the release of drugs.

6.2.2.2.3 Glucomannans

In the solid state, konjac glucomannan has an extended twofold helical structure. It is insoluble in water, but soluble in alkali solutions. If dissolved in water, konjac glucomannan is expected to yield a high viscosity; however, extra effort is required to solubilize the polymers (vigorous stirring, heating, or ultrasonic treatment). Konjac glucomannan solutions exhibit a typical non-Newtonian flow behavior with shear thinning properties. A thermally irreversible gel can be formed when konjac glucomannan solution is heated in alkali conditions. Konjac glucomannan gives synergistic interactions with xanthan gum, and the mixture can form gels at a total polymer concentration as low as 0.1%. Thermally reversible gels can also be prepared by mixing konjac glucomannan with agarose or carrageenan.

Konjac flour is suitable as a thickening, gelling, texturing, and water binding agent. It is a major ingredient in vegetarian meat analogue products. The thermally irreversible gel prepared from konjac flour in alkali conditions is a traditional Japanese food item while the thermally reversible gels formed by the synergistic interactions between konjac glucomannan and other gum applications are found in products such as health jellies.

6.2.2.2.4 Arabinoxylans

Pentosans or arabinoxylans form random coil structures in aqueous solutions. However, arabinoxylans isolated from wheat flour can form thermally irreversible gels upon oxidative gelation and cross-linking of ferulic acid residues. Wheat and rye arabinoxylans are important functional ingredients in baked products: they affect functional properties such as water binding, dough rheology, and starch retrogradation. The presence of arabinoxylans also protects the gas cells in the dough during baking. Due to the lack of commercial products, most of the applications of arabinoxylans or pentosans are in the bakery industry.

6.2.2.2.5 Cereal β-D-Glucan

Solutions of β-D-glucan polymers, isolated from different sources and under various extraction conditions give, diverse rheological properties, ranging from viscoelastic fluids, through weak-gel structures to real gels. High molecular weight β-D-glucans usually exhibit shear thinning flow behavior at a high shear rate region but Newtonian flow behavior at low shear rates. The low molecular weight β-D-glucans, on the other hand, can form gels upon cooling. The gelation of β-D-glucans is affected by two structural features: the molecular weight and structural regularity of polymer chains. The lower molecular weight β-D-glucans favor gel formation, possibly due to the increased mobility of shorter polymer chain to form junction zones. When the molecular weight is fixed, the gel forming capacity of β-D-glucans is affected by the ratio of tri-/tetra saccharides in β-glucans (4.2 > 3.3 > 2.2 for β-D-glucans from wheat, barley, and oats, respectively).[7]

Oat β-D-glucans are components of dietary fiber known for positive health benefits, such as lowering of cholesterol levels and moderating the glycemic response.[8] There are two β-D-glucan-based products commercially available: Oatrim and Glucagel. Oatrim is prepared by treating oat bran or oat flour with a thermo-stable α-amylase at high temperatures. The application of Oatrim includes bakery products, frozen desserts, processed meats, sauces, and beverages. Oatrim is also used as a fat replacer. Glucagel is prepared by partial hydrolysis of barley β-D-glucans. The low molecular weight β-D-glucans (15,000 to 150,000 Daltons) form gels at 2% polymer concentration. The major functionalities of Glucagel are its gelling and fat mimetic properties. Glucagel is used as fat substitutes in bakery, dairy products, dressings, and edible films. β-D-Glucan isolated from oats can also be used in personal care industry as a moisturizer in lotions and hand creams.

6.2.3 Pectins

6.2.3.1 Sources and Structures

Pectins are the major components of most higher plant cell walls; they are particularly prevalent in fruits and vegetables. Commercial pectins are prepared mostly from some by-products of the food industry, such as apple pulp, citrus peels, and sugarbeet pulp. Pectins are the most complex class of plant cell wall polysaccharides. They comprise of two families of covalently linked polymers, galacturonans and rhamnogalacturonans (type I) (Figure 6.8 a).

Galacturonans are segments of pectins with (1→4)-linked α-D-galactosyluronic acid residues in the backbone, such as those in the linear homogalacturonans, in the substituted xylogalacturonans and in rhamnogalacturonans type II (RG II) (Figure 6.8b to Figure 6.8d). The carboxylic acid groups in galacturonans may be methyl esterified; the degree of esterification has an important effect on the conformation and solution properties of these polymers. Based on the degree of esterification, pectins are divided into two categories: low methyl (LM) pectins that contain less than 50% methyl esters, and high methyl (HM) pectins with more that 50% methyl esters. Xylogalacturonans are relatively recently discovered subunits of pectic polysaccharides, present in storage tissue of reproductive organs of peas, soybeans, apple fruit, pear fruit, onions, cotton seeds, and watermelon. They have xylopyranosyl residues α-(1→3)-linked to part of the galactosyluronic acid residues in the galacturonan backbone (Figure 6.8c). The rhamnogalacturonans

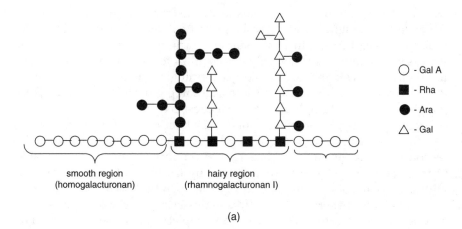

- ○ - Gal A
- ■ - Rha
- ● - Ara
- △ - Gal

smooth region
(homogalacturonan)

hairy region
(rhamnogalacturonan I)

(a)

FIGURE 6.8(a)
Typical structures of pectins: (a) smooth and hairy regions of pectins; (b) homogalacturonan;
(c) xylogalacturonan; (d) rhamnogalacturonan II (RG II); (e) rhamnogalacturonans type I (RG I).

type II have been found in the cell walls of many tissues of edible plants, such as apple (juice), kiwi, grape (wine), carrot, tomato, onion, pea, and raddish. These polysaccharides have a homogalacturonan backbone with very complex side chains with respect to sugar residue content and linkage structure (Figure 6.8d); however, their structure is apparently highly con-served, as identical structural features have been found in the cell walls of various plants. The side chains contain rhamnose and some rare residues, such as apiose, aceric acid (3-C-carboxy-5-deoxy-L-xylose), KDO (3-deoxy-D-manno-octulosic acid), and DHA (3-deoxy-D-lyxo-heptulosaric acid). RG II is a low molecular weight polymer (~4.8 kDa) present in the primary cell walls as a dimer cross-linked by borate diesters.

Rhamnogalacturonans type I (RG I) have a backbone composed of alter-nating (1→2)-linked α-L rhamnosyl and (1→4)-linked α-D-galacturonic acid residues. Depending on the source of pectins, 20 to 80% of rhamnose residues may be branched at O-4 with side chains which vary in length and composi-tion. The side branches may be composed of arabinans, galactans and type I arabinogalactans. Arabinans are branched polysaccharides with (1→5)-linked α-L-arabinofuranosyl units constituting the backbone. Short arabinan chains are linked to the backbone via O-2 and/or O-3 linkages. Type I arabinogalactans have linear chains of (1→4)-linked β-D-galactopyranosyl residues with up to 25% of short chains of (1→5)-linked α-L-arabinofuranosyl residues. Pectins with type I arabinogalactans have been found in potato, soybean, onion, kiwi, tomato, and cabbage.

Pectins are highly heterogeneous with respect to their molecular weight and chemical structure. For isolated and purified pectins, the molecular weight is determined by the extraction modes and conditions used. The weight average molecular weight of pectins from various fruit sources is typically in the order of $10^4 \sim 10^5$ Daltons.[9]

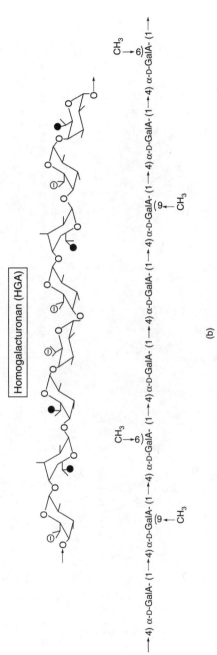

Homogalacturonan (HGA)

$\rightarrow$4) α-D-GalA- (1$\rightarrow$4) α-D-GalA- (1$\rightarrow$4) α-D-GalA- (1$\rightarrow$4) α-D-GalA- (1$\rightarrow$4) α-D-GalA- (1$\rightarrow$4) α-D-GalA- (1$\rightarrow$4) α-D-GalA- (1$\rightarrow$

$CH_3\rightarrow$6)

$CH_3\rightarrow$6)

(9$\leftarrow CH_3$

(9$\leftarrow CH_3$

(b)

FIGURE 6.8(b)

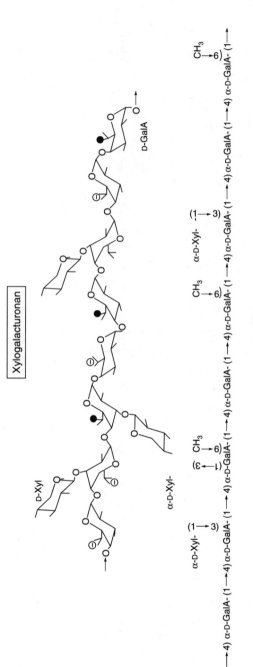

(c)

FIGURE 6.8(c)

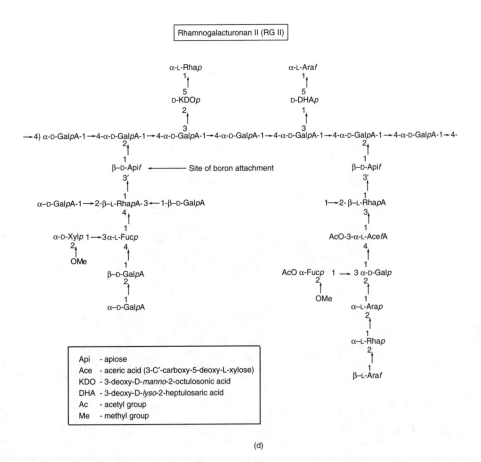

FIGURE 6.8(d)

6.2.3.2 Functional Properties and Applications

Pectins are generally soluble in water. Increases in molecular weight and ionic strength and decreases in the degree of esterification reduce their solubility. Pectic acid and divalent salts of pectic acid are essentially insoluble in water. However, they can be converted into monovalent salt forms, i.e., sodium or potassium pectate, to become soluble. In the solid state pectins exist in right-handed helices stabilized by intramolecular and intermolecular hydrogen bondings. In solution, pectins adopt random coil conformations with some degree of rigidity. The persistence length of pectin chains obtained experimentally or predicted from theoretical calculation is in the range of 2 to 31 nm.[10] The Mark-Houwink exponent is generally in the range of 0.8 ~ 0.9 corresponding to a slightly stiff conformation. Pectin molecules tend to associate to form stiff rods or segmented rods in solution through noncovalent interactions. Recent studies using the dynamic light scattering showed clearly two groups of particles of different sizes in pectin solutions. Presumably the

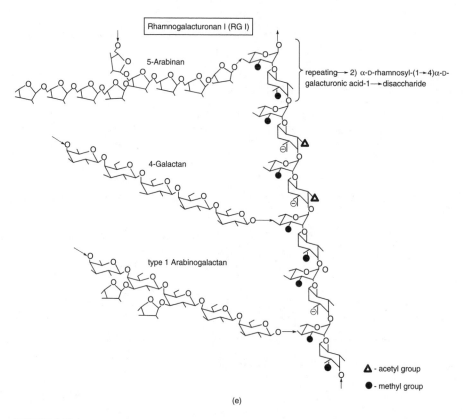

(e)

FIGURE 6.8(e)

smaller particles represent the individual molecules whereas the larger ones correspond to the supramolecular aggregates. Considering the molecular rigidity and aggregation tendency, pectins should be effective thickening agents. However, the molecular weights of pectin isolates are usually not as high as many other viscous polysaccharides. The average molecular weight of pectins from various sources is typically in the order of $10^4 \sim 10^5$. Consequently, pectins are used in food industry mostly as gelling agents rather than thickening additives.

Aqueous pectin solutions show normal random coil type flow behavior. The viscosity decreases with increasing temperature and decreasing concentration and molecular weight of polymers. Chou and Kokini[11] showed that pectins from citrus, apple, and tomato in citrate phosphate buffer solution (pH 4.6) have the critical concentration of $c^* \sim 6$. The specific viscosity (η_{sp}) increases with concentration according to the relationship of $\eta_{sp} \propto c^{1.4}$ when $c < c^*$ and to the relationship of $\eta_{sp} \propto c^{3.3}$ above c^*. In addition, the degree of esterification, electrolyte concentration, pH, and the presence of co-solute also influence solution viscosity. For example, in absence of external salt or at low ionic strength, pectin chains exist in an extended conformation because of the strong intramolecular electrostatic repulsion and the solutions show the highest viscosity. Addition of salts of monovalent cations to pectin solutions suppresses the electrostatic repulsion, hence reduces viscosity. However, addition of calcium or other polyvalent cations may increase the viscosity of pectin solutions.[12] This effect has been attributed to bridging between suitably positioned carboxyl groups leading to the formation of supramolecular structures. The viscosity of pectin solutions usually increases with decreasing pH in a calcium-free solution. In this case, a decrease in pH suppresses the dissociation of carboxyl groups, thus promoting chain to chain association. Addition of co-solutes such as sucrose, dextrose, and maltose usually increases the viscosity of aqueous pectin solutions. The viscosity-enhancing effect of these sugars was mainly attributed to their ability to compete for water with pectins.

High methoxyl (HM) and low methoxyl (LM) pectins can form gel at concentration above 0.5 ~ 1%. Both high molecular weight and high polymer concentration favor gel formation and enhance gel strength. Other intrinsic and extrinsic factors such as the pH, amount and type of co-solutes, and the degree of esterification (DE) are also important factors affecting the gelation mechanisms of HM and LM pectins. HM pectins form thermally irreversible gels at sufficiently low pH (pH <~ 3.6) and in the presence of sugars or other co-solutes at a concentration greater than ~50% by weight. The DE and overall distribution of hydrophilic and hydrophobic groups have major effects on gelation. Commercial HM pectins are classified as rapid-set, medium-set, and slow-set types. Sucrose and other co-solutes also affect gelation although to a lesser extent. The amount of co-solute required increases with increasing DE. LM pectins require calcium or other divalent cations for gelation, and the reactivity to calcium is governed by the proportion and distribution of carboxyl groups and DE. The relative amount of calcium considerably affects the gelling properties of LM pectins. Gelation is favored by increased soluble solids and decreased pH. LM pectin gels prepared at low pH are thermally reversible whereas those prepared at neutral pH are thermally irreversible.

The main uses of pectins are as gelling agents in various food applications including dairy, bakery, and fruit products. HM pectins have long been used in traditional jams and jellies, whereas LM pectins are used in low-calorie, low-sugar jams, and jellies. Pectin gels can be used as an alternative to gelatin in fruit deserts and amidated LM pectins are used to prepare milk gels and desserts. Pectins are also used as a protein dispersion stabiliser in acidified dairy products such as yogurt and milk-based fruit drinks, and other protein drinks prepared from soya and whey.

6.2.4 Exudate Gums

Exudate gums are polysaccharides produced by plants as a result of stress, including physical injury and/or fungal attack. Gum arabic, gum tragacanth, gum karaya, and gum ghatti have been used by humans for many thousands of years in various food and pharmaceutical applications. Generally, these gums are structurally related to arabinogalactans, galacturonans, or glucuronomannans (Table 6.1). They all contain a high proportion of glucuronic or galacturonic acid residues (up to 40%).

6.2.4.1 Sources and Structures

6.2.4.1.1 Gum Arabic

Gum arabic is exuded from the bark of *Acacia* trees that grow primarily in Africa. The thorny trees grow to a height of 7 to 8 meters, and the gum is obtained by cutting sections of the bark from the tree. The structure of gum arabic is relatively complex. The main chain of this polysaccharide is built from (1→3) and (1→6)-linked β-D-galactopyranosyl units along with (1→6)-linked β-D-glucopyranosyl uronic acid units. Side branches may contain α-L-rhamnopyranose, β-D-glucuronic acid, β-D-galactopyranose, and α-L-arabinofuranosyl units with (1→3), (1→4), and (1→6) glycosidic linkages (Figure 6.9). Gum arabic has a high water solubility (up to 50% w/v) and relatively low viscosity compared to other exudate gums (Table 6.2). The highly branched molecular structure and relatively low molecular weight of this polymer are responsible for these properties. Another unique feature of gum arabic is its covalent association with a protein moiety. It is thought that the protein moiety rich in hydroxyproline (Hyp), serine (Ser), and proline (Pro) constitutes a core to which polysaccharide subunits are attached via Ara-Hyp linkages (the wattle blossom model). The protein moiety of gum arabic is responsible for the surface activity, foaming, and emulsifying properties of this polymer.[13]

6.2.4.1.2 Tragacanth Gum

Tragacanth gum is dried exudates from branches and trunks of *Astragalus gummifer* Labillardiere or other species of *Astragalus* grown in West Asia (mostly in Iran, some in Turkey). It is collected by hand, then graded, milled, and sifted to remove impurities. Tragacanth gum contains a water-soluble fraction and a water-insoluble fraction and the water-soluble fraction is accounted for 30 to 40% of the total gum. The water soluble fraction is a highly branched neutral polysaccharide composed of 1→6-linked D-galactosyl backbones with L-arabinose side chains joined by 1→2-, 1→3- and/or 1→5-linkages. The water-insoluble fraction (~60 to 70%), is tragacanthic acid which is consisted of D-galacturonic acid, D-galactose, L-fucose, D-xylose, L-arabinose and L-rhamnose. It has a (1→4)-linked α-D-galacturonopyranosyl backbone chain with randomly substituted xylosyl branches linked at the 3 position of the galacturonic acid residues.

6.2.4.1.3 Gum Karaya

Gum karaya is a branched acidic polysaccharide obtained from the exudates of the *Sterculia urens* tree of the Sterculiaceae family grown in India. The

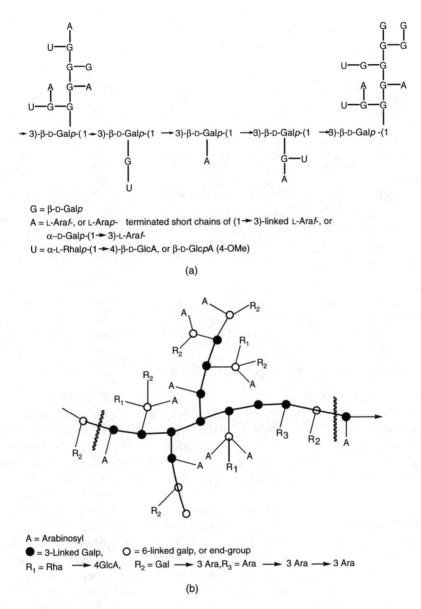

G = β-D-Galp
A = L-Araf-, or L-Arap- terminated short chains of (1→3)-linked L-Araf-, or
   α-D-Galp-(1→3)-L-Araf-
U = α-L-Rhap-(1→4)-β-D-GlcA, or β-D-GlcpA (4-OMe)

(a)

A = Arabinosyl
● = 3-Linked Galp,   ○ = 6-linked galp, or end-group
R₁ = Rha →4GlcA,   R₂ = Gal → 3 Ara,R₃ = Ara → 3 Ara → 3 Ara

(b)

FIGURE 6.9
Typical structural features of gum arabic.

backbone chain is a rhamnogalacturonan consisting of α-(1→4)-linked
D-galacturonic acid and α-(1→2)-linked-L-rhamnosyl residues. The side chain
is made of (1→3)-linked β-D-glucuronic acid, or (1→2)-linked β-D-galactose
on the galacturonic acid unit where one half of the rhamnose is substituted
by (1→4) linked β-D-galactose.

TABLE 6.2

Main Botanical and Structural Characteristics of Exudate gums

Gum	Species of Origin	General structure	Viscosity Concentration %	Viscosity (Pas × 10⁻³)ᵃ
Gum arabic	*Acacia senegal*	Substituted acidic arabinogalactan	5.01	717
Gum tragacanth	*Astragalus gummifer*	Mixture of arabinogalactan and glycano-rhamnogalacturonan	1.0 3.0	1,000 >10,000
Gum karaya	*Sterculia urens*	Glycano-rhamnogalacturonan	1.0 5.0	300,045,000
Gum ghatti	*Anogeissus latifolia*	Glycano-glucuronomannoglycan	5.01	2,882,440

ᵃ Viscosity obtained at shear rates ≤ 10 s⁻¹.

6.2.4.1.4   Gum Ghatti

Gum ghatti is an amorphous translucent exudate of the *Anogeissus latifolia* tree of the Combretaceae family grown in India. The monosaccharide constituents of gum ghatti are L-arabinose, D-galactose, D-mannose, D-xylose, and D-glucuronic acid in the ratio of 10:6:2:1:2, with traces of 6-deoxyhexose.

6.2.4.2   Functional Properties and Applications

6.2.4.2.1   Gum Arabic

Gum arabic is a low viscosity gum and its solutions exhibit Newtonian flow behavior even at high concentrations. It is extremely soluble in water and the solubility can be as high as 55%. Gum arabic is a surface active gum that is able to stabilize oil-in-water emulsions. The protein-rich high molecular weight fraction of gum arabic is preferentially adsorbed onto the surface of oil droplets while the carbohydrate portion inhibits flocculation and coalescence by electrostatic repulsions and steric forces (see Chapter 4 for details).

Gum arabic found broad applications in the confectionary and beverage industries as an emulsifier and stabilizer and flavor encapsulation agent. For example, it is used as an emulsifier in the production of concentrated citrus juices and cola flavor soft drinks. Gum arabic is also used in candies to prevent sugar crystallization and to emulsify the fatty components in products such as pastilles, caramel, and toffee. Gum arabic is an ingredient in chewing gums and cough drops. Spray-drying of flavor oils with gum arabic solutions produces microencapsulated powders that can be easily incorporated into convenient dry food products such as soup and dessert mixes.

6.2.4.2.2   Tragacanth Gum

Tragacanth gum swells rapidly in both cold and hot water to form a viscous colloidal suspension rather than a true solution. When added to water, the soluble tragacanthin fraction dissolves to form a viscous solution while the insoluble tragacanthic acid fraction swells to a gel-like state, which is soft and adhesive. When more water is added, the gum first forms a uniform mixture; after 1 or 2 days, the suspension will separate into two layers with dissolved tragacanthin in the upper layer and insoluble bassorin in the lower layer. The viscosity of the suspension reaches a maximum after 24 hours at room temperature, and hydration can be accelerated by an increase in temperature. The suspension typically exhibits shear thinning behavior. The ability to swell in water, forming thick viscous dispersions or pastes makes it an important gum in the food, pharmaceutical, and other industries. It is the most viscous natural water-soluble gum and is an excellent emulsifying agent with good stability to heat, acidity, and aging.

Although the functional properties of tragacanth gum have not been studied in detail, it has been used extensively by the industry. It is a good surface design thickener in the textile and printing industry as it can be easily mixed with natural dyes and can convey controlled design and color onto fabric. Tragacanth gum has been used in the pharmaceutical and personal care industries as an emulsifier and stabilizer in medicinal emulsions, jellies, syrups, ointments, lotions, and creams. Food applications of tragacanth gum include salad dressings, oil and flavor emulsions, ice creams, bakery fillings, icings, and confectionaries.

6.2.4.2.3 Gum Karaya

The solubility of gum karaya in water is poor. However, the gum swells up to many times its own weight to give dispersions. Homogenous dispersion can be prepared from fine powders whereas coarse granulated gum will produce a discontinuous, grainy dispersion. Dispersions of gum karaya exhibit Newtonian flow behavior at low concentrations ($<0.5\%$) and shear thinning flow behavior at semi-dilute concentrations ($0.5\% < c < 2\%$). Further increases in gum concentration produce pastes or spreadable gels.

Gum karaya is used as a stabilizer in dairy products and dressings, such as packaged whipped creams, cheese spreads, frozen desserts, and salad dressings. It is a water binding agent in processed meats and pasta. Gum karaya has also been used as dental adhesives, bulk laxatives, and adhesives for ostomy rings. Gum karaya also found applications in manufacturing long-fibered light weight papers and in the textile industry to help to convey the dye onto fabrics.

6.2.4.2.4 Gum Ghatti

Gum ghatti does not dissolve completely in water to give clear solutions, instead, it forms dispersions. The dispersions exhibit non-Newtonian flow behavior. Gum ghatti is also an excellent emulsifier and has been used to replace gum arabic in more complex systems. For example, gum ghatti is used as an emulsifier and stabilizer in beverages and butter-containing table syrups. Gum ghatti is used to encapsulate and stabilize oil-soluble vitamins. It is also used as a binder in making long-fibered, light weight papers.

6.2.5   Mucilage Gums

Mucilage gums are very viscous polysaccharides extracted from seeds or soft stems of plants; examples are okra mucilage (from *Hibiscus esculentus*), psyllium (from *Plantago* species), yellow mustard (from *Sinapis alba*), and flax mucilage (from *Linum usitatissimum*). All of them are acidic polysaccharides with structures somewhat related to some of the exudate gums. Their utilization in certain food products is increasing due to their functional properties (viscosity, gelation, water binding) as well as to their bio-active role in prevention and/or treatment of certain diseases.

6.2.5.1   *Yellow Mustard Mucilage*

Yellow mustard mucilage can be extracted from whole mustard seeds or from the bran. The mucilage contains a mixture of a neutral polysaccharide, composed mainly of glucose, and an acidic polysaccharide, containing galacturonic and glucuronic acids, galactose, and rhamnose residues. Detailed analysis of the neutral fraction of yellow mustard mucilage showed that it contains mainly (1→4)-linked β-D-glucose residues. The O-2, O-3, and O-6 atoms of the (1→4)-β-D-glucan backbone may carry ether groups (ethyl or propyl) (Figure 6.10a). These groups interfere with extensive chain to chain interactions (as seen in cellulose) and contribute to the solubility of this polysaccharide. The acidic fraction of yellow mustard mucilage has a rhamnogalacturonan backbone consisting of (1→4)-linked β-D-galacturonopyranosyl and (1→2)-linked α-L-rhamnopyranosyl residues (Figure 6.10b). Side branches, attached mainly to the O-4 of the α-L-Rha*p* in the backbone, contain such sugar residues as β-D-Gal*p* and 4-O-Me-β-D-Glc*p*A. Depending on the polymer concentration, yellow mustard mucilage can form either viscous solution of weak gels. When it is mixed with locust bean gum, however, the gel rigidity can be increased substantially. It has been shown that the neutral (1→4)-β-D-glucan fraction of yellow mustard mucilage synergistically interacts with galactomannans. Yellow mustard mucilage is produced in Canada for food and cosmetic purposes. It is used in processed meat formulations and salad dressing as a stabilizer and bulking agent.

6.2.5.2   *Flaxseed Mucilage*

Flaxseed mucilage can be easily extracted from the seeds by soaking them in warm water. The mucilage constitutes the secondary wall material in the outermost layer of the seed. Upon hydration of the seeds, it expands, breaks the mucilage cells, and exudes on the surface of the seeds. Flaxseed mucilage contains 50 to 80% carbohydrates and 4 to 20% proteins and ash. The carbohydrate portion of flaxseed mucilage contains many sugar residues, which

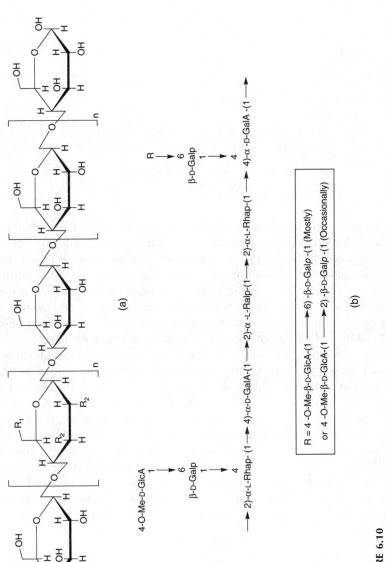

FIGURE 6.10

Two polysaccharide structures from yellow mustard mucilage: (a) structure of a cellulose-like polysaccharide (R_1 and R_2 are ethyl or propyl groups); (b) structure of a pectic polysaccharide.

TABLE 6.3

Relative Amount of Neutral and Acidic Sugar Residues
in Water-Soluble Polysaccharides from Flaxseed

Sugar Residue	Minimum	Maximum	Mean
L-Rhamnose[a] (%)	8.2	23.8	14.2
L-Fucose[a]	1.9	9.1	3.5
L-Arabinose[a]	3.2	15.4	10.6
D-Xylose[a]	10.9	32	23.2
D-Galactose[a]	12.7	26.9	19.1
D-Glucose[a]	21.3	40	28.9
Rhamnose/Xylose Ratio	0.3	2.2	0.7
Galacturonic acid[b]	13.8	25.1	

[a] Relative amount of neutral sugar residues. Results based on
examination of 109 accessions of flaxseed from 12 geograph-
ical regions.

[b] Galacturonic acid content (%) based on the gum weight
basis. Results based on examination of 12 different genotypes of flaxseed.

may vary in proportions depending on the source and methods of extraction
(Table 6.3).[14] Flaxseed mucilage contains a mixture of neutral polysaccharides, composed mainly of xylose, arabinose and galactose residues, and
acidic polysaccharides, containing galactose, rhamnose, and galacturonic
acid residues. The neutral fraction of flaxseed mucilage has a backbone of
(1→4)-linked β-D-xylopyranosyl residues, to which arabinose and galactose-
containing side chains are linked at O-2 and/or O-3. The acidic fraction of
flaxseed mucilage has a rhamnogalacturonan backbone with (1→4)-linked
α-D-galacturonopyranosyl and (1→2)-linked α-L-rhamnopyranosyl residues.
The side chains contain L-fucose and D-galactose residues. The ratio of neutral
to acidic polysaccharides in flaxseed may vary substantially with their origin.
As seen in the Table 6.3, the ratio of rhamnose to xylose, which roughly
indicates the ratio of acidic to neutral polysaccharides, may range from 0.3 to
2.2. Unfractionated flaxseed mucilage forms a viscous solution, but it is the
neutral fraction that mainly contributes to the high viscosity and weak gel-
like properties of this gum. Flaxseed mucilage has not yet been widely
utilized mostly because of limited information about the structure and functional properties of this gum. Similar to other gums, flaxseed mucilage can
be used as a thickener, stabilizer, and water-holding agent.

6.2.5.3 Psyllium Gum

Psyllium gum can be extracted from seeds of the *Plantago* species. The gum
is deposited in the seed coat; it is, therefore, advantageous to mechanically
separate the outer layers from the rest of the seed before extraction. Psyllium
gum can be extracted with hot water or mild alkaline solutions. The molecular
structure of the gum is a highly branched acidic arabinoxylan (Figure 6.11).
The xylan backbone contains β-D-Xyl*p* residues linked mainly via (1→4) and

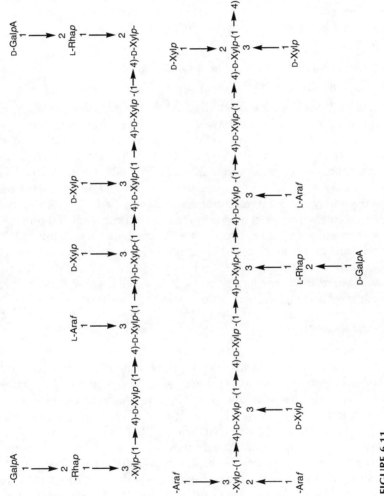

FIGURE 6.11
Structural features of psyllium gum.

glycosidic bonds. The xylose units are branched at O-2 and O-3 with single L-Araf and D-Xylp residues or disaccharides (aldobiouronic acid) containing L-Rhap and D-GalpA. D-Glucuronic acid residues have also been found in this gum. Psyllium gum has a very high molecular weight (~1500 kDa) and does not completely dissolve in water. When dispersed in water, it swells and forms a mucilageous dispersion with gel-like properties. It is used primarily as a laxative and dietary fiber supplement in pharmaceutical and food industries. It has been reported that psyllium gum can lower plasma low-density cholesterol and can be used to delay and reduce allergic reactions by entrapping and holding allergens and toxins in the gel structure.

6.2.5.4 Functional Properties and Applications

6.2.5.4.1 Yellow Mustard Mucilage

Yellow mustard mucilage solution/dispersion is a non-Newtonian system exhibiting shear thinning flow behavior and weak gel structures. Synergistic interactions can be observed when yellow mustard gum is mixed with galactomannans, and the extent of synergism follows the order of locust bean gum > guar gum > fenugreek gum, which is coincident with the order of mannose to galactose ratios in these gums (4:1, 4:2 and 4:4 for locust bean gum, guar gum, and fenugreek gum, respectively; Table 6.1). A thermally reversible gel can be prepared by one part of locust bean gum and nine parts of yellow mustard mucilage at a total polymer concentration as low as 0.1%. Yellow mustard mucilage also exhibits strong surface activity and capacity to stabilize oil/water emulsions.

Yellow mustard mucilage has been used to stabilize salad dressings and as a moisturizer and stabilizer in lotions and hand creams.

6.2.5.4.2 Flaxseed Gum

Flaxseed gum is soluble in cold water and exhibits Newtonian flow behavior at low concentrations and shear thinning flow at high polymer concentrations. Flaxseed gum is a low to medium viscosity gum and the viscosity is markedly influenced by pH. For example, higher viscosity can be obtained at pH 6 to 8 than at pH 2 to 6. Flaxseed gum exhibits surface activity and the ability to stabilize oil/water emulsions and foams.

Flaxseed gum has been traditionally used as an egg white substitute in bakery and ice cream products. Recently, flaxseed gum has been used in cosmetics to give a smooth feeling to creams and lotions. A saliva substitute is also prepared using flaxseed gum using its lubricating and moisturizing characteristics. Flaxseed gum has also been used as a bulk laxative, stabilizer, and suspending agent in barium sulphate suspensions for x-ray testing. The fast drying and film-forming properties of flaxseed gum make it a useful ingredient in hairdressing products. In addition, flaxseed gum found applications in printing, textile, and papers industries.

6.2.5.4.3 Psyllium Gum

Psyllium husk has been the most popular laxative agent in the pharmaceutical industry. It does not dissolve but forms mucilagenous dispersions with a similar appearance of wallpaper paste. Freshly prepared psyllium gum dispersion (1%) exhibits similar flow behavior to those of random coil polysaccharides in which a Newtonian plateau is observed at low shear rates and shear thinning flow behavior is observed at higher shear rates. However, upon aging, psyllium gum dispersions form cohesive gels and show observable syneresis. The gel strength increases with the increase of the gum concentration. Upon prolonged storage, psyllium gum gels continue to contract and the contraction process can be accelerated by freezing and thawing cycles.

Psyllium has also been explored to replace wheat gluten in gluten-free bread. For example, adding psyllium gum and hydroxypropylmethylcellulose (HPMC) (at 2 and 1%, respectively) to rice flour can make a bread with a loaf volume close to that of hard wheat control.[15] Recently, psyllium husk powders are used as an environmentally friendly binding and stabilizing agent in the landscaping and cottage industry to bind sands and crushed stones.

6.2.6 Fructans

6.2.6.1 Sources and Structures

Fructans are reserve polysaccharides in certain plants, either complementing or replacing starch. They can also be produced by certain species of bacteria. The two main kinds of fructans are inulin and levan. Inulins are found in roots or tubers of the family of plants known as Compositae, including dandelions, chicory, lettuce, and Jerusalem artichoke. They can also be extracted from the Liliacae family, including lily bulbs, onion, tulips, and hyacinth. Inulin is a low molecular weight polysaccharide containing $(2\rightarrow1)$ linked β-D-Fru*p* residues (Figure 6.12). It has a D-glucopyranose nonreducing end unit linked to the O-2 position of the β-D-Fru*p* residues. The degree of polymarization (DP) ranges usually from 20 to 40 units with the exception of inulin from artichoke globe, which has DP up to 200.

Levans are found mainly in grasses. They are higher molecular weight polysaccharides than inulins (with average DP from 100 to 200), containing a backbone of $(2\rightarrow6)$ linked β-D-Fru*p* residues with $(2\rightarrow1)$ linked branches of one to four D-frupyranosyl units (Figure 6.12).

6.2.6.2 Functional Properties and Applications

Fructan is a dietary fiber, bulking, and laxative agent, though the principal mechanism of its action is not through providing bulk. Like other fibers, fructans from inulin or levan are undigested in the human upper intestine, but both are fermented in the colon to give lactate and short-chain fatty acids (acetate, propionate, and butyrate) and stimulate the growth of beneficial to

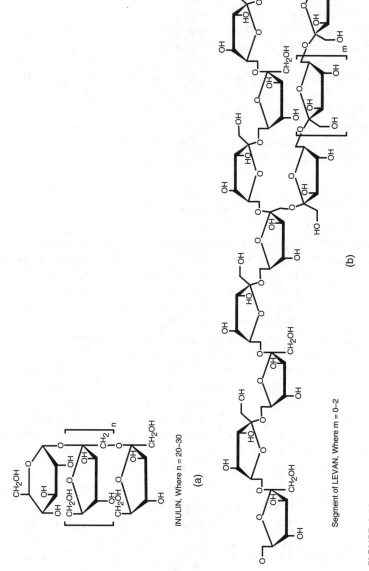

FIGURE 6.12
Structural features of two fructans: (a) inulin; (b) levan.

colon bifidobacteria. The selective stimulation of the growth of certain populations of intestinal bacteria helps to protect the colon from infection.

6.3 Seaweed Polysaccharides

6.3.1 Alginates

6.3.1.1 Sources and Structures

Alginates constitute the primary structural polysaccharides of brown seaweeds (Phaeophyceae). The major species of seaweeds that produce alginates are *Macrocystis pyrifera*, grown primarily along the California coast of the USA, south- and north-western coasts of South America, and coasts of Australia and New Zealand. Other good sources of alginates are *Laminaria hyperborea*, *Laminaria digitata*, and *Laminaria japonica*, grown along the north Atlantic coast of the USA, Canada, France, and Norway. Alginates can also be synthesized by bacteria, *Pseudomonas aeruginosa* and *Azobacter vinelandii*.

Alginates are unbranched copolymers of $(1\rightarrow4)$-linked β-D-mannuronic acid (M) and α-L-guluronic acid (G) residues (Figure 6.13). If the uronic acid groups are in the acid form (–COOH), the polysaccharide, called alginic acid, is water insoluble. The sodium salts of alginic acid (–COONa), sodium alginates, are water soluble. The sequence of mannuronic and guluronic residues significantly affects the physicochemical properties of alginates. It has been shown that alginates contain homopolymeric blocks of contiguously linked mannuronic acid (M-blocks), alternated by guluronic acid (G-blocks) sequences. Mixed GM-blocks can also occur along the chain. The ratio of β-D-mannuronic acid to α-L-guluronic acid residues is usually 2:1, although it may vary with the algal species, the age of the plant as well as the type of tissue the alginates are extracted from. For example, the highest amount of G is found in *L. hyperborea* (the ratio of M:G is 1:2.3). In bacterial alginates, the content of G may vary widely, ranging from 15 to 90% (the ratio of G:M may vary from 0.4 to 2.4). The unique properties of alginates originate from their conformational features. The D-Man$p$A residues are in $^4C_1$, whereas the L-Gul$p$A residues are in $_4C^1$ conformation. Two axially linked L-Gul$p$A residues, being in the $_4C^1$ conformation, form characteristic cavities, which act as binding sites for divalent ions (e.g., Ca^{2+}). The consequence of this binding capacity of G-blocks is that alginates form gel immediately upon contact with the divalent ions. The gel networks are formed by cross-linking alginate chains via calcium ions (Figure 6.13).

6.3.1.2 Functional Properties and Applications

Alginates form thermally stable cold-setting gels in the presence of calcium ions. The gel strengths depend on the type of ions and the method of their

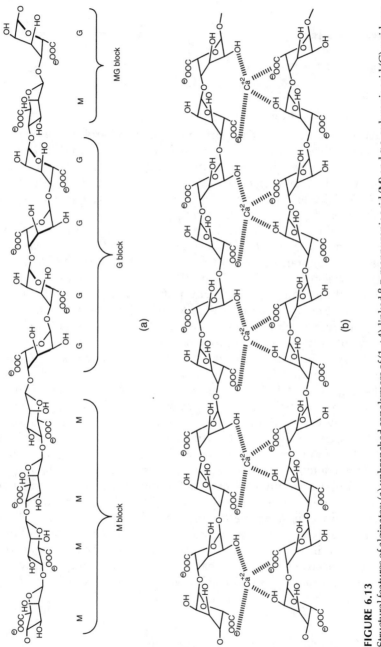

FIGURE 6.13

Structural features of alginates: (a) unbranched copolymers of (1→4)-linked β-D-mannuronic acid (M) and α-L-guluronic acid (G) residues; (b) demonstration of cross-linkings of alginate chains via calcium ions.

introduction. The effect of ions on the gelation potential of alginates generally follows the order of $Mg^{2+} \ll Ca^{2+} < Sr^{2+} < Ba^{2+}$.[13] High **G** alginates produce strong brittle gels with good heat stability (except for the low molecular weight molecules) but syneresis is observed after freeze-thaw treatments. High **M** alginates produce weaker, however, more elastic gels with good freeze-thaw stability.

Two methods have been used to introduce the cross-linking ions: diffusion and internal setting. In the diffusion method, cross-linking ions (e.g., Ca^{2+}) diffuse from an outer reservoir into an alginate solution. Products such as artificial berries, pimento strips, and onion rings can be prepared by the diffusing method. In the internal setting method, an inert calcium is converted into an active cation by changing the pH of the alginate solution. Cold water dessert gel, instant imitation bakery jelly, and facial masque are prepared by this method.

Alginates have been chemically modified to improve their functionalities. For example, propylene glycol esters of alginic acid (PGAs) are industrially produced; they tolerate calcium ions and are suitable for applications in acidic environments, such as in fermented milk-based products and salad dressings. PGAs also exhibit surface activities due to the presence of hydrophobic group (ester) and, therefore, can be used as emulsifiers and stabilizers for foams.

6.3.2 Carrageenans

6.3.2.1 Sources and Structures

Carrageenans are structural polysaccharides of marine red algae of the Rhodophyceae class. They are extracted mainly from *Chondrus crispus, Euchema cottoni, Euchema spinosum, Gigartina skottsbergi,* and *Iradaea laminarioides.* These red seaweeds grow mostly along the Atlantic coasts of North America, Europe, and the western Pacific coasts of Korea and Japan. κ-Carrageenans, ι-carrageenans, and furcellarans are linear polysaccharides whose backbone structure is based on a repeating disaccharide sequence of sulphate esters of (1→3) linked β-D-galactose and (1→4) linked 3,6-anhydro-α-D-galactose (Figure 6.14a). They differ from each other in the number and position of sulphate groups. κ-Carrageenans have one sulphate group per repeating disaccharide unit, positioned at C-4 of the β-D-galactopyranosyl residue, whereas ι-carrageenans have two sulphate groups, positioned at C-4 of the β-D-galactopyranosyl residue and C-2 of the 3,6-anhydro-α-D-galactopyranosyl residue (Figure 6.14c). Furcellaran has a similar structure to κ-carrageenan, but it is less sulphated; only 40% of the β-D-galactopyranosyl residues carry the sulphate group at C-4. Interestingly, the β-D-galactose units in carrageenans are in the $^4C_1$ conformation, whereas the 3,6-anhydro-α-D-galactose units are in the $_4C^1$ conformation. These two types of monosaccharide conformations, along with the presence of axial and equatorial glycosidic linkages, allow κ- and ι-carrageenans to assume a helical conformation. In solution, they exist as parallel, three-fold double helices, stabilized by the

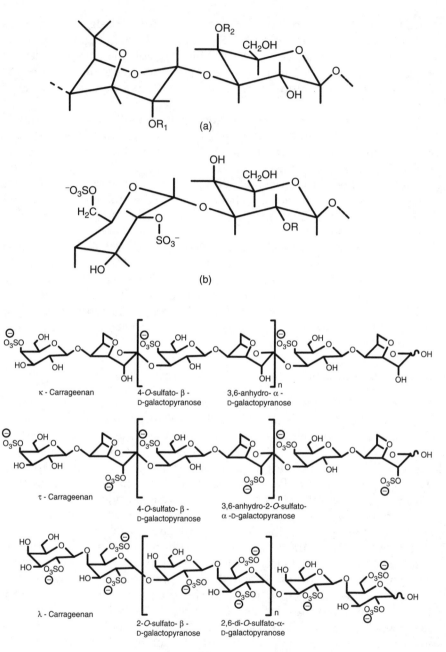

FIGURE 6.14

Structural features of carrageenans: (a) disaccharide repeating sequence of sulphate esters of 1→3 linked β-D-galactose and 1→4 linked 3,6-anhydro-α-D-galactose; (b) repeating disaccharide unit of β-D-galactopyranosyl residue sulphated at C-2 and 2,6-di-O-sulfato-α-D-galactopyrano-syl units in λ-carrageenans; (c) comparison of structure features of κ, ι and λ-Carrageenans.

H-bonds between the hydroxyl groups on O-2 and O-6 of the β-D-galactopyr-anosyl residue. The ordered helical structures are interrupted by the helix-breaking units, or kinks, lacking the anhydro sugars, thus depriving the chain of the helix compatible $_4C^1$ conformation. In the presence of some cations (K$^+$, Rb$^+$, Ca^{++}), the double helices of furcellaran, κ- and ι-carrageenans can aggregate and form gel. The functional properties of carrageenan gels, such as rigidity, turbidity, and tendency to syneresis (separation of water from gel upon aging), generally decrease with the increasing degree of sulphation in these polymers.

λ-Carrageenans constitute another group of the red seaweed polysaccha-rides. The repeating disaccharide unit in λ-carrageenans consists of β-D-galacto-pyranosyl residue sulphated at C-2 (instead of C-4 as in ι- and κ-carrageenans) and 2,6-di-O-sulfato-α-D-galactopyranosyl units (instead of 3,6-anhydro-α-D-galactopyranosyl residue) (Figure 6.14b, c). λ-Carrageenans are the most sulphated polysaccharides, carrying three sulphate groups per repeating disaccharide unit. Also, all sugar residues in λ-carrageenans are in the $^4C_1$ conformation. As a consequences of these structural differences, and in con-trast to ι- and κ-carrageenans, λ-carrageenans assume a flat, ribbon-like conformation rather than helical structures. λ-Carrageenans are nongelling polysaccharides used as cold soluble thickeners in syrups, fruit drinks, pizza sauces, and salad dressings.

6.3.2.2   *Functional Properties and Applications*

Carrageenans are used mainly as gelling, thickening, and suspending agents. κ-Carrageenan gives firm, clear, and brittle gels with poor freeze-thaw sta-bility. These gels may be softened by adding locust bean gum through synergistic interactions. ι-Carrageenan forms soft elastic gels with good freeze-thaw stability. The gelation of ι-carrageenan is ionic strength dependent: higher ionic strength promotes formation of junction zones and favors the formation of gels. λ-Carrageenan is a nongelling polysaccharide. There is a lack of 3,6-anhydro group in the (1→4)-linked α-D-galactopyranosyl residues which is required for the initial formation of double helix. λ-Carrageenan is used as a cryoprotectant. A combination of λ-carrageenan with locust bean gum can improve the freeze-thaw behavior of frozen products.

κ-Carrageenan has been used widely in dairy products to prevent whey separation on account of its interactions with casein micelles. It is also used in ice creams as a second stabilizer to prevent the phase separation caused by the incompatibility of locust bean gum and β-casein. κ-Carrageenan is also used as a water binder in cooked meats and as a thickener in toothpaste and puddings.

6.3.3   Agar

6.3.3.1   *Sources and Structures*

Agar constitutes another group of polysaccharides from red-purple algae of the Rhodophyceae class. The agar-yielding species of *Gracilaria* and *Gelidium*

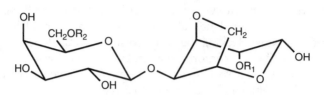

FIGURE 6.15

Structural features of agarose: repeating disaccharide unit of 1→3 linked β-D-galactose and 1→4- linked 3,6-anhydro-α-L-galactose residues.

grow in the waters along the coast of Japan, New Zealand, South Africa, Southern California, Mexico, Chile, Morocco, and Portugal. Agar differs from ι- and κ-carrageenans in that the 3,6-anhydro sugar in agar is the L-enantiomer, whereas in carrageens it is the D-enantiomer. Agar is, therefore, a linear polysaccharide built up of the repeating disaccharide unit of (1→3)-linked β-D-galactose and (1→4)-linked 3,6-anhydro-α-L-galactose residues (Figure 6.15). Also, in contrast to carrageenans, agar is only lightly sulphated and may contain methyl groups. Methyl groups, when present, occur at C-6 of the (1→3)-linked β-D-galactose or C-2 of (1→4)-linked 3,6-anhydro-α-L-galac-tose residues. Carrageenans containing 3,6-anhydro-α-D-galactose residue form three-fold right-handed helices, whereas agar, containing 3,6-anhydro-α-L-galactose residues, forms three-fold left-handed helices; the reversal is related to the enantiomeric change. The agar helix is more compact due to the smaller amount of sulphate groups. Agar is a well known thermo-revers-ible gelling polysaccharide, which sets at 30 to 40°C. Being less sulphated than furcellaran, and κ- and ι-carrageenans, agar can form strong gels, which are, subject to pronounced syneresis, attributed to strong aggregation of double helices (not weakened by the sulphate groups).

6.3.3.2   *Functional Properties and Applications*

Among all the gums available agar is the best gelling agent. Its gelling power arise from strong hydrogen bond of the three equatorial hydrogen atoms on the 3,6-anhydro-L-galactose residues in the agarose. Agar gives cold setting gels at ~38°C. However, the melting temperature is much higher, about 85°C. This gives agar gel a very large gelling/melting hysteresis. Agar gel found applications in the baking industry because of its heat resistant characteris-tics, which help to prevent chipping, cracking, or sweating of icings, toppings, and glazes in baked goods. Agar can form a stiff gel at low concentrations and it is compatible with most of the other gums. Agar gel has a mouth-feel and other attributes similar to gelatin; it is therefore used in kosher gel desserts, pie fillings, and jelly candies.

Adding a small portion of locust bean gum to some type of agar (e.g., *Gelidium* agar in the ratio of 1:9) can enhance gel strength and improve gel texture. This is attributed to a synergistic interaction between locust bean gum and agar. The synergistic interactions found applications in the food industry to produce less brittle gels. In addition, agar is a very important microbiological culture medium due to its excellent gelling characteristics and high resistance to microorganisms.

6.4 Microbial Polysaccharides

6.4.1 Xanthan Gum

6.4.1.1 Sources and Structures

Xanthan gum is an extracellular polysaccharide produced by the bacterium *Xanthomonas campestris*. It was first produced by Kelco-AIL, and after rigorous toxicological and safety testings, it received the FDA approval in 1969 for food uses in the United States. Now xanthan is approved worldwide and

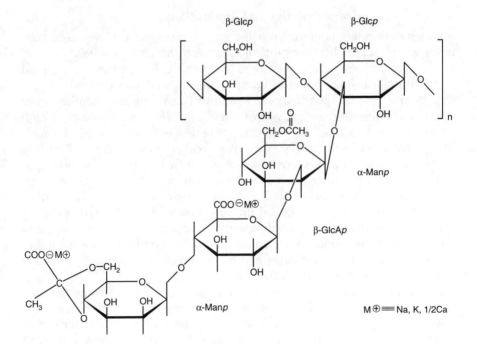

FIGURE 6.16
Structure of xanthan gum: a cellulose-like backbone of (1→4)-linked β-D-Glc*p* residues and the trisaccharide branch consisting of the β-D-Man*p*-(1→4)-β-D-Glc*p*A-(1→2)-α-D-Man*p*-(1→ unit substituted at O-3 of alternate glucose residues.

it is one of the most extensively studied polysaccharides. The primary structure of xanthan consists of the cellulose-like backbone of (1→4)-linked β-D-Glc$p$ residues substituted, at O-3 of alternate glucose residues, with a trisaccharide. The trisaccharide consists of the β-D-Man$p$-(1→4)-β-D-Glc$p$A-(1→2)-α-D-Man$p$-(1→ unit (Figure 6.16). Noncarbohydrate substituents include the acetyl group at O-6 of the inner Man$p$ residue and the pyruvate group at O-4,6 of the terminal Man$p$. The pyruvic acid content of xanthan can vary substantially depending on the strain of *X. campestris*, resulting in different viscosities of xanthan solutions. Xanthan gum is a high molecular weight polysaccharide (3×10^5 to 8×10^6 Da). Molecular modelling studies suggest that xanthan gum can assume a helical structure, with the side branches positioned almost parallel to the helix axis and stabilizing the structure. Xanthan gum forms very viscous solutions, and, at sufficiently high polymer concentration, it exhibits weak gel-like properties. It can form thermo-reversible gels when mixed with certain galactomannans (e.g., locust bean gum) or konjac glucomannan. Xanthan is widely used in foods because of its good solubility in either hot or cold solutions, high viscosity even at very low concentrations, and excellent thermal stability.

6.4.1.2 Functional Properties and Applications

Xanthan gum is a nongelling gum and has been the most widely used gum in the food industry because of its unique shear thinning flow behavior and weak gel structures. Xanthan gum can hydrate in cold water, however, good hydration and solubilization of the gum depends on particle size, solvent quality and rate of agitation, and amount of heat applied. Xanthan gum exhibits very high low-shear viscosity and relatively low viscosity at high shear rate; these unique characteristics provide the advantages of easy mixing, pouring, and swallowing, yet gives excellent suspension and coating properties to colloidal suspensions at rest. Xanthan gum also found applications in dairy products, such as ice creams, sour cream, and sterile whipping cream, to provide optimal viscosity, long-term stability, heat shock protection, ice crystal control, and improved transfer characteristics during processing. Xanthan gum contributes to smoothness, air incorporation, and retention of baked goods. Xanthan gum is also mixed with dry cake ingredients to give uniform hydration and batter mixing, which is important for the overall quality and shelf life of finished cakes.

Strong synergism is observed between xanthan gum and some mannan containing polysaccharides, such as galactomannans and glucomannans. The synergistic interactions result in increased viscosity and/or formation gels. Maximum synergism is achieved after heating the samples at 90°C. However, the extent of interactions between xanthan gum and galactomannans could be reduced by changing the pHs and salt concentrations. The optimum ratios of synergistic interactions are 1:1 for xanthan–LBG system and 1:4 for xanthan–guar system.

6.4.2 Pullulan

6.4.2.1 Sources and Structures

Pullulan is an extracellular homopolysaccharide of glucose produced by many species of the fungus *Aureobasidium*, specifically *A. pullulans*. Pullulan contains (1→4) and (1→6)-linked α-D-glucopyranosyl residues. The ratio of (1→4) to (1→6) linkages is 2:1. Pullulan is generally built up of maltotriose units linked by (1→6) with much smaller amount of maltotetraose units (Figure 6.17). The molecular weight of pullulans ranges from 1.5×10^5 to 2×10^6. The presence of (1→6) glycosidic linkages increases flexibility of pullulan chains and resulted in their good solubility in water compared with other linear polysaccharides (e.g., amylose).

6.4.2.2 Functional Properties and Applications

Pullulan forms viscous solutions, but does not gel. It is only partially digestible by human salivary α-amylase, producing hydrolysis products too short for further hydrolysis. The viscosity is stable with pH, therefore, pullulan applications are found in sauces and other food systems. Pullulan is also used as an adhesive for papers, wood products, and metals. The low digestibility of pullulan make it a good ingredient for manufacturing low-caloric foods or beverages by replacing starches. Pullulan is also a good film-forming agent for manufacturing edible films. It can be used as a coating agent to give food product glossy surface.

6.4.3 Gellan Gum

6.4.3.1 Sources and Structures

Gellan gum is a deacetylated form of the extracellular bacterial polysaccharide from *Auromonas elodea*. Gellan gum was discovered in 1977, and it is now approved for food use in many countries including Australia, Canada, United States, Mexico, Chile, Japan, South Korea, and Philippines. It has a repeating tetrasaccharide sequence of →3)-β-D-Glc*p*-(1→4)-β-D-Glc*p*A-(1→4)-β-D-Glc*p*-(1→4)-α-L-Rha*p*-(1→ (Figure 6.18). In the native form gellan has an L-glyceryl substituent on O-2 of the (1→3)-linked glucose and an acetyl group on some of the O-6 atoms of the same residue; both groups are normally lost during commercial extraction. A few other polysaccharides originating from different bacterial species share the same linear backbone, and have been grouped in the gellan gum family. Welan, S-657, and rhamsan, not yet approved for use in food, differ from gellan in having side branches (Figure 6.18). Upon cooling of gellan solutions, the polysaccharide chains can assume double helices, which aggregate into weak gel structures (supported by van der Waals attractions). In the presence of appropriate cations (Na^+ or Ca^{++}), the double helices form cation-mediated aggregates, which leads to formation of strong gel networks. Acyl substituents present in native gellan interfere with the aggregation process, giving much weaker gels. In the branched

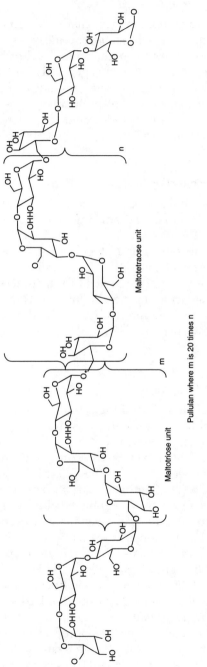

FIGURE 6.17
Two building blocks of pullulan: a major repeating unit — maltotriose and a minor repeating unit — maltotetraose.

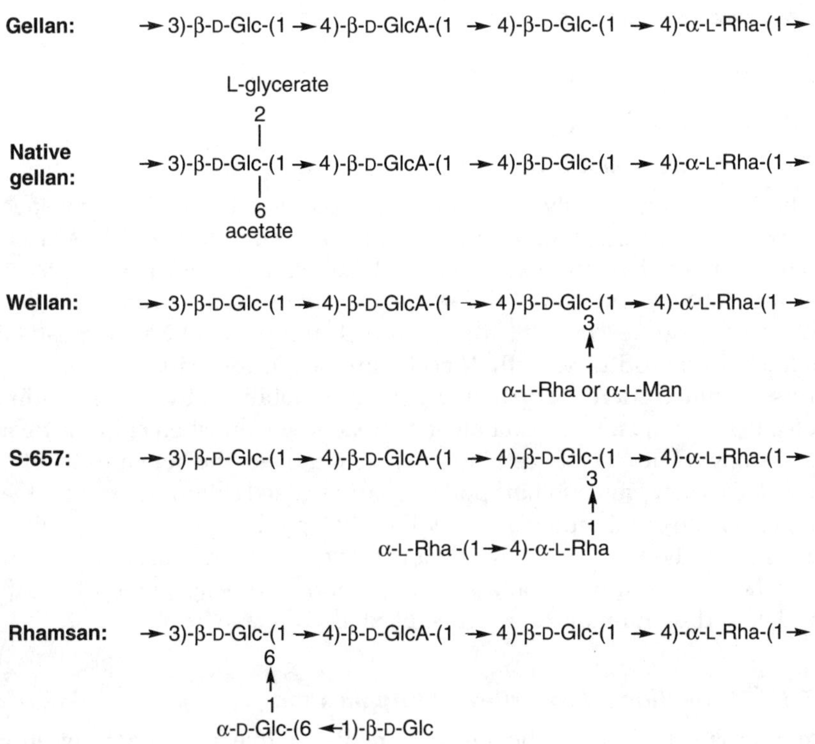

Gellan: → 3)-β-D-Glc-(1 → 4)-β-D-GlcA-(1 → 4)-β-D-Glc-(1 → 4)-α-L-Rha-(1 →

L-glycerate
2
|
Native   → 3)-β-D-Glc-(1 → 4)-β-D-GlcA-(1 → 4)-β-D-Glc-(1 → 4)-α-L-Rha-(1 →
gellan: |
6
acetate

Wellan: → 3)-β-D-Glc-(1 → 4)-β-D-GlcA-(1 → 4)-β-D-Glc-(1 → 4)-α-L-Rha-(1 →
3
↑
1
α-L-Rha or α-L-Man

S-657:   → 3)-β-D-Glc-(1 → 4)-β-D-GlcA-(1 → 4)-β-D-Glc-(1 → 4)-α-L-Rha-(1 →
3
↑
1
α-L-Rha -(1 → 4)-α-L-Rha

Rhamsan: → 3)-β-D-Glc-(1 → 4)-β-D-GlcA-(1 → 4)-β-D-Glc-(1 → 4)-α-L-Rha-(1 →
6
↑
1
α-D-Glc-(6 ← 1)-β-D-Glc

FIGURE 6.18
Structural features of gellan gum and related polysaccharides.

variants of gellan, the side chains also interfere with the cation-induced aggregation, allowing only 'weak gel' formation.

6.4.3.2   *Functional Properties and Applications*

Gellan gum is used as a gelling agent, and the gel characteristics depend on the degree of acylation and presence of counter ions. For example, high acyl (HA) gellan gum gives soft, elastic, transparent, and flexible gels at polymer concentrations higher than 0.2%. HA gels set and melt at ~70 to 80°C with no thermal hysteresis. However, low acyl (LA) gellan gum can form hard, nonelastic and brittle gels in the presence of cations, including Ca^{2+}, Mg^{2+}, Na^+, K^+ and H^+. The gel strength of LA gellan gum increases with increasing ion concentration. Cation concentration can also affect gel setting and melting temperatures. LA gellan gels also exhibit significant thermal hysteresis, which is opposite to the gels formed by HA gellan gum.

Gellan gums are used as gelling agents in dessert jellies, dairy products, and sugar confectionery. LA gellan gum can be used to improve the properties of traditional gelatin dessert jellies. Gellan gum can also be used to prepare structured liquids.[16]

6.5　Animal Polysaccharides

6.5.1　Chitin and Chitosan

6.5.1.1　Sources and Structures

Chitin is a structural polysaccharide that replaces cellulose in many species of lower plants, e.g., fungi, yeast, green, brown, and red algae. It is also the main component of the exoskeleton of insects and shells of crustaceans (shrimp, lobster, and crab). The molecular structure of chitin is similar to that of cellulose, except that the hydroxyl groups at O-2 of the β-D-Glc*p* residues are substituted with *N*-acetylamino groups (Figure 6.19). Chitin forms a highly ordered, crystalline structure, stabilized by numerous intermolecular H-bonds. It is insoluble in water. However, when chitin is treated with strong alkali, the *N*-acetyl groups are removed and replaced by amino groups. This new water-soluble polysaccharide, called chitosan, contains, therefore, (1→4)-linked 2-amino-2-deoxy-β-D-glucopyranosyl residues. Chitosan is the only polysaccharide carrying a positive charge. It is used in a number of medical applications (contact lenses, wound dressings, etc). Chitosan is not digested by humans and can be used as a dietary fiber.

6.5.1.2　Functional Properties and Applications

Chitosan is soluble in acidic aqueous media to give a unique polycationic structure. Chitosan can form a thermo-irreversible gel by chemical and enzymatic reactions. Chitosan gel can also be prepared by introducing large organic counter ions, such as 1-naphthol-4-sulphonic acid or 1-naphthylamine-4-sulphonic acid.

Chitin, chitosan and their derivatives exhibit unique thermal and gelling properties that can be widely applied in pharmaceutics, cosmetics, foods, and membranes technologies. Chitosan is a good filming agent which could be used in textile finishing, paper sheet formation, glass fiber coating, dye application, and shrink proofing of wool. Chitosan is also used as a dietary fiber for weight control.

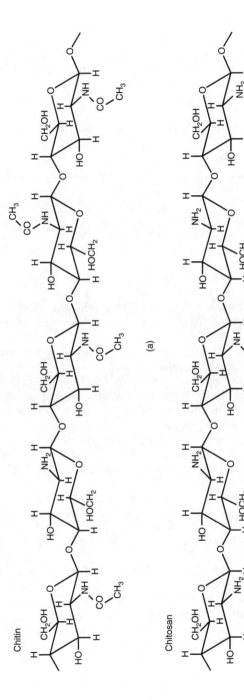

FIGURE 6.19
Structural features of (a) chitin and (b) chitosan.

Suggested Reading

Cui, W. *Polysaccharide Gums from Agriculture Products: Processing, Structure and Functional Properties*. Lancaster, PA: Technomic Publishing Company Inc., 2001.

Glicksman, M. *Food Hydrocolloids Volume I*. Boca Raton, FL: CRC Press, 1982.

Glicksman, M. *Food Hydrocolloids Volume II*. Boca Raton, FL: CRC Press, 1983.

Phillips, G.O. and Williams, P.A. (eds.) *Handbook of Hydrocolloids*. Boca Raton, FL: CRC Press, 2000.

Whistler, R.L. and BeMiller, J.N. (eds.) *Industrial Gums*. San Diego, CA: Academic Press, 1993.

References

1. Nishinari, K., Yamatoya, K., and Shirakawa, M. Xyloglucan. In: *Handbook of Hydrocolloids*. G.O. Phillips, and P.A. Williams, eds. Boca Raton, FL: CRC Press, 247–267. 2000.

2. Shirakawa, M., Yamatoya, K., and Nishinari, K. Tailoring of xyloglucan properties using an enzyme. *Food Hydrocolloids*, 12, 25–28. 1998.

3. Wang, Q., Ellis, P. R., Ross-Murphy, S. B., and Burchard, W. Solution characteristics of the xyloglucan extracted from Detarium senefalense Gmelin. *Carbohydrate Polymers*, 33, 115–124. 1997.

4. Yoshimura, M., Takaya, T., and Nishinari, K. Effects of xyloglucan on the gelatinization and retrogradation of corn starch as studied by rheology and differential scanning calorimetry. *Food Hydrocolloids*, 13, 101–111. 1999.

5. Yamanaka, S., Yuguchi, Y., Urakawa, H., Kajiwara, K., Shirakawa, M., and Yamatoya, K. Gelation of tamarind seed polysaccharide xyloglucan in the presence of ethanol. *Food Hydrocolloids*, 14, 125–128. 2000.

6. Salazar-Montoya, J. A., Ramos-Ramirez, E. G., and Delgado-Reyes, V. A. Changes of the dynamic properties of tamarind (*Tamaindus indica*) gel with different saccharose and polysaccharide concentrations. *Carbohydrate Polymers*, 49, 387–391. 2002.

7. Cui, W. *Polysaccharide Gums from Agriculture Products: Processing, Structure and Functional Properties*. Lancaster, PA: Technomic Publishing Company Inc. 2001.

8. Wood, P. Relationship between solution properties of cereal b-glucans and physiological effects — a review. *Trends in Food Science and Technology*, 15, 313–320. 2004.

9. Corredig, M., Kerr, W., and Wicker, L. Molecular characterization of commercial pectins by separation with linear mix gel permeation columns in-line with multi-angle light scattering detection. *Food Hydrocolloids*, 14, 41–47. 2000.

10. Rinaudo, M. Physicochemical properties of pectins in solution and gel states, In: *Pectin and Pectinase*. Visser, J. and Voragen, A. G. J. eds. Elsevier, New York, 21–34. 1996.

11. Chou, T. and Kokini, J. Rheological properties and conformation of tomato paste pectins, citrus and apple pectins. *Journal of Food Science*, 52, 1658–1664. 1987.

12. Lotzkar, H., Schultz, T., Owens, H., and Maclay, W. Effect of salts on the viscosity of pectinic acid solutions. *Journal. Physical Chemistry,* 50, 200–210. 1946.
13. Draget, K. Alginates, In: *Handbook of Hydrocolloids.* G. O. Phillips and P. A. Williams, eds. Boca Raton, FL: CRC Press, 379–395. 2000.
14. Oomah, B. D., Kenaschuk, E. O., Cui, W. W., and Mazza, G. Variation in the Composition of Water-Soluble Polysaccharides in Flaxseed. *Journal of Agricultural and Food Chemistry,* 43, 1484–1488. 1995.
15. Haque, A., Morris, E., and Richardson, R. Polysaccharide substitutes for gluten in non-wheat bread. *Carbohydrate Polymers,* 25, 337–344. 1994.
16. Sworn, G. Gellan gum, In: *Handbook of Hydrocolloids.* G. O. Phillips, P. A. Williams, eds. Boca Raton, FL: CRC Press, 115–135. 2000.

7

Understanding Starches and Their Role in Foods

Qiang Liu

CONTENTS

7.1   Introduction

Among food carbohydrates, starch occupies a unique position. It is the major carbohydrate storage material in many higher plants and is considered the second largest natural biopolymer next to cellulose. Starch is deposited in plant organs in the form of granules that are relatively dense, insoluble in cold water, and range from 1 to 100 μm in size depending on the plant species. Starch contributes to the physicochemical properties of food products made from cereals, tubers, roots, legumes, and fruits. It is the basic source of energy for the majority of the world's population. In human nutrition, starch plays a major part in supplying the metabolic energy that enables the body to perform its different functions. Recent studies suggest that slowly digested starch and enzyme resistant starch have significant implications for human health. Unlike some carbohydrates and digestible starches, resistant starch resists enzymatic hydrolysis in the upper gastrointestinal tract, thus resulting in little or no direct glucose absorption. In addition, resistant starch causes increased microbial fermentation in the large intestine to produce short-chain fatty acids, a similar physiological effect to dietary fiber.

Starch is also one of the most important raw materials for industrial use. In its native granular form, starch has limited applications. However, using chemical and physical modifications, starch has been applied in a wide variety of industrial products including food ingredients, sizing agents for paper, textiles, and starch-based plastics (Table 7.1). More information on modified starch is given in Chapter 8.

Starch occurs throughout the plant world. Cereal grains, legume seeds, and tubers are the most important sources of starch. Table 7.2 shows starch contents (% dry matter basis) in some plant materials for food use. Some

TABLE 7.1

Industrial Uses of Starch

Industry	Use of Starch/Modified Starch
Food	Viscosity modifier, edible film, glazing agent
Adhesive	Binding
Paper and board	Binding, sizing, coating
Textile	Sizing, finishing, and printing
Pharmaceuticals	Diluent, binder, drug delivery (encapsulation)
Oil drilling	Viscosity modifier
Detergent	Surfactants, suspending agent, bleaching agents and bleaching activators
Agrochemical	Mulches, pesticide delivery, seed coatings
Plastics	Food packaging, biodegradable filler
Cosmetics	Face and talcum powders
Purification	Flocculant
Medical	Plasma extender/replacers, transplant organ preservation (scaffold), absorbent sanitary products

Source: Adapted from Ellis et al., 1998.[1]

TABLE 7.2

Major Components (% Dry Matter Basis) of Some Plant Materials for Food Use

Plant Source	Starch (%)	Protein (%)	Lipids (%)
Wheat, grain	67	15	2
Rice, polished	89	8	1
Maize, kernel	57	12	7
Sorghum, grain	72	12	4
Potato	75	8	<1
Yam	90	7	<1
Cassava (tapioca)	90	<1	<1
Bean (common)*	42	23	2

Source: Adapted from Galliard, 1987 and Hedley, 2000.[2,3]

fresh plant crops, such as corn, potato, and sweet potato, contain about 15% or more starch. Nuts such as chestnuts contain up to 33% starch. Starch also exists in the stem-pith (e.g., sago) and fruits (e.g., banana). The starch content increases with the degree of refinement in milled products; it is about 70% in white wheat flour compared to about 60% in whole grain. The increase in starch is accompanied by a parallel decrease in cellulose and hemicellulose.

Starch is synthesized in the plastid compartment of plant cell. It is accumulated during the day in plant leaf cells and broken down at night to achieve a more or less constant supply of sucrose to the nonphotosynthetic tissues. Starch is formed by a complex biological pathway involving photosynthesis. However, the pathway of starch synthesis is not completely understood. Variations in structures and properties of starch may be associated with different species, growth conditions, environments, and genetic mutations of plants.[4] By manipulating the expression of one or more starch synthesizing enzyme genes, it is possible to make starches from amylose-free to high amylose content and to alter the type and amount of starch produced in a plant.[5]

Most starches are composed of a mixture of two molecular entities (polysaccharides), a linear fraction, amylose, and a highly branched fraction, amylopectin. The content of amylose is between 15 and 25% for most starches. The ratio of amylose and amylopectin in starch varies from one starch to another. The two polysaccharides are homoglucans with only two types of chain linkages, an α-(1→4) of the main chain and an α-(1→6) of the branch chains.

Although starch consists mainly of a homopolymer of α-D-glucopyranosyl units, it is one of nature's most complex materials. Starch molecules are hydrogen bonded and aligned radially in the granules that have a semicrystalline characteristic. Like many other complex materials, starch granule organization and molecular structure have been the subject of countless studies by researchers from many different disciplines.

In this chapter, starch chemistry, granular and molecular structure, functionality and roles of starch in food are discussed.

7.2 Starch Isolation and Chemistry

Starch exists in the form of granules and is composed essentially of homopolymer of α-D-glucopyranosyl units and small amounts of noncarbohydrate components, particularly lipids, proteins, and phosphorus. To understand the role of starch in plant foods and to determine the properties of starch, starch must first be extracted from plant materials. The functional properties of starch are affected by its two constituent polysaccharides, amylose and amylopectin, the physical organization of these biopolymers in granules, as well as interactions between starch polymers and other components.

7.2.1 Starch Isolation

Starch can be extracted using various processes, depending on the plant source and end use of the starch. Wet milling processing is widely used in cereal starch production. The purpose of wet milling is to separate the kernel (grain) into its constituent chemical components. Figure 7.1 presents the major procedures for extracting starches from cereal grains and tubers. As shown in the schematic, the major steps of cereal starch isolation include steeping, milling, and separation. Steeping is the first critical processing step in wet milling. It usually involves using an aqueous solution of sulfur dioxide and lactic acid at certain pH and temperature for certain periods of time. The main purposes of the steeping process are:

- To loosen the granules in the packed cell structure
- To isolate starch with minimum damage and maximum purity
- To reduce or inhibit the activity of undesirable microorganisms

After steeping, the softened grain is degerminated by coarse grinding. A hydroclone separator is used to separate the germs which are less dense and contain high levels of oil. The remaining materials are reground to liberate most of the remaining starch and other components. Starch can be separated from other components by centrifugation, based on density difference between starch and other components such as protein.

For tuber starch, such as potato starch, the main extraction processes include tuber soaking, disintegration, and centrifugation. Soaking is carried out in an aqueous solution of sodium bisulfite at controlled pH to prevent biochemical reaction such as discoloration of tuber. Disintegration and centrifugation are used to separate starch from other components. Starch granules

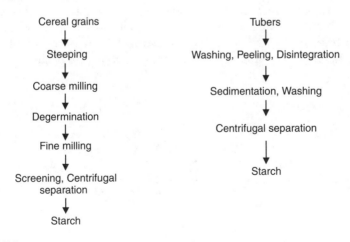

FIGURE 7.1
Major steps in starch isolation from cereal grains and tubers.

are unevenly distributed in the cell walls of tubers. They can be liberated from tuber by disruption of the cell walls. This is done during tuber disintegration by a cylindrical drum containing rotary saw blades on its circumference or a juice extractor for extraction at a small scale. Starch can be purified by washing, sedimentation, and centrifugation.

The wet starch is then either dried, or chemically modified or enzymatically hydrolyzed to manufacture new products (starch derivatives) and to improve their functional properties (for more detail on starch modification, see Chapter 8).

The isolation of starches from legume seeds is difficult due to the presence of insoluble flocculent protein and fine fiber, which decreases sedimentation and co-settles with the starch to give a brownish deposit.[6] Legume starches are isolated using aqueous techniques as well as pin milling and air classification.[7]

7.2.2 Starch Chemistry

As mentioned earlier, most starches are composed primarily of amylose and amylopectin. In some mutant species, starch granules may contain nearly 100% amylopectin. In addition to amylose and amylopectin, granules also contain some minor components such as proteins, lipids, inorganic substances, and nonstarch polysaccharides.

7.2.2.1   Amylose

7.2.2.1.1   Molecular Structure

Amylose is essentially a linear macromolecule consisting of ᴅ-glucopyranose residues linked together by (1→4) bonds (Figure 7.2). The degree of polymerization (DP) is between 100 and 10,000. Each macromolecule bears one reducing end and one nonreducing end. Amylose from some starch sources contains about 2 to 8 branch points per molecule. The chain length of these branch chains varies from 4 to 100 DP.[8] In some plant species, amylose has a few phosphate groups, probably at the C-6 position of glucose residues.[9]

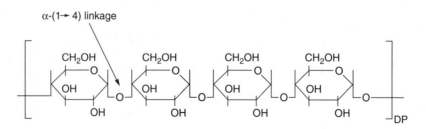

FIGURE 7.2
The glucopyranose unit linkage in amylase.

7.2.2.1.2 Physicochemical Properties

The abundance of hydroxyl groups along the amylose molecules (Figure 7.2) imparts hydrophilic properties to the polymer, giving it an affinity for moisture. Because of their linear nature, mobility, and the presence of many hydroxyl groups along the polymer chains, amylose molecules have a tendency to orient themselves in a parallel fashion and approach each other closely enough to permit hydrogen bonding between adjacent chains. As a result, the affinity of the polymer for water is reduced and the solution becomes opaque.

The interaction of amylose and iodine results in formation of complexes with characteristic color. Teitelbaum et al[10,11] postulated that the principal chromophore was the pentaiodide ion (I_5^-). The color of the amylose-iodine complex has been shown to vary with amylose chain length. John et al.[12] reported that the color of the complexes changed from brown (DP 21-24), to red (DP 25-29), red violet (DP 30-38), blue violet (DP 39-40), and finally blue (DP > 47). When DP was lower than 20, no color was formed. Since amylose assumes a helical structure, in the complex, iodine molecules occupy the central cavity of the helical molecule. Many physicochemical properties of amylose, such as iodine binding capacity and DP, depend on the botanical origins of the starch and the conditions used for its fractionation and purification.

Polar lipids and surfactants (e.g., monoglycerides and fatty acids) can also form a helical inclusion complex with amylose. Long and saturated monoacyl chains form more stable complexes. However, unsaturated monoglycerides have a poor complex-forming ability.[13] The inclusion complex gives V-type x-ray diffraction pattern (see Section 7.3.2.1 for more detail). At high or intermediate water content, differential scanning calorimetry (DSC) measurements show that these complexes melt at 100 to 120°C.[14] The transition is reversible; the complex is reformed on cooling. The complex formation is dependent on the length of the carbon chains of the lipids and surfactants. Lipids and surfactants are required to have a minimum of four to eight carbons in the chain and optimal chain lengths are between 12 and 18 carbons. Less than nine carbons in the chain give a complex of low stability.[15]

7.2.2.1.3 Analysis

Physicochemical properties of starch are greatly influenced by the content, molecular weight, and molecular weight distribution of amylose. To determine the proportion of amylose in starch, the iodine reaction has been most commonly used because it is specific, sensitive, and easy to analyze qualitatively and quantitatively. The methods such as blue value (absorbance at 680 nm for starch-iodine complex using amylose and amylopectin standards), potentiometric, and amperometric titration have been used for more than 50 years. These procedures are based on the capacity of amylose to form helical inclusion complexes with iodine, which display a blue value color characterized by a maximum absorption wavelength (λ_{max}) above 620 nm. Iodometric methods give a measure of medium to long chain α-(1→4)-glucan content, and the small amount of iodine bound by normal amylopectin is compensated for in calculations or by calibrating colorimetric assays with standard amylose-amylopectin mixtures. During the titration of starch with iodine solution, the amount (mg) of iodine bound to 100 mg of polysaccharide is determined. The value is defined as iodine binding capacity or iodine affinity (IA). The amylose content is based on the iodine affinity of starch vs. purified linear fraction from the standard: 100 mg pure linear amylose fraction has an iodine affinity of 19.5 to 21.0 mg depending on amylose sources. Amylopectin binds 0 to 1.2 mg iodine per 100 mg.[16] The amylose content determined by potentiometric method is considered an absolute amylose content of starch if the sample is defatted before potentiometric titration. The amylose content estimated by all of the procedures based on iodine complex formation might be considered as apparent amylose content. Table 7.3 gives the apparent and absolute amylose content of starch from various sources.

There are many factors that influence the determination of amylose content of starch, such as starch source, sample preparation, and molecular structure of starch. The intermediate materials (see Section 7.2.2.3) in starch and long linear chains in amylopectin could also influence the results of amylose content. The measurement of amylose content in starch can be affected by the existence of lipid. If monoacyl lipids are present in starch, they will form

TABLE 7.3

The Apparent and Absolute Amylose Content of Starch from Various Crops

Starch	Iodine Affinity	Amylose content (%) Apparent	Absolute
Rice (Japonica)	4.00	20.0	17.5
Wheat (Asw)	4.86	24.3	21.7
Barley (Bomi)	6.08	30.4	27.5
Maize (Normal)	5.18	25.9	21.5
Maize (Hylon 7)	13.4	67.0	58.6
Water Chestnut	4.94	24.7	23.3
Chestnut	4.32	21.6	19.6
Sago	5.16	25.8	24.3
Lotus	3.37	16.9	15.9
Kuzu	4.06	20.3	21.0
Sweet Potato	4.18	20.7	18.9
Yam	4.29	21.5	22.0
Lentil*	6.97-9.09	29-45.5	—
Tapioca	3.41	17.1	16.7
Arrowhead	5.20	26.0	25.6
Edible Canna	4.80	24.0	22.2
Potato	4.44	22.2	21.0

Source: Adapted from Hizukuri, 1996 and *Hoover and Sosulski, 1991.[17,18]

complexes with amylose when starch is gelatinized. This interaction competes with iodine binding, hence results in a lower apparent amylose content. Lipid interference can be prevented by prior extraction using solvents such as hot *n*-propanol-water (3:1 v/v)[18] to remove internal starch lipids.

Amylose content of starch may also be estimated measuring the melting enthalpy of amylose-lipid complex using DSC.[19] In this technique, lipid is added to starch during starch gelatinization. Starch-lipid complex is formed during cooling and storage. The complex is then melted by heating near or slightly above 100°C. Based on the value of enthalpy of melting complex, amylose content can be calculated. Amylose content of starch can be determined using size exclusion chromatography based on elution times and peak area for different molecular sizes of amylose and amylopectin.[20] In addition to the above methods, Concanavalin A method (Con A) has been employed for amylose determination.[21] Con A is also used to precipitate amylopectin from a starch solution.

Pure amylose obtained from various starches contains limited numbers of branch linkages and is a mixture of branched and unbranched molecules. The branched molecules of amylose from various origins have their own characteristic structures such as molecular size, inner chain length, and the number of side chains. The molar fractions of unbranched and branched molecules vary with the origin of the amylose. For example, wheat amylose probably contains only a small number of very large branched molecules, whereas sweet potato amylose has a small number of relatively large unbranched molecules.[22]

7.2.2.2 Amylopectin

7.2.2.2.1 Molecular Structure

Amylopectin is a highly branched polysaccharide. The structure consists of α-D-glucopyranose residues linked mainly by (1→4)-linkages (as in amylose) but with a greater proportion of nonrandom α-(1→6)-linkages, which gives a highly branched structure (Figure 7.3). Amylopectin is one of the largest biological molecules and its molecular weight (M_w) ranges from 10^6 to 10^9 g × mol^{-1}, depending on botanical origin of the starch, fractionation of starch, and method used to determine the molecular weight.

7.2.2.2.2 Physicochemical Properties

The large size and the branched nature of amylopectin reduce its mobility in solution and eliminate the possibility of significant levels of interchain hydrogen bonding. On average, amylopectin has one branch point every 20 to 25 residues. The branch points are not randomly located.[23]

The amylopectin chains can be classified into three types according to their length and branching points. The shortest A chains carry no branch points. The B chains are branched by A chain or other B chains (e.g., B1, B2, and B3

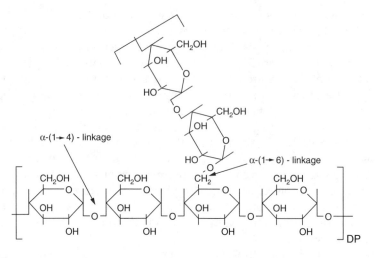

FIGURE 7.3
The glucopyranose unit linkage in amylopectin.

as shown in Figure 7.4). The C chain carries B chains but contains the sole reducing terminal residue.

It is now widely accepted that linear branched chains with DP ~15 in amylopectin are the crystalline regions present in the granules.[24] These short chains form double helical ordered structure; part of the double helices can pack together in organized arrays in cluster form. This concept is compatible with cluster models of amylopectin, which during the last 20 years have received general acceptance. Modifications have been proposed since the model was introduced.[25-28] Figure 7.4 shows some acceptable models for amylopectin structure.

Another unique feature of amylopectin is the presence of covalently linked phosphate monoesters. They can be linked to either the C3 or C6 position of the glucose monomers, and occur to a greater extent in starch from tuberous species, especially potato starch. The gelatinization enthalpy decreases as the degree of phosphorylation at the C6 position increases.[29] Due to the charged nature of phosphate monoesters, electrostatic repulsion between molecules increases, which results in the change of gelatinization and pasting properties of starch.

7.2.2.2.3 *Analysis*

Structural analysis of isolated amylose and amylopectin components has been carried out by standard methods based on methylation, periodate oxidation, and partial acid hydrolysis studies. Methylation and periodate oxidation studies established the linkage types and frequency of branching of amylopectin, together with the characterization of oligosaccharides from partial hydrolysis with acid and/or with enzymes, provided evidence that

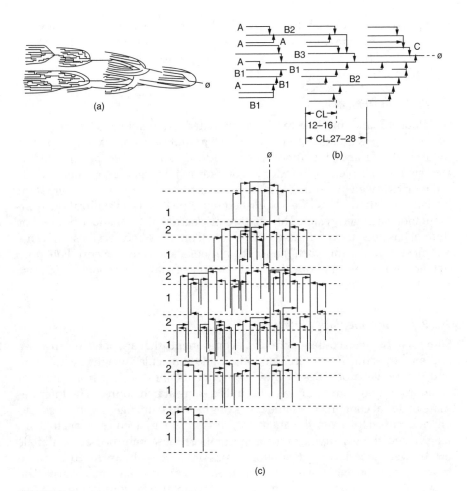

FIGURE 7.4
Schematic models of amylopectin molecule: (a) adapted from French, 1972;[26] (b) adapted from Hizukuri, 1986;[25] and (c) adapted from Robin et al., 1974.[27] In models Ø is the reducing end. (A), (B) (B1, B2 and B3), and (C) chains are defined in the text. CL represents chain length. Crystalline and amorphous granular regions are represented by area (1) and (2), respectively.

the α-D-glucopyranose residues are joined mainly by (1→4) linkage with 4 to 5% by (1→6) linkage.[9,32]

The amylopectin chain profile can be obtained by size exclusion chromatography or ion-exchange chromatography of enzymically debranched amylopectin. The A:B chain ratio in amylopectin can be determined using debranching enzymes. The debranching enzymes, isoamylase and pullulanase, specifically hydrolyze the branch linkages to produce linear chains. An important difference between the debranching enzymes is that pullulanase debranches side chains of two or more glucose units, whereas isoamylase requires at least three glucose units to work with. The A:B chain ratio is typically from 0.8 to 2.2 on molar basis or from 0.4 to 1.0 on weight basis.[25]

Size exclusion chromatography (SEC) and high performance anion-exchange chromatography with pulsed amperometric detection (HPAEC–PAD) are the two basic techniques used to estimate the chain length and chain length distribution of amylopectin.[26-31]

7.2.2.2.4   Fractionation

Amylopectin and amylose can be fractionated using aqueous leaching, dispersion, and precipitation processes. Starch granules are completely dispersed in hot water or aqueous dimethyl sulfoxide, and amylose is precipitated as a crystalline complex by the addition of hydrophilic organic solvents, such as butanol, followed by cooling. Amylopectin is recovered from the supernatant by lyophilization. Amylose is purified by repeated recrystallization from 1-butanol-saturated water. The mixture of isoamyl alcohol and 1-butanol or thymol has been used for the precipitation of amylose because these solvents are most efficient for fractionation of amylose and amylopectin. For more detailed information about the fractionation of starch components, please refer to cited literature.[16]

7.2.2.3   Intermediate Materials

Some starches also contain a group of components that are between amylose and amylopectin. The molecular weight and molecular weight distribution and functional properties of these components are different from those of standard amylose and amylopectin in the same starch source. The intermediate material contains the same types of glucosidic linkages as amylose and amylopectin. However, the chain length of linear molecules, branch chain length, and the density of branching points of these intermediate materials are between amylose and amylopectin. Amylose with up to 20 or more branch points on average may be considered as intermediate materials.[17] The amylopectin fraction of high amylose starch can also be considered as intermediate material. The amount and structural features of these intermediate materials vary with starch sources.

For example, wrinkled pea starch contains about 29% intermediate material. These intermediate materials of wrinkled pea starch had increased proportions of long chains with decreasing molecular weight and their iodine affinity (IA) is between 8 and 10.[33] In oat starch, the IA values of intermediate material are larger than that of amylopectin and the molecular weight is less than that of amylose.[34] The average chain length (branched) of intermediate material is about 29, compared to 21 for amylopectin.[17]

7.2.2.4   Minor Components

Starch granules usually contain 0.5 to 2% (w/w) noncarbohydrates, including 0.05% to 0.5% (w/w) proteins, 0.1 to 1.7% (w/w) lipids, and 0.1 to 0.3% (w/w) ash. Although these components are present at very low levels in starch granules, they play an important role in the physicochemical properties of starch. For example, nonstarch polysaccharides in starch tend to bind water and develop viscosity. The presence of proteins can lead to unwanted flavors or colors in starch and starch hydrolysis products, via reaction between free amino acid groups and reducing sugars (Maillard reactions). In addition, proteins may affect surface charge, rates of hydration and may also interfere with the interactions between starch granules and hydrolytic enzymes.

Lipids can form an amylose-lipid complex that enhances the resistance of starch to enzyme hydrolysis. Lipid may be present on the surface or inside of the starch granule, depending on the plant source. The main components of surface lipids are triglycerides, free fatty acids, glycolipids, and phospholipids. Monoacyl lipids (containing a single fatty acid residue which has a chain length of 16 or 18 carbon atoms) are the main components of internal lipids of some maize and wheat starches.[35]

The basic composition of starch ash are: phosphorus, calcium (CaO), potassium (K_2O), sodium (Na_2O), and silicon (SiO_2).[36] Phosphorus in potato starch is bound as esterified phosphate at carbon 6 and 3 of various D-glucose units; C6 bound phosphate constitutes about two thirds of the total starch bound phosphate.[29] The phosphorus content of starch can be determined by colorimetry and ^{31}P-NMR methods.[37]

Starches from different sources vary in their content of minor components. Cereal starches in general contain lipids that appear to be associated with the amylose fraction, whereas tuber starches have very low lipid contents. Cereal starches have more protein than potato starch. On the other hand, potato starch has higher phosphorus than cereal starches.

7.3 Granular Structure

Starch is found in nature as particles (granules). A pound of corn starch contains about 750 billion granules[38] and a typical 15 μm diameter corn starch granule consists of over a billion molecules.[39] In molecular terms, the granule is an immense and highly organized structure. Starch granules consist of amorphous and crystalline regions. The crystalline regions or the crystallites are formed from the short branch chains of amylopectin molecules arranged in clusters. The areas of branching points are believed to be amorphous. The granular structure of starch is essentially determined by genetic factors that govern starch biosynthesis.[32]

Water is an integral component of the granule structure and participates in the important hydration process that takes place during gelatinization and subsequent granule swelling and dissolution.

The structure of the starch granule also depends on the way amylose and amylopectin are associated by intermolecular hydrogen bonds. The presence of α-(1→6)-D bonds in amylopectin is responsible for the alternation between amorphous and crystalline zones.[24] It is inferred that the molecules of amylose and amylopectin are distributed throughout the granule. The degree of mutual binding by hydrogen bonds between amylose and amylopectin and amylopectin itself is responsible for structural heterogeneity of starch granule. When these bonds are strong, numerous, and regular, the chains associate as crystalline networks. In contrast, in the amorphous areas, hydrogen bonding is weaker, and this portion of the granule is easily distinguishable from the crystalline zones (the presence of α-$(1\rightarrow6)$-linkages).

7.3.1 Starch Morphology

In nature, starch exists in the form of granules, which can differ in size and shape. The origin of starch granules can be inferred from their size, shape, and, the hilum position (the original growing point of granule) (Table 7.4).[38,40,41]

Tuber starch granules are generally voluminous and oval shaped with an eccentric hilum (Figure 7.5). Cereal starch granules such as maize, oat, and rice have polygonal or round shapes. High amylose maize starch exhibits filamentous granules (budlike protrusions). Legume seed starch granules are bean-like with a central elongated or starred hilum. The hilum is not always distinguishable, especially in very small granules.

TABLE 7.4

The Size and Shape of Starch Granules Derived from Major Botanic Sources

Starch Sources	Granule Size (μm)	Granule Shape
Cereals		
Maize[b]	5–25	Round, polygonal
High amylose[a]	~15	Round, filamentous
Waxy maize[b]	5–25	Round, oval indentations
Wheat[a]	2–38	Round, lenticular
Oats[a]	5–15	Round
Barley[a]	2–5	Round, lenticular
Sorghum[a]	4–24	Round, polygonal
Millet[a]	4–12	Round, polygonal
Rye[a]	12–40	Lenticular
Rice[a]	3–8	Polygonal
Pulses		
Horsebean[a]	17–31	Spherical
Smooth pea[a]	5–10	Reniform (simple)
Wrinkled pea[a]	30–40	Reniform
Roots and Tubers		
Potato[b]	15–100	Voluminous, oval, oyster
Waxy potato[c]	14–44	Round, oval
Sweet potato[c]	2–42	Round, oval and polygonal
Tapioca (cassava)[b]	5–35	Round-oval, truncated on side

Source: From [a]Blanshard, 1987;[40] [b]Pomeranz, 1985;[38] [c]Hoover, 2001.[41]

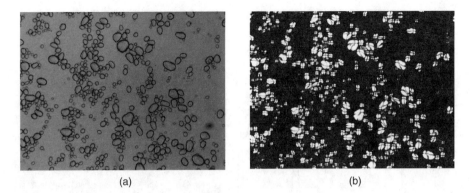

FIGURE 7.5
The potato starch granules observed under microscope without (a) polarized light and with
(b) polarized light.

The size of starch granules vary from 2 to 100 μm in diameter. The size of
starch granules is usually expressed as a range or as an average of the length
of the longest axis. Potato starch has the largest granules among all the
starches. The size of most cereal starch granules is smaller than that of tuber
and legume starches.

Under polarized light in a microscope, a typical birefringence cross is
observed as two intersecting bands (the Maltese cross) (Figure 7.5b). It indicates that the starch granule has a radial orientation of crystallites or there
exists a high degree of molecular order within the granule.

Various methods are used to study granule morphology. Microscopy
shows the distribution of various structures in the granule and some of their
details. The complementary methods of transmission electron microscopy
(TEM), scanning electron microscopy (SEM), and atomic force microscopy
(AFM) have been widely used. The external surface area of starch granule
can be determined by the shape and size of the particle which is usually
determined using optical microscopy or light scattering.

Starches isolated from corn, sorghum, millet, large granules of wheat, rye,
and barley were found to have pores on their surfaces by SEM.[42] However,
some granules contain many pores, others a few. For some starch granules,
there are no surface pores observed. The presence of pores in starch granules
would result in a macroporous structure whose available surface area would
be much greater than the boundary surface area. Provision for surface open-
ings and interior channels is accommodated in the model of Oostergetel and
van Bruggen,[43] a channel runs through the center of the superhelical structure. The channels are naturally occurring features of the granule and are
likely to predispose the initial sites of enzymic attack during germination.
Granules of different ages and sizes vary in the amount of loose material on
their surface.[44]

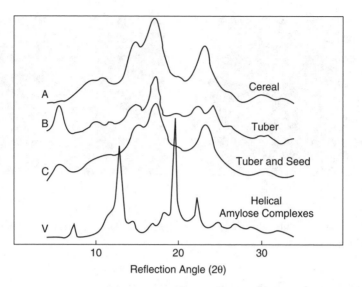

FIGURE 7.6
X-ray diffractometer tracings from different starches. Labeling refers to (A) A-type from cereal starches, (B) B-type from tuber starch, (C) C-type from seed starches, and (V) V-type from helical amylose complexes. (Adapted from Zobel, 1988.)[45]

7.3.2   Ordered Structure of Starch Granules

7.3.2.1   X-ray Diffraction Pattern

Starch is a semi-crystalline material. The semi-crystalline structure can be identified at the light microscope level (Figure 7.5) and through characteristic x-ray diffraction patterns. Figure 7.6 presents four major types of x-ray diffraction patterns of native starches, A, B, C, and V.[45] The A type is characteristic of most starches of cereal origin; the B type of potato, other root starches, amylomaize starches, and retrograded starch; the C type of smooth pea and various bean starches. The V type can only be found in amylose helical complex starches after starch gelatinization and complexing with lipid or related compounds. The x-ray diffraction pattern of starch could be altered by heat/moisture treatment. For example, B type of potato starch can be converted to A or C type using heat/moisture treatment.[46]

7.3.2.2   The Characteristics of Starch Crystallites

The x-ray diffraction pattern reflects different packing of amylopectin side-chain double helices. The A type structure has a closely packed arrangement of double helices, whereas the B type structure consists of a more open packing of helices with a correspondingly greater amount of inter-helical water (Figure 7.7).[47,48] Individual double helices in the two structures are very similar. The tightly packed A type structure would be expected to be more stable. There is some evidence from calorimetric studies indicating the

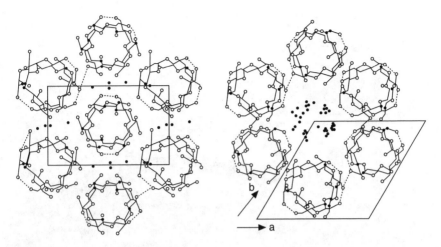

FIGURE 7.7
The helix packing projection in (a), (b) plane and unit cells in (A) type (left) and (B) type (right) amylose. Water molecules are indicated by dots. H-bonds are indicated by dashed lines. (Adapted from Wu and Sarko, 1978.)[47]

A type starch has a higher melting temperature and hence is more stable than the B type.[49]

The crystalline regions are predominantly located in the hard layers of the granule and are composed of stacks of crystalline lamellae which form the backbone of the starch granule as shown in Figure 7.8. For amylopectin, crystalline and amorphous lamellae are organized into larger, more or less spherical structures as blocklets. The blocklet concept was proposed to describe the organization of amylopectin lamellae into effectively spherical blocklets, with diameters from 20 to 500 nm depending on the botanic origin of the starch and their location within the granule.[50] SEM studies of wheat starch found small blocklets approximately 25 nm in size in the semi-crystalline shells, and larger 80 to 120 nm blocklets in the hard crystalline layers. Potato starch shows much larger blocklets of 200 to 500 nm.[50]

In recent years, amylopectin was proposed to be viewed as a structural analogue to a synthetic side-chain liquid crystal polymer.[51] The model of side-chain liquid crystals has been employed to interpret the mechanism of starch gelatinization at different moisture content as shown in Section 7.4.1.4.

7.3.2.3   Structure Model of Unit Cell Packing in A and B Type Starches

A number of structure models have been proposed for the A and B type starches.[47,48,52,53] Structures proposed by Wu and Sarko,[47,48] which postulate right-handed, parallel-stranded double helices packed antiparallel to one another (12 residues), received wide acceptance. The schematic packing projection of double helix arrangement in A and B type starch are shown in Figure 7.7. The unit cell packing for A type starch is orthorhombic (unit cell dimensions of a = 1.17 nm, b = 1.77 nm, and c = 1.05 nm) with 8 molecules of water

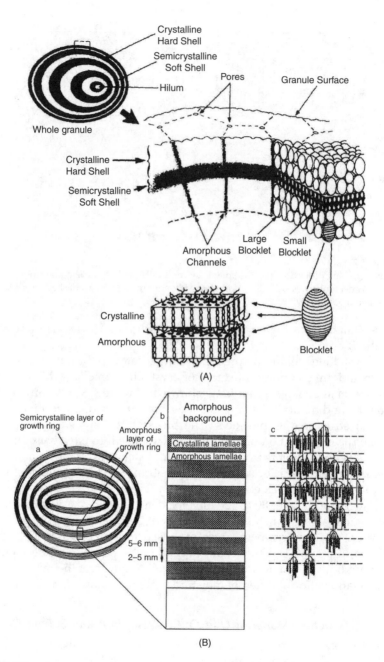

FIGURE 7.8

Overview of starch granule structure. (A) At the lowest level of granule organization (upper left), the alternating crystalline (hard) and semi-crystalline (soft) shells are shown. The shells are thinner towards the granule exterior and the hilum is shown off center. At a higher level of structure the blocklet structure is shown, in association with amorphous radial channels. Blocklet size is smaller in the semi-crystalline shells than in the crystalline shells. At the next highest level of structure one blocklet is shown containing several amorphous crystalline lamellae.

per cell. For B type starch, packing is hexagonal (unit cell dimensions of a = b = 1.85 nm, c = 1.04 nm) and the cell contains 36 molecules of water. In addition, the helices in B type starches are clustered around a central column of water. In A type starches, the centers are filled with a starch helix.[47,48]

Revisions have been made since the A and B types were originally proposed.[52,53] Revised A starch is a face-centered monoclinic lattice with unit cell dimensions a = 2.124 nm, b = 1.172 nm, c = 1.069 nm (the fiber axis), and γ = 123.5°.[52] The cell contains 12 glucose residues in two left-handed double helices with four water molecules located between the helices. The c-axis is parallel to the helix axis. Parallel-stranded helices are packed in a parallel mode. Such packing results in all chains having the same polarity and is in agreement with biochemical data that argue for a parallel molecular arrangement. The repeating molecular chain unit in the cell is a maltotriose moiety with α-(1→4)-linked glucose residues in the $^4C_1$ pyranose conformation. For B type starch, the revised version considers the strands to be left-handed. The wide variety of proposed models is at least partly due to the variable amount of water ranging from 10% to 50% in B type starch.[54] At low humidity (27% moisture content) and high temperature, B type starch may be irreversibly converted to A type starch[55] while remaining in the solid state as fiber or granules.[54] Because of this solid state transition, and because the fiber repeat distances are equal, it has long been thought that the chain conformation of B type starch is very similar to that of A type starch. Thus, the chains in B type starch are also arranged in double helices, with the structures differing in crystal packing and water content.[24]

Based on these revised structures, one of the low energy conformations of a flexible amylosic chain leads to single strands that readily form rigid double helices. These double helices associate in pairs and can pack together in organized arrays → cluster form, stabilized by hydrogen bonding and van der Waals forces. These pairs associate with one another to give the A or B type structures, depending on their chain length and water content.

7.3.2.4 Requirements for Formation of A and B Type Starch Crystallites

Whether A or B starch crystallites are formed is thought to be related to the average chain lengths of chains within the amylopectin clusters. A type starch crystallites are formed from shorter chains and B type starch crystallites from longer chains. From the study of waxy (*wx*)-type starch, A type starch crystallites are believed to be formed in warmer and dry conditions with denser crystalline packing, whereas B type starch crystallites are formed

---

FIGURE 7.8 (continued)
In the next diagram the starch amylopectin polymer in the lamellae is shown. (B) Schematic representation of starch granule structure from Jenkins et al.: (a) a single granule with alternating amorphous and semicrystalline layers, representing growth rings; (b) expanded view of the semicrystalline layer of a growth ring, consisting of alternating crystalline and amorphous lamellae; (c) the cluster structure of amylopectin within the semicrystalline layer of the growth ring. (Adapted from Gallant et al., 1997 and Jenkins et al., 1994.)[50,122]

in wet and cold conditions with less dense crystalline packing. In the C type starch granules, B type crystallite is enhanced in the inner part of granules and A type crystallite is enhanced in the outer part of granules.[56]

7.3.3 Amorphous Structure of Starch Granules

The noncrystalline material in starch granules has been considered amorphous. In the schematic model of amylopectin cluster (Figure 7.4), the branching regions are believed to be amorphous region. It has been suggested that some amylose molecules are located in this amorphous region with some interaction with the branch chain of amylopectin. Amorphous starch has been defined as material not exhibiting an x-ray diffraction pattern.[57] These amorphous regions of starch are susceptible to chemical reaction. Thus they can be removed by acid treatment or can react with functional groups. Gidley and Bociek[58] have shown that amorphous starch gives a nuclear magnetic resonance (NMR) pattern consistent with a nonrandom relationship among glucose units, suggesting some order exists in the amorphous domains of the granule. The distinction between crystalline and amorphous states in starch is not absolutely clear in terms of molecular order.

The amorphous regions are the main swellable structural elements of the native granule. During swelling, plasticization and stretching of amorphous glassy layers along the direction of chain orientation could, therefore, provide a rapid and limited absorption of water.

The distribution of crystalline and amorphous regions of starch has been approached through acid hydrolysis. A rapid initial attack occurs on the amorphous regions of starch containing branching points with α-1$\rightarrow$6 linkage at the early stage of acid hydrolysis. Then, a slower hydrolysis takes place on the more crystalline areas during the second stage. The crystallinity of starch granules can be determined using the separation and integration of the areas under the crystalline peak and amorphous region in x-ray diffraction pattern.[59] In general, the crystallinity of A type starch is higher than that of B type starch. The values for starch crystallinity vary from 15 to 45% (Table 7.5)[59,60] depending on starch source and methods used for calculating the crystallinity. The crystallinity of starch is greatly influenced by the moisture content in the granules.[46] Amylose content has little effect on crystallinity for A type starches. However, a higher amylose content (e.g., amylomaize) results in lower crystallinity for B type starches. This may be due to the molecular structure of high amylose starch, such as longer chain length and unique chain length distribution.[61] There is no trend on crystallinity within the C type starch series.

7.3.4 Role of Water in Starch Granules

Starch is biosynthesized in a water medium; as a result, starch granules contain much water. The amount and distribution of water within starch

TABLE 7.5

Crystallinity of Starch with A-, B-, and C-Type
Structure and Different Amylose Content

Starch	Crystallinity (%)	Amylose (%)
Starches with A Structure		
Oat	33	23
Barley*	22–27	22–26
Rye	34	26
Wheat	36	23
Waxy rice	37	—
Sorghum	37	25
Rice	38	17
Maize	40	27
Waxy maize	40	0
Dasheen	45	16
Naegeli amylodextrin	48	—
Starches with B Structure		
Amylomaize	15–22	55–75
Edible canna	26	25
Potato	28	22
Starches with C Structure		
Sweet potato	38	20
Horse chestnut	37	25
Tapioca	38	18

Source: From Zobel, H.F. 1988 and *Tang et al., 2001.[59,60]

granules play a very important role in the granular structure of starch as discussed earlier. Water strongly absorbs on polar groups at lower hydration levels, swells and plasticizes unordered regions at humidities > 5%. It also intercalates stoichiometrically in the crystalline hydrated structures stabilized by hydrogen bonds. The amount of water bound by starch is determined by measuring the water uptake or loss when starch is exposed to various relative humidities and temperatures (moisture sorption isotherm). In this section, only bound water in the starch granules will be discussed.

At room temperature, starch granules absorb water to a limited extent. The water absorption in starch is related to amorphous regions since the crystallites have a defined water content. During dehydration, the granules shrink and crack, which indicates that the changes in starch granules accompanying drying are due to loss of water. It also indicates that the crystal structure of starch has a significant amount of water, particularly in the case of B type starch where Wu and Sarko[47] proposed a channel of water as large as the double helix. In the A type structure, the double helices are packed in a monoclinic unit cell with eight water molecules per unit cell. In the B type structure, double helices are packed in a hexagonal unit cell with 36 water molecules per unit cell.[24]

Since starch is a heterogeneous and semi-crystalline biopolymer, the binding of water depends on the density and regularity of packing of the starch chains. Starch chains are found to be packed rather inefficiently in dry starch compared with crystalline cellulose. There are very small voids into which water can penetrate with a relatively small volume increase. After uptake of about 10% water, all voids are filled.[40] The amounts of water are considered as bound water or structural water. The bound water contributes to a semi-crystalline structure of starch granules. In other words, the crystallites are not able to form without the presence of water. Increasing water content in starch results in an increase in molecular mobility of the chain in the amorphous regions. It could also result in changes in the proportions of A and B types of starch crystallites. Recent research indicates that the amount of double helices is significantly increased when the water content in starch granules is increased.[62] It suggests that increasing water content in starch results in formation of additional double helices inside the granules from both amylopectin and amylose.

It has been found that the total crystallinity of starch granules is always lower than the proportion of the double helices present.[63] In the dry powder only about half of the double helices were found to be arranged within crystalline order. Crystalline order arrangement cannot be made by those double helices at limited water content. Double helices lose their long-range order at the edge of crystallites during starch extraction from plant sources. This could result in reduced crystalline size and total crystallinity. Thus, differences in granular structure can be attributed to changes in the proportion of amylopectin forming double helices and in the proportion of these double helices that form crystallites. Granular structure can also be changed by the distribution of water in the starch granules.

7.4   Starch Functionality

Starch has numerous useful functional properties for food and nonfood applications. These include thickening, coating, gelling, adhesion, and encapsulation. Some of these functionalities are unique to the polymer due to the structure of amylose and amylopectin and their organization.

Gelatinization, retrogradation, and pasting, which underlie starch functionality, are the three most important phenomena in starch applications. In this section, the mechanism, analytical techniques, and factors influencing starch gelatinization, retrogradation, and pasting will be discussed. The study of starch gelatinization and retrogradation will help us to understand the relationship between the structure and properties of starch.

7.4.1 Starch Gelatinization

7.4.1.1 Definition

Native starch granules are insoluble in cold water but swell in warm water. When starch granules are heated in the presence of water, an order-to-disorder phase transition occurs. Starch gelatinization is the collapse (disruption) of molecular orderliness within the starch granule along with concomitant and irreversible changes in properties such as granular swelling, crystallite melting, loss of birefringence, viscosity development, and solubilization. The point of initial gelatinization and the range over which it occurs are governed by starch concentration, method of observation, granule type, and heterogeneity within the granule population under observation.[64] It should be noted that the term gelatinization does not have the same meaning to all starch researchers, as different methods have been used to monitor different events occurring during this phenomenon. New information on starch gelatinization is often the result of using new methods of characterization and investigation.[57]

7.4.1.2 Analytical Techniques to Probe Starch Gelatinization

Starch gelatinization occurs during the processing of starchy food. Product quality is greatly influenced by starch gelatinization. Because gelatinization is of such great importance in food processing, a variety of analytical techniques have been employed to probe the phenomenon and to understand its mechanism. These include: viscometry, optical microscopy,[65] electron microscopy,[46] differential scanning calorimetry (DSC),[66] x-ray diffraction,[67] nuclear magnetic resonance (NMR) spectroscopy,[68] Fourier transform infrared (FTIR) spectroscopy,[69] and most recently simultaneous x-ray scattering.[70] Microscopic examination of the granules undergoing gelatinization allows the observation of the degree and duration of swelling, as well as the integrity and size of the swollen granules. The calorimetry technique (e.g., DSC) is able to detect the heat flow changes associated with both first-order and second-order transitions of starch polymer. Thus, the enthalpy for the phase transition and melting temperature of starch crystallites can be determined. DSC has been widely employed to study gelatinization of cereal, tuber, and legume starches. DSC provides valuable insight into the order-to-disorder transition phenomenon of granular starches. X-ray diffraction has been used to study the crystallinity change and to characterize the transition of crystal structure during starch gelatinization. The molecular structure information of starch during gelatinization can also be obtained using FTIR and NMR techniques. The FTIR technique detects the absorption of different bond vibrations in starch molecules during gelatinization and it is sensitive to changes in molecular structure (short-range order), such as starch chain conformation, helicity, crystallinity, and moisture content. The NMR technique provides information of the loss of structural order within granules during gelatinization. Using ^{13}C CPMAS NMR technique, Cooke and

Gidley[68] suggested that double helix melting, rather than loss of crystallinity, could be primarily responsible for gelatinization enthalpy. X-ray scattering is used to study the structure of semi-crystalline carbohydrate polymers. Using the high intensity of a synchrotron of small angle x-ray scattering (SAXS) and wide angle x-ray scattering (WAXS), real time experiments during starch gelatinization can be carried out. The technique provides useful new insights into the organization of the structure within the granule at length scales characteristic of the lamellar structure.[70]

However, some of these methods are experimentally limited by certain parameters such as starch/water ratio and the temperature range over which gelatinization can be studied. For example, calorimetry is particularly well suited to investigate the phase transition of starch/water systems because it permits the study of starch transitions over a wide range of moisture content, the determination of transition temperatures over 100°C, and the enthalpy changes during transitions.

7.4.1.3   Flory-Huggins Equation

Figure 7.9 shows the DSC thermograms of potato starch as a function of starch contents and temperature. For most starches, at high water content, gelatinization is accompanied by the endothermic transition (usually called the **G** transition) which occurs essentially in a rather narrow temperature range (from about 65°C to 75°C). At lower water contents (<14% w/w), the crystallites melt (referred to as the **M** transition) at a higher temperature as a function of water content. Endotherms occurring at even higher temperatures have been attributed to dissociation of amylose-lipid complexes for some cereal starches.

This observation has led to a thermodynamic treatment of melting data for starches using the Flory-Huggins equation,[71] which describes phase transition behavior of polymer/diluent mixtures.[72-74] According to this theory, which is based on assumptions of equilibrium thermodynamics, the equilibrium melting point, T_m^0, of a polymer is depressed to a new value, T_m in the presence of a diluent. The depression of T_m is dependent on the volume fraction of the diluent, v_1, by:

$$\frac{1}{T_m} - \frac{1}{T_m^0} = \frac{R}{\Delta H_u} \frac{V_u}{V_1} \left(v_1 - \chi_1 v_1^2 \right)$$

where R is the gas constant, ΔH_u is the enthalpy of fusion per repeating unit, V_u/V_1 is the ratio of molar volumes of the repeating unit (anhydroglucose in the case of starch) and the diluent, and χ_1 is the polymer/diluent interaction parameter. Although the Flory theory was found to describe the behavior of starch/water mixtures reasonably well at low water content, its application to starch systems has been questioned, because true equilibrium conditions are not met during starch gelatinization.[75,76] The approach of Flory-Huggins equation can be applied to explain certain modifications in

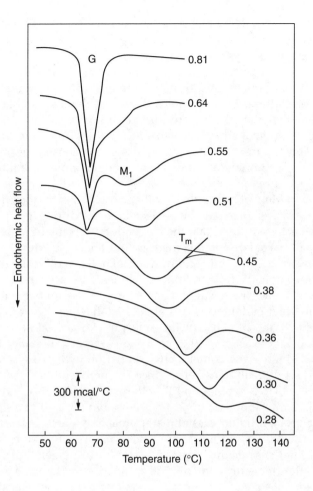

FIGURE 7.9
Differential scanning calorimetry (DSC) thermograms of potato starch at various water fractions upon heating. (Adapted from Donovan, 1979.)[66]

the structure from a partly crystalline to an amorphous form[77] and to compare the thermal responses of different starches.

7.4.1.4 Mechanism of Starch Gelatinization

The mechanism of starch gelatinization as a function of water content has been studied by many researchers based on DSC study. The following mechanisms or hypotheses were proposed for starch gelatinization at various moisture contents. Donovan[66] suggested that there are two distinct mechanisms by which ordered regions of starch undergo phase transition over a wide range of moisture contents. The low temperature endotherm reflects stripping and disorganization of polymer chains from crystallites — processes that are facilitated by the swelling action of water on the amorphous regions. As the water content decreases and becomes insufficient for the above process of complete melting, the partially hydrated crystallites tend to melt at a higher temperature whose value depends on the volume fraction of the diluent (water), as predicted by Flory theory.

Evans and Haisman[78] suggested that the biphasic endothermic transition of starch at intermediate moisture content as shown in Figure 7.9 reflects two types of melting. First, granules that have the least stable crystallites melt at a low temperature. As a result, the disordered starch chains absorb more water, thus making it unavailable for the remaining ungelatinized granules. This means that the effective solvent concentration is further reduced by repartitioning of water. Hence, the remaining granules melt at a higher temperature.

Colonna and Mercier[74] provided another explanation for the physicochemical changes taking place at the molecular level during gelatinization of starch. They have proposed that the two endotherms might reflect a sequential process of partial disentanglement within the crystallites (low temperature) followed by intramolecular double helix→coil transitions (high temperature), to yield a fully hydrated/swollen polysaccharide gel network.

Slade and Levine[76] have suggested that a glass → rubber transition precedes the melting endotherms of starch crystallites and that gelatinization is controlled by the mobility of the amorphous material surrounding the crystallites. Water plasticizes the amorphous regions in starch to facilitate annealing. At high water content, any correlation with glass transition temperature (T_g) is difficult to establish since T_g approaches 0°C. With moisture contents in the 13 to 20% range and T_g in the order of 60 to 30°C,[79] relationships between glass transition and annealing temperatures could be significant.

In recent years, a liquid crystalline approach to starch gelatinization has been taken by Donald's group.[80] The model (Figure 7.10) serves as a tool for understanding the structure changes in starch during gelatinization. In dry starch granules (moisture content <5% w/w), the amylopectin helices are in a glassy nematic state and the rigid crystalline parts (mesogens) are somewhat disordered (Figure 7.10a). At intermediate water contents (>5%, <40% w/w) there are two endothermic transitions observed in DSC analysis of amylopectin.[81] The first is thought to be due to the rearrangement of dislocations between constituent amylopectin helices leading to a smectic-nematic transition (Figure 7.10b). The second is the helix-coil transition as the amylopectin helices unwind in an irreversible transition. According to their results,[80] the double helices can only unwind if they are dissociated from their crystallites. Thus, even if the temperature required to unwind the helices is lower than that required to dissociate the crystallites, unwinding is prohibited at temperatures below the endothermic transition. It would suggest that the unwinding (fast) appears only as the helices dissociate side-by-side (slow) making the endotherms merge more or less together. This is, however, also dependent on the moisture content.[80] In excess water (>40% w/w) lamellar break up and the helix-coil transition occur at the same point, since free unassociated helices are unstable (Figure 7.10c).

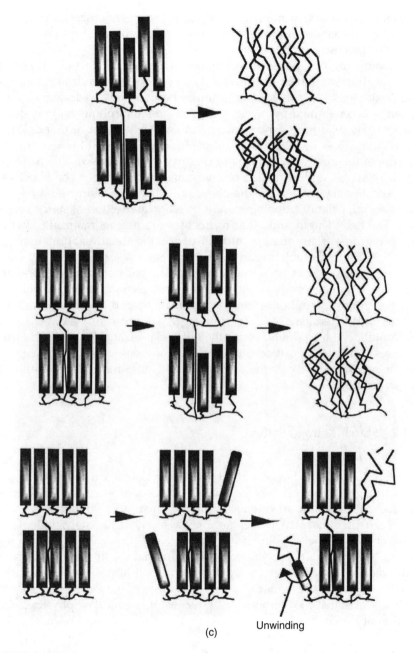

(c)

Unwinding

FIGURE 7.10

Starch gelatinization according to the side-chain liquid crystal approach: (a) the single stage process at low water contents; (b) the two stage process involved in limiting water; (c) the two stage process involved in excess water. (Adapted from Waigh et al., 2000.)[81]

Another view is that the occurrence of biphasic endothermic transitions in DSC measurement of starch is thought to reflect disorganization (melting) and rearrangement of crystallites.[82]

A combination of physical techniques was also used to determine the gelatinization properties of starch/water systems.[65] At high moisture content, water acts as a solvent and granules become fully hydrated. Gelatinization is accompanied by swelling in both the amorphous and crystalline phases, loss of birefringence (loss of molecular orientation), and loss of crystallinity (loss of long range order and double helical structure). The rupture of the inter- and intra-molecular hydrogen bonds and hydrophobic bonds of starch, and formation of intermolecular hydrogen bonds between starch and water accompany the phase transition of gelatinization.[46]

In general, gelatinization temperatures reflect perfection of starch crystallites. The helix length and other forms of molecular reorganization within micro-regions of granules could influence the gelatinization temperatures.[83–85] Thus, the gelatinization characteristics are influenced by starch sources, moisture content, and environmental conditions such as pressure, mechanical damage, presence of small molecular solutes (nonionic and electrolytes), physical modification (e.g., annealing and heat-moisture treatment), and chemical modification (e.g., substitution) as well as by hydrophilic hydrocolloids. Nevertheless, gelatinization is a process that requires further study. It seems to be critically based upon the degree of plasticization, mobility of the amorphous regions, and rupture within the starch granule.

7.4.2 Starch Retrogradation

7.4.2.1 Definition

Starch retrogradation has been used to describe changes in physical behavior following gelatinization. It is the process that occurs when starch molecules reassociate and form an ordered structure such as double helices during storage. In an initial step, two chains may associate. Ultimately, under favorable conditions, a crystalline order appears and physical phase separation occurs.[64] Retrogradation is important in industrial use of starch, as it can be a desired end point in certain applications but it also causes instability in starch pastes. Structural modification, either by genetic means to change the pathway of starch biosynthesis or by means of chemical or physical modification of starch, has been employed to alter the process of retrogradation.

7.4.2.2 Analytical Techniques to Probe Starch Retrogradation

To study starch retrogradation, several modern analytical techniques have been used to understand and control this important phenomenon during food processing and storage. These techniques include differential scanning calorimetry (DSC),[86] x-ray diffraction,[87] nuclear magnetic resonance (NMR),[63] rheological analysis,[87] Fourier transform infrared spectroscopy (FTIR),[69] and Raman spectroscopy and microscopy.[88]

For example, the helices and variably ordered semi-crystalline arrays of these helices in retrograded starch, can been monitored and determined by x-ray diffraction for the crystal pattern and crystallinity, NMR for the double helix content, or differential scanning calorimetry (DSC) to observe endotherm changes when retrograded starch structures are lost on heating.

7.4.2.3   Factors Influencing Starch Retrogradation

Starch retrogradation is influenced by the botanical source (e.g., cereal starch vs. tuber starch) and the fine structure of amylopectin (e.g., chain length and distribution). The amylose:amylopectin ratio affects the kinetics of retrogradation. In nonmutant genotype starches, the amylose is responsible for short term (<1 day) changes.[69] The amylopectin molecule is responsible for longer term rheological and structural changes of starch gels.[89] X-ray diffraction and shear modulus studies on various starches showed that the initial rates of development and stiffness of gels followed the order of smooth pea > maize > wheat > potato, while the long term increase followed the order of smooth pea > potato > maize > wheat.[90] In addition, molecular size and size distribution of starch affect the rate of retrogradation. For example, retrogradation is faster for chain lengths in the DP range of 75 to 100.[91] Water content in the starch gel and storage temperature can affect the rate and extent of starch retrogradation. Certain polar lipids, surfactants, and sugars can retard or reduce the extent of refrigeration.

The retrograded starch shows a B type x-ray diffraction pattern.[45] Because starch retrogradation is a kinetically controlled process,[76] the alteration of time, temperature, and water content during processing can produce a variety of end products. At lower water contents, water acts as a plasticizer which will affect the T_g (glass transition temperature) of a partially crystalline polymer.[92] Therefore, the amount of water will affect the glass transition of starch based foods, hence the properties, processing, and stability of many starch-based food products.[76]

From the study of wheat starch gels,[93,94] maximum crystallization has been shown to occur at 40 to 50% moisture. The retrogradation of the *wx*, *du wx* and *su2 wx* maize starches is strongly influenced by moisture content (in the range of 50 to 80%) and initial heating temperature, whereas retrogradation of *ae wx* maize starch is minimally influenced by either.[95] The difference in retrogradation behaviors from these mutant starches is likely related to the different amylopectin structures.

7.4.2.4   Mechanism of Starch Retrogradation and Avrami Equation

In general, retrogradation takes place in two stages. The first and fastest stage is the formation of crystalline regions from retrograded amylose. The second stage involves the formation of an ordered structure within amylopectin.

During the retrogradation, the molecular interactions (mainly hydrogen bonding between starch chains) occur. These interactions are found to be time and temperature dependent. Amylose is able to form double helical association of 40 to 70 glucose units,[96] whereas amylopectin crystallization occurs by association of the outermost short branches (e.g., DP = 15).[87]

The rate of retrogradation of amylopectin differs from one starch variety to another due to the different lengths of the external chains of amylopectin. Partial beta-amylolysis produced a significant fraction of chains having 2 to 6 glucose units, which hinders the reassociation of the long external chains, therefore reducing the retrogradation of amylopectin.[97]

Retrogradation of starch has been widely studied by fitting an Avrami equation:

$$\varphi = (A_L - A_t)/(A_L - A_0) = \exp(-kt^n)$$

where φ is the fraction of the total change in retrogradation endotherm enthalpy still to occur, and A_0, A_t and A_L are experimental values of retrogradation enthalpy at time zero (0), different time (t), and infinity (complete retrogradation) (L), respectively. k is the rate constant and n is the Avrami exponent.

The Avrami equation has been used to model crystallization of starch gels,[94,98] starch retrogradation, and bread staling. Some authors determined the crystallization rate constant (k) when the modes of nucleation and growth were assumed.[99,100] Others used Avrami plots to determine the Avrami index (n) and gain insight into the mechanism of crystallization.[101,102] The kinetics of crystallization of starch depend on the storage temperature,[98] starch concentration, and initial heating temperatures.

7.4.3   Starch Pasting

7.4.3.1   *Definition*

When starch is cooked, the flow behavior of a granule slurry changes markedly as the suspension becomes a dispersion of swollen granules, partially disintegrated granules, then molecularly dispersed granules. The cooked product is called a starch paste.

Pasting is defined as the state following gelatinization of starch. In general, a starch paste can be described as a two-phase system composed of a dispersed phase of swollen granules and a continuous phase of leached amylose. It can be regarded as a polymer composite in which swollen granules are embedded in and reinforce a continuous matrix of entangled amylose molecules.[103] If the amylose phase is continuous, aggregation with linear segments of amylopectin on cooling will result in the formation of a strong gel.

7.4.3.2   *Rheological Measurement*

Starch rheology is to study the stress-deformation relationships of starch in aqueous systems. The rheological properties of starch are important to both food and industrial processing applications. During processing, starch dispersions will be subjected to combined high heating and shearing that affect their rheological change as well as the final characteristics of the product. Starch gelatinization, especially granular swelling, changes the rheolgical properties of starch. The subsequent retrogradation will further modify the rheolgical properties of starch. Depending on the starch source and concentration, the final structure of starchy products will give a thickened solution or a gelled structure.

Numerous techniques have been used to characterize the rheological properties of starch during the heating process. Most rheological studies were conducted at temperatures lower than 95°C and in a range of shear rates sometimes irrelevant to processing conditions. The dynamic rheometer allows the continuous assessment of dynamic moduli during temperature and frequency sweep testing of the starch suspensions. The storage moduli (G') (elastic response) is a measure of the energy stored in the material. The loss modulus (G'') (viscous response) is a measure of the energy dissipated or lost per cycle of sinusoidal deformation.[104] The ratio of the energy lost to the energy stored for each cycle can be defined by tan δ, which is used to indicate the degree of elasticity of a system. The G', G'', and tan δ are used to evaluate the rheological properties of starch and starch products. The G' values depend on the swelling power of starch.[105]

Rheological properties of starch pasting can also be determined by rapid visco-analyzer (RVA) (see Section 7.4.3.3). Rheology is directly related to the microstructure of starch. The rheological properties of starch are influenced by many factors such as the amylose/amylopectin ratio, minor components, the chain length of amylose and amylopectin molecules, the concentration of starch, shear and strain, and temperature. Rheological properties of starch are also influenced by hydrothermal treatments.[105]

7.4.3.3 Pasting Profiles of Starches from RVA

Rheological properties of starch have been widely investigated using a Brabender Visco Amylograph and Rapid Visco Analyzer (RVA). The RVA provides information on starch characteristics similar to the Brabender Visco Amylograph with additional versatility of testing parameters. The RVA has the advantages of using a small sample size, short testing time, and the ability to modify testing conditions. A typical Rapid Visco™ Analyzer profile from normal maize starch is shown in Figure 7.11. Other commercial starches give different values with a similar profile as normal maize starch. In this RVA profile, native starch granules are generally insoluble in water below 50°C. Thus, the viscosity is low. When starch granules are heated, the granules absorb a large amount of water and swell to many times their original size. The viscosity increases on shearing when these swollen granules have to squeeze past each other. The temperature at the onset of this rise in viscosity is known as the pasting temperature as shown in Figure 7.11. The pasting temperature provides an indication of the minimum temperature required

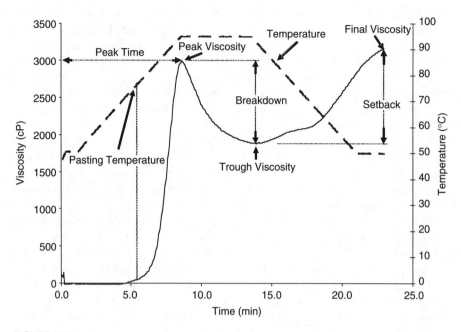

FIGURE 7.11
Typical RVA pasting profile of normal maize starch for viscosity (—) and temperature (---) as a function of time.

to cook a given sample. When a sufficient number of granules become swollen, a rapid increase in viscosity occurs. Granules swell over a range of temperature, indicating their heterogeneity of behavior. The peak viscosity occurs at the equilibrium point between swelling and polymer leaching. Peak viscosity and temperature indicate the water binding capacity of the starch. As the temperature increases further and holds at a high temperature for a period of time, granules rupture and subsequent polymer alignment occurs, which decreases the apparent viscosity of the paste. This process is defined as breakdown. The viscosity at this stage gives an indication of paste stability. It is important to stress that only intact swollen granules can give paste viscosity, and not fragmented granules or solubilized starch substance. As the system is subsequently cooled, re-association between starch molecules, especially amylose, occurs to various degrees. In sufficient concentration this usually causes the formation of a gel, and the viscosity will be increased to a final viscosity. This phase of the pasting curve is commonly referred to as the setback region, and involves retrogradation of the starch molecules. The final viscosity gives an indication of the stability of the cooled, cooked paste under low shear.

The starch pasting properties of some commercial starches by RVA are presented in Table 7.6.[106] Among these commercial starches, potato starch shows the highest peak viscosity and the lowest pasting temperature with moderate final viscosity and lower setback viscosity. Waxy maize starch has

TABLE 7.6

The Pasting Properties of Some Commercial Starches

Starch	Peak Viscosity (cP)	Final Viscosity (cP)	Setback Viscosity (cP)	Pasting Temperature (°C)
Normal maize	2974 ± 17	3157 ± 5	1308 ± 30	81.1 ± 0.5
Waxy maize	2437 ± 6	1053 ± 13	226 ± 7	70.9 ± 0.3
Potato	5523 ± 8	2352 ± 19	480 ± 5	65.6 ± 0.1
Tapioca	2249 ± 13	1437 ± 11	551 ± 6	69.5 ± 0.5
Wheat	2499 ± 25	3272 ± 21	1302 ± 13	85.5 ± 0.5
Rice	2000 ± 13	2380 ± 13	767 ± 16	89.2 ± 0.1

Source: From Liu, Q., 2003.[106]

the lowest final viscosity and setback viscosity with moderate peak viscosity and pasting temperature due to lack of amylose. Rice starch shows the highest pasting temperature and lowest peak viscosity with moderate final viscosity and setback viscosity. Wheat starch has the highest final viscosity and setback viscosity with moderate peak viscosity and higher pasting temperature. Normal maize starch shows the highest setback viscosity and higher final viscosity with moderate peak viscosity and higher pasting temperature.

7.4.3.4 The Characteristics of Starch Pastes

As described earlier, it is believed that viscosity of starch dispersions is strongly influenced by the swelling of starch granules. Starch granules swell radially in the first stage of swelling. Solubilization of amylose starts in the amorphous region of the granules. Water diffuses first into the center of the granules and the degree of leaching of amylose out of the granules is low. If there is lipid in the system, amylose-lipid complexes are believed to restrain swelling and amylose leaching.[107] Once the amylose-lipid complexes dissolve, the rate of amylose leaching out of the granules increases substantially. When the temperature increases, the amylopectin-rich granules swell tangentially. The granule deforms and loses its original shape. The presence of amylose in the continuous phase surrounding the swollen granules will result in the formation of a strong gel on cooling.

However, the swelling of starch granules and the rate of gel formation primarily depends on the variety and source of the starch. This is because amylose content and molecular weight of amylose varies with the starch origin. Different granule structure also results in different degrees of amylose leaching. For example, tuber starches form gels much more slowly than cereal starches at similar conditions.[108] If starch granules tend to hydrate with ease, swell rapidly, and rupture to a great extent, the starch paste loses viscosity relatively easily, producing a weak bodied, stringy, and cohesive paste. For some starches, the paste may remain fluid or form a semi-solid or solid gel demonstrating considerable strength.

Starches have been classified into four types based on their gelatinized paste viscosity profiles.[109] Type I is the group of high swelling starches (e.g., potato, tapioca, waxy cereal), which are characterized by a high peak viscosity followed by rapid thinning during cooking. Type II is moderately swelling starch, which shows a lower peak viscosity, and much less thinning during cooking (e.g., normal cereal starches). Type III consists of restricted swelling starches (e.g., chemically cross-linked starches), which show a relatively less pronounced peak viscosity and exhibit high viscosity that remains constant or increases during cooking. Type IV is highly restricted starch (e.g., high amylose maize starch), which does not swell sufficiently to give a viscous solution.

7.5 Role of Starches in Foods

As a natural component, starch contributes to the characteristic properties of food products made from wheat, rice, maize, potato, and legume seed. It is also widely used as a functional ingredient in many food products such as sauces, puddings, confectionery, comminuted meat, fish products, and low fat dairy products. In order to meet the requirement of some food products, starch is chemically and/or biotechnologically modified as an ingredient.

The role of starch in some foods is well known. In a food system, the roles of starch are to stabilize the structure and interact with other components to deliver or maintain nutrient and flavor. For example, starch can serve as thickening agents for sauces, cream soups, and pie fillings; colloidal stabilizers for salad dressing; moisture retention for cake toppings; gel-forming agents for gum confections; binders for wafers and ice cream cones; and coating and glazing agents for nut meats and candies. An increased understanding of starch functionality and how it is affected by other ingredients and methods could lead to improvements in the quality of food products. The properties of foods are not governed simply by the starch and its molecular structure. Processing conditions and the amount of water available are crucial determinants of the final characteristics of the product.

Starch molecules differ from other carbohydrate hydrocolloids because they are made functionally useful only by altering the granule package. When starch is heated in the presence of water or other components during processing, swelling of the granule, loss of crystallinity, and a largely irreversible uptake of water are the results. In general, gelatinized starch begins to change (retrograde) when the system is cooled. Thus, starch retrogradation must be considered as a potential destabilizing influence on the physicochemical properties generated by starch gelatinization for the development of starch related food products. Based on the requirement of food products such as clarity and storage stability, starch from different sources with different physicochemical properties can be selected to meet specific needs. For example, maize starch retrogrades and becomes increasingly opaque much more rapidly at low temperature (e.g., 4°C) than potato starch. In this case, potato starch would provide greater clarity to the food product.

7.5.1 Starchy Food Products

7.5.1.1 Baked Products

Starch is the main component in many baked products such as bread. The contribution of starch is related to its three important properties: water absorption, gelatinization, and retrogradation. The rheological behavior of wheat starch dispersions over a period of 30 days is dominated by the changes in the amylose fraction.[110] The swollen granules and partially solubilized starch act as essential structural elements of bread.[111] Protein is responsible for the formation of the viscoelastic gluten in dough. Upon heating, the gluten transforms from a gel to a coagel by polymerization (the gel loses its cohesiveness).[112] The transformation from dough to bread involves changes both in the starch and the protein fraction. Baking includes the solidification of dough and a change from a foam type system with gas cells to an open pore system.[112] On cooling and aging of bread, rearrangements in the starch fraction lead to a series of changes including gelation and crystallization. This transformation is thought to be the major cause of bread crumb firming on aging of bread.[113] The branched molecules within the swollen granules undergo a slow association to rigidify the swollen granules. It is possible that the intra-granular amylose fraction enhances the rigidity of starch granules on bread staling.[114] The transformation also involves amylopectin retrogradation. The association of branched molecules is relatively weak, and is readily reversed merely by heating to 50°C. Heating restores the elastic gel structure of the original fresh bread.[38] However, bread staling is a complicated process that involves a number of physicochemical changes. Although retrogradation is mainly considered as a problem for deterioration of product quality, controlling starch retrogradation would open new possibilities to optimize the nutritional properties of starchy food.

7.5.1.2 Snack Foods

Many snack foods are high in starch. Starches and starch derivatives have a long history of use in snack foods to help achieve various textural attributes. One stage in the production of snacks is extrusion cooking, in which a slurry of starch-rich cereal or potato is subjected to conditions of high temperature, high pressure, high moisture, and high shear forces. Some highly processed snacks are double extruded, gelatinized, expanded, and fried.

Specialty starches can provide a number of functional benefits in making snack products, including different expansion, crispness, oil pickup, and overall eating quality. In expanded or puffed snacks, the target texture can be obtained by selecting combinations of high amylose and high amylopectin starch or specific starches with suitable amylose to amylopectin ratios. Normal maize and high amylose maize starches are used when increased crunchiness and strength are required for the snacks. Waxy maize starches can be used to increase the expansion of expanded or puffed snacks and increase the crispness of snacks because the starch is composed of almost 99% amylopectin. Due to the difficulty in rolling into thin sheets of uniform thickness, potato starch is not suitable for such applications since it has the greatest swelling capacity.[115]

Native starches, especially the waxy type, generally are not resistant to the high temperature and shear processing conditions experienced in producing extruded snacks, as we can see in rheological properties of starch. Thus, modified starches, such as cross-linked starch, are selected to provide resistance to shear and excessive heat during processing. The action of cross-linking lowers the swelling capacity of the starch, resulting in a snack that has reduced expansion with nonuniform texture.

Snacks with enhanced resistant starch can be produced through effective processing or by selecting resistant starch sources. It provides the opportunity for snack manufacturers to develop high-quality fiber-fortified snacks.

7.5.2   Resistant Starch

7.5.2.1   Definition

As mentioned earlier, resistant starch is impervious to the effects of digestive enzymes, is not digested in the small intestine, and undergoes some degree of fermentation in the large intestine, which allows it to function likes dietary fiber.

Resistant starch has been categorized into four main types: type I is physically entrapped starch within whole plant cells and food material (e.g., in partly milled grains and seeds). The presence of intact cell walls contributes to the resistant starch content of legumes. More extensive milling and chewing can make these starches more accessible and less resistant. Type II resistant starch includes those granules from certain plant sources (e.g., the native crystalline starch granules in raw potatoes, green bananas, and high amylose corns). Type III resistant starch comprises retrograded starch. Type IV resistant starch includes chemically modified starches that are used by food manufacturers to improve the functional characteristics of the starch. Although these resistant starches are found very widely in processed foods, their physiological effects may not be the same.

7.5.2.2   Characteristics

Resistant starch has a low calorie profile and can be used as a bulking agent in reduced sugar or reduced fat food formulations. It holds significantly less

TABLE 7.7

The Resistant Starch Content of Some
Common Foods

Food	% of total resistant starch
Bread	2.2–4.3
Special breads	7.2–9.5
Breakfast cereals	0.0–9.0
Wholegrain cereals	1.7–12.3
Potato	0.6–9.0
Pasta	1.3–4.2
Legume seeds	12–20

Source: From Würsch, 1999.[116]

water than traditional dietary fiber. Resistant starch does not compete for the water needed by other ingredients and allows for easier processing because it does not contribute to stickiness. This may be advantageous in the production of low moisture products such as cookies and crackers. Products ranging from bread, crackers to muffins have better taste, mouth feel, and appearance when resistant starch is used to replace traditional fiber sources. In most applications, it does not alter the taste, texture, or appearance of the food. Table 7.7[116] presents the resistant starch content of some common foods.

7.5.2.3   Sources and Formation

The major commercial source of resistant starch is high amylose starch. High amylose maize and legume starch granules (type II) have unique properties imparting resistance to digestive enzymes. Resistance is probably related to the crystalline order or packing of the glucan chains of amylose and amylopectin. Resistant starch type III can be formed in starch gels, flours, dough, and baked goods and mostly consists of retrograded amylose, and recrystallized amylose fragments released during starch degradation. Resistant starch content can be increased by storing the wet cooked food in the cold, by heating and cooling, or by freezing. Storage at low temperatures results in starch retrogradation and formation of resistant starch fractions. The highest yield of resistant starch is obtained by controlled retrogradation of high amylose starch, debranched amylopectin, and amylose after autoclaving, or by chemical modification. Figure 7.12 presents the major steps for the production of resistant starch type III from normal maize starch. After starch is debranched by enzymes, more linear fractions are produced. Retrogradation is favored by the presence of linear fraction. It increases the amount of ordered structure that resists enzyme hydrolysis. The amount of resistant starch type III is influenced by the starch source, the dose of debranching enzyme, heating and cooling cycle times, and storage conditions.

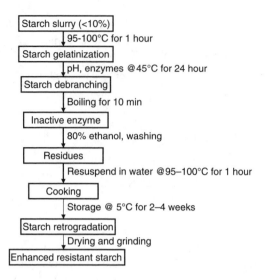

FIGURE 7.12
Production of resistant starch type III.

7.5.2.4 Measurement

Several procedures for determination of resistant starch have been developed since resistant starch was first recognized by Englyst et al. in 1982.[117] Resistant starch is measured chemically as the starch not hydrolyzed after 2 hours of incubation at 37°C with pancreatin containing α-amylase plus proteolytic and lipolytic enzymes, amyloglucosidase, and invertase.[117] Another procedure was developed for the measurement of resistant starch incorporating the α-amylase/pullulanase treatment and omitting the initial heating step at 100°C,[118] so as to more closely mimic physiological conditions. Under these conditions, the measured resistant starch contents of samples were much higher.

More recently, the methods for measuring resistant starch have been modified to provide a reliable method (as much as feasible) to reflect *in vivo* conditions. These modifications included changes in enzyme concentrations employed, types of enzymes used (all used pancreatic α-amylase, but pullulanase was removed, and in some cases replaced by amyloglucosidase), sample pre-treatment (chewing), pH of incubation and the addition (or not) of ethanol after the α-amylase incubation step, and shaking and stirring during incubation. All of these modifications will have some effect on the determined level of RS in a sample. For more detailed information about resistant starch measurement, refer to AACC method 32-40 and AOAC method 2002.02.

7.5.2.5 Physiological Effects

Resistant starch is slowly fermented in the large intestine by the colonic flora. The metabolites formed include volatile fatty acids such as acetic, propionic, and butyric acids. Resistant starch seems to yield more butyrate than some highly fermentable soluble fibers. Thus, resistant starches have been shown to have equivalent and/or superior impacts on human health similar to the conventional fiber enriched food ingredients. These impacts include the following:

- Decreasing dietary caloric values for body fat deposition, important for the prevention of obesity
- Lowering glycemic index, important for diabetes
- Reducing blood cholesterol levels to prevent and control cardiovascular diseases
- Decreasing the risk of colon cancer through enhancing short chain fatty acid production, especially butyrate

Resistant starch appears to have a unique combination of physiological and functional properties compared to traditional types of fiber. It can be used as a functional food ingredient for making breads, buns, crumpets, muffins, cakes, biscuits, breakfast cereals, snacks, confectionery, pasta, drinks, yogurts, and ice cream. These foods with improved nutritional profiles (high level of resistant starch) have met consumer demands for appearance, taste, and texture. Their use will in turn improve the health status of consumers.

7.5.3   Starch and Health

7.5.3.1   *Glycemic Index of Starchy Foods*

The nutritional properties of starch in foods are to a large extent related to its availability for digestion and/or absorption in the gastrointestinal tract. A number of food properties have been identified as being of prime importance for the glycemic response, and starch foods cover the entire range from slow to rapid glucose release and thus, insulin response. The nutritional properties of starch in foods vary considerably, and may be of importance for the treatment, as well as the prevention, of several diseases commonly found in our societies, such as diabetes and cardiovascular disease. Starch in cooked potatoes, most types of bread, and breakfast cereals is rapidly digested and absorbed. In contrast, starch in legumes, pasta, and certain rice or cereal products is slowly digested and absorbed. These starchy foods show significant differences in their glycemic response and are defined as high and low glycemic index (GI) foods. The glycemic index is defined as the incremental blood glucose area after ingestion of the test product as a percentage of the corresponding area following an equicarbohydrate load of a reference product (pure glucose or white bread).[119] The suggested advantageous effects of low GI starchy foods include:[120]

- Improved metabolic control in individuals with diabetes
- Reduced blood lipid levels in individuals with hyperlipidemia and in healthy individuals
- Improved glucose tolerance
- Protection against cardiovascular disease
- Protection against diabetes in genetically susceptible populations
- Prolonged satiety
- Prolonged physical performance during endurance exercise
- Reduced cariogenic potential

7.5.3.2   Digestion Process

The digestibility of starch affects the nutritional (caloric) value of starch present in grain or tubers for food uses. Digestion of starch consists of breaking up the glycosidic bonds by glycosidases in order to liberate glucose. When food that has been thoroughly chewed reaches the stomach, acidity inactivates the salivary α-amylase, but by then the large starch molecules have been reduced from several thousand to a few glucose units. Little hydrolysis of the carbohydrate occurs in the stomach. As the stomach empties, the hydrochloric acid in the material entering the small intestine is neutralized by secretions from the pancreatic ducts, bile, and pancreatic juices. The digestion of starch dextrins is continued by the action of pancreatic α-amylase. The products of digestion by α-amylase are mainly maltose and maltotriose, and α-limit dextrins containing about eight glucose units with one or more α-1→6 glucosidic bonds. Finally hydrolysis of di- and oligosaccharides is carried out by surface enzymes of the small intestinal epithelial cells. Di-, oligo-, and polysaccharides that are not hydrolyzed by α-amylase and/or intestinal surface enzymes cannot be absorbed, and they reach the lower tract of the intestine, which from the lower ileum onward, contain bacteria.

7.5.3.3   Factors Influencing Starch Digestion

Modern food processing techniques, especially milling, may damage the plant cell structure and starch may be completely gelatinized. This could lead to rapid digestion of starch in the small intestine. The starch in wheat flour is 48% rapidly digestible starch and 49% slowly digestible as measured *in vitro*. The ratio is changed from 1:1 in the flour to 10:1 in bread, where the values are 90% rapidly digestible starch and 10% slowly digestible starch because the starch granules are gelatinized during the bread making process.[121]

Starches with B and C type x-ray diffraction patterns, such as potato starch and banana starch, suffer poor enzyme digestibility compared with A type starch, such as maize, wheat, and rice starches. Most enzymes hydrolyze

TABLE 7.8

Nutritional Classification of Starch

Class	Example of Occurrence	Site of Digestion and Absorption	Glycemic Response
Rapidly digestible starch	Processed foods	Small intestine	Large
Slowly digestible starch	Legume, Pasta	Small intestine	Small
Resistant starch			
Physically inaccessible	Whole grains	Large intestine	None
Resistant granules	Unripe banana	Large intestine	None
Retrograded starch	Processed foods	Large intestine	None

Source: From Englyst and Hudson, 1997.[121]

maize, wheat, rice, and other A type starches by boring holes into the granule and then hydrolyzing the starch from the inside out, but they attack potato and other B type starches and banana starch (C type) mainly by hydrolyzing the lamella on the surface of the starch granule. It has long been known that the physicochemical form of starch affects both the rate and the extent of its hydrolysis by amylolytic enzymes. Thus, there are corresponding differences in the digestibility of starch in foods. A relatively slow rate of starch hydrolysis is a characteristic of food which provokes a low glycemic response in human subjects. Furthermore, starch which escapes digestion in the small intestine is likely to have physiological effects similar to some of the components of dietary fiber. It is possible to manipulate the digestibility of starch in food by the use of appropriate processing techniques, such as preservation of the cell wall structure of plant foods, or introduction of a secondary structure that will hinder access of amylase.

Table 7.8[121] presents the nutritional classification of starch. Rapidly digestible starch is starch that is rapidly and completely digested in the small intestine; slowly digestible starch is completely but more slowly digested in the small intestine; resistant starch is defined as the sum of starch and starch degradation products that, on average, reach the human large intestine.

The digestibility of starches in cereal products is very much dependent on the intactness of tissue structures, the degree of swelling of starch granules, and the amount of retrograded amylose outside the starch granules.

References

1. Ellis, R.P., Cochrane, M.P., Dale, M.F.B., Duffus, C.M., Lunn, A., Morrison, I.M., Prentice, R.D.M., Swanston, J.S., and Tiller, S.A.: Starch production and industrial use. *Journal of Agricultural and Food Chemistry* 77, 289–311, 1998.
2. Galliard, T.: Starch availability and utilization, In: *Starch: Properties and Potential.* Ed. T. Galliard. John Wiley & Sons, Chichester, U.K., 1987, p. 4.
3. Hedley, C.L.: Introduction, In: *Carbohydrate in Grain Legume Seeds.* Ed. C.L. Hedley. CABI Publishing, Oxon, U.K., 2000, p. 12.

4. Shannon, J. and Garwood, D.L.: Genetics and physiology of starch development, In: *Starch: Chemistry and Technology.* Eds. R.L. Whistler, J.N. BeMiller, and E.F. Paschall. Academic Press, Inc., New York, 1984, p. 25.
5. Keeling, P.L.: Plant biotechnology: Technical barriers to starch improvement, In: *Starch: Structure and Functionality.* Eds. P.J. Frazier, A.M. Donald, and P. Richmond. The Royal Society of Chemistry, Cambridge, U.K., 1997.
6. Hoover, R. and Sosulski, F.W.: Studies on the functional characteristics and digestibility of starches from *Phaselous vulgaris* biotypes. *Starch* 37, 181–191, 1985.
7. Reichert, R.D. and Youngs, C.G.: Nature of the residual protein associated with starch fractions from air-classified field peas. *Cereal Chemistry* 55, 469–480, 1978.
8. Hizukuri, S., Takeda, Y., and Yasuda, M.: Multi-branched nature of amylose and the action of debranching enzymes. *Carbohydrate Research* 94, 205–213, 1981.
9. Morrison, W.R. and Karkalas, J.: Starch, In: *Methods in Plant Biochemistry.* Academic Press, Inc., New York, 1990, p. 323.
10. Teitelbaum, R.C., Ruby, S.L., and Marks, T.J.: On the structure of starch-iodine. *Journal of the American Chemical Society* 100, 3215–3217, 1978.
11. Teitelbaum, R.C., Ruby, S.L., and Marks, T.J.: A resonance Raman/Iodine Mossbauer investigation of starch-iodine structure. Aqueous solution and iodine vapor preparations. *Journal of the American Chemical Society* 102, 3322–3328, 1980.
12. John, M., Schmidt, J., and Kneifel, H.: Iodine-maltosaccharide complexes: relation between chain-length and color. *Carbohydrate Research* 119, 254–257, 1983.
13. Eliasson, A.-C.: Interactions between starch and lipids studied by DSC. *Thermochimica Acta* 246, 343–356, 1994.
14. Biliaderis, C.G., Page, C.M., Slade, L., and Sirett, R.R.: Thermal behavior of amylose-lipid complexes. *Carbohydrate Polymers* 5, 367, 1985.
15. Kowblansky, M.: Calorimetric investigation of inclusion complexes of amylose with long-chain aliphatic compounds containing different functional groups. *Macromolecules* 18, 1776, 1985.
16. Banks, W. and Greenwood, C.T.: *Starch and Its Components.* Edinburgh University Press, Edinburgh, 1975.
17. Hizukuri, S.: Starch: analytical aspects, In: *Carbohydrates in Food.* Ed. A.-C. Eliasson. Marcel Dekker Inc., New York, 1996, p. 147.
18. Hoover, R. and Sosulski, F.W.: Composition, structure, functionality, and chemical modification of legume starches: a review. *Canadian Journal of Physiology & Pharmacology* 69, 79–92, 1991.
19. Kugimiya, M. and Donovan, J.W.: Calorimetric determination of the amylose-lysolecithin complex. *Journal of Food Science* 46, 765, 1981.
20. Colonna, P. and Mercier, C.: Macromolecular structure of wrinkled- and smooth-pea starch components. *Carbohydrate Research* 126, 233, 1984.
21. Gibson, T.S., Solah, V., and McCleary, B.V.: A procedure to measure amylose in cereal starches and flours with concanavalin A. *Journal of Cereal Science* 25, 111–119, 1997.
22. Takeda, Y., Hizukuri, S., Takeda, C., and Suzuki, A.: Structures of branched molecules of amyloses of various origins, and molar fractions of branched and unbranched molecules. *Carbohydrate Research* 165, 139–145, 1987.
23. Thompson, D.B.: On the non-random nature of amylopectin branching. *Carbohydrate Polymers* 43, 223–239, 2000.
24. Imberty, A., Buléon, A., Tran, V., and Pérez, S.: Recent advances in knowledge of starch structure. *Starch* 43, 375–384, 1991.

25. Hizukuri, S.: Polymodal distribution of the chain length of amylopectins, and its significance. *Carbohydrate Research* 147, 342–347, 1986.
26. French, A.D.: Fine structure of starch and its relationship to the organization of the granules. *Journal of Japanese Society Starch Science* 19, 8–25, 1972.
27. Robin, J.P., Mercier, C., Charbonniere, R., and Guilbot, A.: Lintnerized starches. Gel filtration and enzymatic studies of insoluble residues from prolonged acid treatment of potato starch. *Cereal Chemistry* 51, 389–406, 1974.
28. Nikuni, Z.: Granule and amylopectin structure. *Starch* 30, 105–111, 1978.
29. Hizukuri, S. , Tabata, S., and Nikuni, Z.: Studies on starch phosphate. Part 1. Estimation of glucose-6-phosphate residues in starch and the presence of other bound phosphates. *Starch* 22, 338–343, 1970.
30. Wong, K.S. and Jane, J.-L. Quantitative analysis of debranched amylopectin by HPAEC-PAD with a postcolumn enzyme reactor. *Journal Liquid Chromatography* 20, 297–310, 1997.
31. Yu, L.-P. and Rollangs, J.E.: Low-angle laser light scattering-aqueous size exclusion chromatography of polysaccharides: molecular weight distribution and polymer branching determination. *Journal of Applied Polymer Science* 33, 1909–1921, 1987.
32. Guilbot, A. and Mercier, C.: Starch, In: *Polysaccharide*. Ed. G.O. Aspinall. Academic Press, Inc., New York, 1985, p. 209.
33. Bertoft, E.: Studies on structure of pea starches, Part 4, Intermediate material of wrinkled pea starch. *Starch* 45, 215–220, 1993.
34. Paton, D.: Oat starch: Some recent developments. *Starch* 31, 184–187, 1979.
35. Galliard, T. and Bowler, P.: Morphology and composition of starch, In: *Starch: Properties and Potential*. Ed. T. Galliard. John Wiley & Sons, New York, 1987, p. 55.
36. Leszczynski, W.: Potato starch processing, In: *Potato Science and Technology*. Eds. G. Lisinska and W. Leszczynski. Elsevier Applied Science, New York, 1989, p. 281.
37. Bay-Smidt, A.M., Wischmann, B., Olsen, C.E., and Nielsen, T.H.: Starch bound phosphate in potato as studied by a simple method from determination of organic phosphate and ^{31}P-NMR. *Starch* 46, 167–172, 1994.
38. Pomeranz, Y.: *Functional Properties of Food Components*. Academic Press, Inc., Orlando, FL., 1985.
39. Thompson, D.B.: Structure and functionality of carbohydrate hydrocolloids in food systems, In: *Development in Carbohydrate Chemistry*. Eds. R.J. Alexander and H.F. Zobel. The American Association of Cereal Chemists, St. Paul, MN, 1992, p. 315.
40. Blanshard, J.M.V.: Starch granule structure and function: a physicochemical approach, In: *Starch: Properties and Potential*. Ed. T. Galliard. John Wiley & Sons, Chichester, U.K., 1987, pp. 17–54.
41. Hoover, R.: Composition, molecular structure, and physicochemical properties of tuber and root starches: a review. *Carbohydrate Polymers* 45, 253–267, 2001.
42. Fanon, J.E., Hauber, R.J., and BeMiller, J.N.: Surface pores of starch granules. *Cereal Chemistry* 69, 284–288, 1992.
43. Oostergetel, G.T. and van Bruggen, F.J.: The crystalline domains in potato starch granules are arranged in a helical fashion. *Carbohydrate Polymers* 21, 7–12, 1993.
44. Oates, C.G.: Towards an understanding of starch granule structure and hydrolysis. *Trends in Food Science and Technology* 8, 375–382, 1997.
45. Zobel, H.F.: Starch crystal transformations and their industrial importance. *Starch* 40, 1–7, 1988.

46. Liu, Q. *Characterization of Physico-Chemical Properties of Starch from Various Potatoes and Other Sources.* University Laval, Quebec City, Canada, 1997.
47. Wu, H.C.H. and Sarko, A.: The double-helical molecular structure of crystalline B-amylose. *Carbohydrate Research* 61, 7–26, 1978.
48. Wu, H.C.H. and Sarko, A.: The double-helical molecular structure of crystalline A-amylose. *Carbohydrate Research* 61, 27–40, 1978.
49. Gidley, M.J.: Factors affecting the crystalline type (A-C) of native starches and model compounds: a rationalisation of observed effects in terms of polymorphic structures. *Carbohydrate Polymers* 161, 301–304, 1987.
50. Gallant, D.J., Bouchet, B., and Baldwin, P.M.: Microscopy of starch: evidence of a new level of granule organization. *Carbohydrate Polymers* 32, 177–191, 1997.
51. Waigh, T.A., Perry, P., Riekel, C., Gidley, M.J., and Donald, A.M.: Communications to the editor: chiral side-chain liquid-crystalline polymeric properties of starch. *Macromolecules* 31, 7980–7984, 1998.
52. Imberty, A., Chanzy, H., Pérez, S., Buléon, A., and Tran, V.: The double-helical nature of the crystalline part of A-starch. *Journal of Molecular Biology* 201, 365–378, 1988.
53. Imberty, A. and Pérez, S.: A revisit to the three-dimensional structure of B-type starch. *Biopolymer* 27, 1205–1221, 1988.
54. Cleven, B., Van den Berg, C., and Van den Plas, L.: Crystal structure of hydrated potato starch. *Starch* 30, 223–238, 1978.
55. Sair, L.: Heat-moisture treatment, In: *Methods in Carbohydrate Chemistry.* Ed. R.C. Whistler. Academic Press, New York, 1964, pp. 283–285.
56. Bogracheva, T.Y., Morris, V.J., Ring, S.G., and Hedley, C.L.: The granular structure of C-type pea starch and its role in gelatinization. *Biopolymers* 45, 323–332, 1998.
57. Zobel, H.F.: Starch granule structure, In: *Developments in Carbohydrate Chemistry.* Eds. R. Alexander and H.F. Zobel. The American Association of Cereal Chemistry, St. Paul, MN, 1992, pp. 1–36.
58. Gidley, M.J. and Bociek, S.M.: ^{13}C CP/MAS NMR studies of amylose inclusion complexes, cyclodextrins, and the amorphous phase of starch granules: relationship between glycosidic linkage conformation and solid-state ^{13}C chemical shifts. *Journal of the American Chemical Society* 110, 3821–3829, 1988.
59. Zobel, H.F.: Molecules to granules: A comprehensive starch review. *Starch* 40, 44–50, 1988.
60. Tang, H., Ando, H., Watanabe, K., Takeda, Y., and Mitsunaga, T.: Physicochemical properties and structure of large, medium and small granule starches in fractions of normal barley endosperm. *Carbohydrate Research* 330, 241–248, 2001.
61. Shi, Y., Capitani, T., Trzasko, P., and Jeffcoat, R.: Molecular structure of a low-amylopectin starch and other high-amylose maize starches. *Journal of Cereal Science* 27, 289–299, 1998.
62. Paris, M., Bizot, H., Emery, J., Buzaré, J.Y., and Buléon, A.: Crystallinity and structuring role of water in native and recrystallized starches by ^{13}C CP-MAS NMR spectroscopy: 1: Spectral decomposition. *Carbohydrate Polymers* 39, 327–339.
63. Gidley, M.J. and Bociek, S.M.: Molecular organization in starches: A ^{13}C CP/MAS NMR study. *Journal of the American Chemical Society* 107, 7040–7044, 1985.
64. Atwell, W.A., Hood, L.F., Lineback, D.R., Varriano-Marston, E., and Zobel, H.F.: The terminology and methodology associated with basic starch phenomena. *Cereal Foods World* 33, 306–311, 1988.

65. Liu, Q., Charlet, G., Yelle, S., and Arul, J.: A study of phase transition in the starch-water system I. Starch gelatinization at high moisture level. *Food Research International* 35, 397–407, 2002.

66. Donovan, J.W.: Phase transitions of the starch-water system. *Biopolymers* 18, 263–275, 1979.

67. Nara, S., Mori, A., and Komiya, T.: Study on the relative crystallinity of moist potato starch. *Starch* 30, 111–114, 1978.

68. Cooke, D. and Gidley, M.J.: Loss of crystalline and molecular order during starch gelatinisation: origin of the enthalpic transition. *Carbohydrate Research* 227, 103–112, 1992.

69. Goodfellow, B.J. and Wilson, R.H.: A Fourier transform IR study of the gelation of amylose and amylopectin. *Biopolymers* 30, 1183–1189, 1990.

70. Jenkins, P.J. and Donald, A.M.: Gelatinisation of starch: a combined SAXS/WAXS/DSC and SANS study. *Carbohydrate Research* 308, 133–147, 1998.

71. Flory, P.J.: *Principles of Polymer Chemistry.* Cornell Univ. Press, Ithaca, NY, 1953.

72. Biliaderis, C.M., Maurice, T.J., and Vose, J.R.: Starch gelatinization phenomena studied by differential scanning calorimetry. *Journal of Food Science* 45, 1669–1674, 1980.

73. Liu, H. and Lelièvre, J.: A differential scanning calorimetry study of melting transitions in aqueous suspensions containing blends of wheat and rice starch. *Carbohydrate Polymers* 17, 145–149, 1992.

74. Colonna, P. and Mercier, C.: Gelatinization and melting of maize and pea starches with normal and high-amylose genotypes. *Phytochemistry* 24, 1667–1674, 1985.

75. Biliaderis, C.G., Page, C.M., Maurice, T.J., and Juliano, B.O.: Thermal characterization of rice starches: A polymeric approach to phase transitions of granular starch. *Journal of Agricultural Food and Chemistry* 34, 6–14, 1986.

76. Slade, L. and Levine, H.: Recent advances in starch retrogradation, In: *Industrial Polysaccharides.* Eds. S.S. Stilva, V. Crescenzi, and I.C.M. Dea. Gordon and Breach Sci., New York, 1987, pp. 387–430.

77. Lelièvre, J.: Starch gelatinization. *Journal of Applied Polymer Science* 18, 293–296, 1973.

78. Evans, I.D. and Haisman, D.R.: The effect of solutes on the gelatinization temperature range of potato starch. *Starch* 34, 224–231, 1982.

79. Zeleznak, K.J. and Hoseney, R.C.: The glass transition in starch. *Cereal Chemistry* 64, 121–124, 1987.

80. Waigh, T.A., Gidley, M.J., Komanshek, B.U., and Donald, A.M.: The phase transformations in starch during gelatinisation: a liquid crystalline approach. *Carbohydrate Research* 328, 165–176, 2000.

81. Waigh, T.A., Kato, K.L., Donald, A.M., Gidley, M.J., Clarke, C.J., and Riekel, C.: Side-chain liquid-crystalline model for starch. *Starch* 52, 450–460, 2000.

82. Biliaderis, C.G.: The structure and interactions of starch with food constituents. *Canadian Journal of Physiological Pharmacology* 69, 60–78, 1991.

83. Tester, R.F. , Debon, S.J.J., Davies, H.V., and Gidley, M.J.: Effect of temperature on the synthesis, composition and physical properties of potato starch. *Journal of the Science of Food and Agriculture* 79, 2045–2051, 1999.

84. Moates, G.K., Noel, T.R., Parker, R., and Ring, S.G.: The effect of chain length and solvent interactions on the dissolution of the B-type crystalline polymorph of amylose in water. *Carbohydrate Research* 298, 327–333, 1997.

85. Safford, R., Jobling, S.A., Sidebottom, C.M., Westcott, R.J., Cooke, D., Tober, K.J., Strongitharm, B.H., Russell, A.L., and Gidley, M.J.: Consequences of antisense RNA inhibition of starch branching enzyme activity on properties of potato starch. *Carbohydrate Polymers* 35, 155–168, 1998.

86. Fisher, D.K. and Thompson, D.B.: Retrogradation of maize starch after thermal treatment within and above the gelatinization temperature range. *Cereal Chemistry* 74, 344–351, 1997.

87. Ring, S.G., Colonna, P., I'anson, K., Kalichevsky, M.T., Miles, M.J., Morris, V.J., and Orford, P.D.: The gelation and crystallization of amylopectin. *Carbohydrate Research* 162, 277–293, 1987.

88. Bulkin, B.J. and Kwak, Y.: Retrogradation kinetics of waxy-corn and potato starches: a rapid, raman-spectroscopic study. *Carbohydrate Research* 160, 95–112, 1987.

89. Gudmundsson, M.: Retrogradation of starch and the role of its components. *Thermochimica Acta* 246, 329–341, 1994.

90. Orford, P.D., Ring, S.G., Carroll, V., Miles, M.J., and Morris, V.J.: The effect of concentration and botanical source on the gelation and retrogradation of starch. *Journal of the Science of Food and Agriculture* 39, 169–173, 1987.

91. Zhang, W. and Jackson, D.S.: Retrogradation behavior of wheat starch gels with differing molecular profiles. *Journal of Food Science* 57, 1428–1432, 1992.

92. Bizot, H., Le Bail, P., Leroux, B., Davy J., Roger, P., and Buleon, A.: Calorimetric evaluation of the glass transition in hydrated, linear and branched polyanhydroglucose compounds. *Carbohydrate Polymers* 32, 33–50, 1997.

93. Zeleznak, K.J. and Hoseney, R.C.: The role of water in the retrogradation of wheat starch gels and bread crumb. *Cereal Chemistry* 63, 407–411, 1986.

94. Longton, J. and LeGrys, G.A.: Differential scanning calorimetry studies on the crystallinity of ageing wheat starch gels. *Starch* 33, 410–414, 1981.

95. Liu, Q. and Thompson, D.B.: Effects of moisture content with different initial heating temperature on retrogradation from different maize starches. *Carbohydrate Research* 314, 221–235, 1998.

96. Jane, J. and Robyt, J.F.: Structure studies of amylose-v complexes and retrograded amylose by action of alpha amylases, and a new method for preparing amylodextrins. *Carbohydrate Research* 132, 105–118, 1984.

97. Würsch, P. and Gumy, D.: Inhibition of amylopectin retrogradation by partial beta-amylolysis. *Carbohydrate Research* 256, 129–137, 1994.

98. Marsh, R.D.L. and Blanshard, J.M.V.: The application of polymer crystal growth theory to the kinetics of formation of the B-amylose polymorph in a 50% wheat-starch gel. *Carbohydrate Polymers* 9, 301–317, 1988.

99. Russell, P.L.: A kinetic study of bread staling by differential scanning calorimetry. *Starch* 35, 277–281, 1983.

100. van Soest, J.J.G., de Wit, D., Tournois, H., and Vliegenthart, J.F.G.: Retrogradation of potato starch as studied by Fourier transform infrared spectroscopy. *Starch* 46, 453–457, 1994.

101. Wong, R.B.K. and Lelièvre, J.: Effects of storage on dynamic rheological properties of wheat starch pastes. *Starch* 34, 231–233, 1982.

102. Inouchi, N., Glover, D.V., Sugimoto, Y., and Fuwa, H.: DSC characteristics of retrograded starches of single-, double- and triple- mutants and their normal counterpart in the inbred OH 43 maize (*Zea mays* L.) background. *Starch* 43, 473–477, 1991.

103. Ring, S.G.: Some studies on starch gelation. *Starch* 37, 80–83, 1985.

104. Ferry, J.D.: *Viscoelastic properties of polymers.* Wiley, New York, 1980.
105. Eerlingen, R.C., Jacobs, H., Block, K., and Delcour, J.A.: Effects of hydrothermal treatments on the rheological properties of potato starch. *Carbohydrate Research* 297, 347–356, 1997.
106. Liu, Q.: Unpublished data. 2003.
107. Conde-Petit, B., Pfirter, A., and Escher, F.: Influence of xanthan on the rheological properties of aqueous starch-emulsifier systems. *Food Hydrocolloids* 11, 393–399, 1997.
108. Svegmark, K., Kidman, S., and Hermansson, A.-M.: Molecular structures obtained from mixed amylose and potato starch dispersions and their rheological behaviour. *Carbohydrate Polymer* 22, 19–29, 1993.
109. Schoch, T.J. and Maywald, E.C.: Preparation and properties of various legume starches. *Cereal Chemistry* 45, 564–573, 1968.
110. Conde-Petit, B. *Interaktionen von Starch mit Emulgatoren in Wasserhaltigen Lebensmittel-Modellen.* Swiss Federal Institute of Technology, Zurich, 1992.
111. Keetels, C.J.A.M., Visser, K.A., van Vliet, T., Jurgens, A., and Walstra, P.: Structure and mechanics of starch bread. *Journal of Cereal Science* 24, 15–26, 1996.
112. Eliasson, A.C. and Larsson, K.: *Cereals in Breadmaking: A Molecular Colloidal Approach.* Marcel Dekker Inc., New York, 1993.
113. Zobel, H.F. and Kulp, K.: The staling mechanism, In: *Baked Goods Freshness: Technology, Evaluation and Inhibition of Staling.* Eds. R.E. Hebeda and H.F. Zobel. Marcel Dekker, Inc. New York, 1996, pp. 1–64.
114. Hug-Iten, S., Handschin, S., Conde-Petit, B., and Escher, F.: Changes in starch microstructure on baking and staling of wheat bread. *Lebensmittel-Wissenschaft und Technologie* 32, 255–260, 1999.
115. Huang , D.P. and Rooney, L.W.: Starch for snack foods, In: *Snack Foods Processing.* Eds. E.D. Lusas and L.W. Rooney. Technomic Publishing Company, Inc., Lancaster, PA, 2001, pp. 115–136.
116. Würsch, P.: Production of resistant starch, In: *Complex Carbohydrates in Foods* . Eds. S. Cho, L. Prosky, and M.L. Dreher. Marcel Dekker, New York, 1999, p. 385.
117. Englyst, H.N., Kingman, S.M., and Cummings, J.H.: Classification and measurement of nutritionally important starch fractions. *European Journal of Clinical Nutrition* 46, 533–550, 1992.
118. Berry, C.S.: Resistant starch: Formation and measurement of starch that survives exhaustive digestion with amylolytic enzymes during the determination of dietary fiber. *Journal of Cereal Science* 4, 301–314, 1986.
119. Jenkins, D.J.A., Wolever, T.M.S., Taylor, R.H., Ghafari, H., Barker, H.M., Fielden, H., Baldwin, J.M., Bowling, A.C., Newman, H.C., Jenkins, A.L., and Golf, D.V.: Glycemic index of foods: a physiological basis for carbohydrate exchange. *American Journal of Clinical Nutrition* 34, 362–366, 1981.
120. Björck, I. and Asp, N.-G.: Controlling the nutritional properties of starch in foods: a challenge to the food industry. *Trends in Food Science and Technology* 5, 213–218, 1994.
121. Englyst, H. N. and Hudson, G. J. Starch and health. Frazier, P. J., Donald, A. M., and Richmond, P. In: *Starch Structure and Functionality.* The Royal Society of Chemistry, Cambridge, UK, 1997.
122. Jenkins, P.J., Cameron, R.E., Donald, A.M., Bras, W., Derbyshire, G.E., Mant, G.R., and Ryan, A.J.: *In situ* simultaneous small and wide-angle X-ray scattering: a new technique to study starch gelatinization. *Journal of Polymer Science Part B: Polymer Physics* 32, 1579–1583, 1994.

8

Starch Modification and Applications

Sherry X. Xie, Qiang Liu, and Steve W. Cui

CONTENTS

8.1 Introduction

Native starches have many disadvantages for industrial applications such as insolubility in cold water, loss of viscosity, and thickening power after cooking. In addition, retrogradation occurs after loss of ordered structure on starch gelatinization, which results in syneresis or water separation in starchy food systems. However, these shortcomings of native starch could be overcome, for example, by introducing small amounts of ionic or hydrophobic groups onto the molecules. The modifications alter the properties of starch, including solution viscosity, association behavior, and shelf life stability in final products. The functionality of starch can be modified through physical, chemical, and biotechnological means. Another purpose of starch modification is to stabilize starch granules during processing and make starch suitable for many food and industrial applications.

Starch can be physically modified to improve water solubility and to change particle size. The physical modification methods involve the treatment of native starch granules under different temperature/moisture combinations, pressure, shear, and irradiation. Physical modification also includes mechanical attrition to alter the physical size of starch granules. Starch is widely modified by chemical methods. The most common chemical modification processes are acid treatment, cross-linking, oxidation, and substitution, including esterification and etherification.

The development of biotechnology provides another means for starch modification during the growth of the plant. Different amylose levels, amylopectin structure, and phosphorus contents from various plant sources can be produced using antisense reduction of enzyme activity of single or multiple enzymes.[1] New starch functionalities can also be identified in naturally occurring mutants, which have been widely employed in food industry because of their natural and specific functional properties.

In this chapter, starch modification using various techniques, functional properties of modified starches, and their applications in foods will be discussed.

8.2   Chemical Modification

Chemical modification can be carried out on three starch states:

- In suspension, where the starch is dispersed in water, the chemical reaction is carried out in water medium until desired properties are achieved. The suspension is then filtered, washed, and air dried.
- In a paste, where the starch is gelatinized with chemicals in a small amount of water, the paste is stirred, and when the reaction is completed, the starch is air dried.
- In the solid state, where dry starch is moisturized with chemicals in a water solution, air dried, and finally reacted at a high temperature (i.e., $\geq 100°C$).

The most common chemical modification includes: oxidation, esterification, and etherification. The chemical modification of starch results in enhanced molecular stability against mechanical shearing, acidic, and high temperature hydrolysis; obtaining desired viscosity; increasing interaction with ion, electronegative, or electropositive substances; and reducing the retrogradation rate of unmodified starch. The levels of chemicals allowed in food grade modified starches are outlined in Table 8.1.

8.2.1. Oxidation

Starch oxidation has been practiced since the early 1800s, and various oxidizing agents have been introduced, for instance, hypochlorite, hydrogen peroxide, periodate, permanganate, dichromate, persulfate, and chlorite.[2] The main uses for oxidized starch are the paper and textile industries. However, the application of oxidized starches in the food industry is increasing because of their low viscosity, high stability, clarity, and binding properties. Oxidized starch for food use is mainly produced by the reaction of starch with sodium hypochlorite. Bleaching agents, such as hydrogen peroxide, peracetic acid, potassium permanganate, and sodium chlorite, are also permitted by the Food and Drug Administration (FDA) in the United States[3] (Table 8.1), however, their levels used in the reaction are too low to change the behavior of starch, therefore, such chemicals are not included in this section.

8.2.1.1   Chemical Reactions

The major reactions of hypochlorite oxidation of starch include cleavage of polymer chains and oxidation of hydroxyl groups to carbonyl and carboxyl groups (Figure 8.1). Oxidation occurs randomly at primary hydroxyls (C-6), secondary hydroxyls (C-2, C-3, and C-4), aldehydic reducing end groups, and glycol groups (cleavage of C-2–C-3 bonds.)[4] The properties of starches may be mainly affected by the oxidations at C-2, C-3, and C-6 positions because of the large number of hydroxyls at these positions.

The reaction rate of starch with hypochlorite is remarkably affected by pH. The reaction rate is most rapid around pH 7 and very slow pH at 10. A hypothesis has been proposed to explain the difference in reaction rate under acidic, neutral, and alkaline conditions.[2]

1. Under acidic conditions, there are two steps involved as shown below (Equation 8.1 and Equation 8.2). Chlorine, a product from the rapid conversion of hypochlorite, reacts with the hydroxyl groups in starch molecules, and further gives the oxidized starch by formation of a keto group. Protons (hydrogen atoms) are released in both reactions. However, under the acid conditions there are excess protons, which hinder the liberation of protons, hence decrease the reaction rate:

$$H{-}\overset{|}{\underset{|}{C}}{-}OH \;+\; Cl{-}Cl \;\xrightarrow{\text{Fast}}\; H{-}\overset{|}{\underset{|}{C}}{-}O{-}Cl \;+\; HCl \qquad (8.1)$$

TABLE 8.1

Regulations for Food Starch — Modified in the United States

172.892 Food starch-modified.

Food starch-modified as described in this section may be safely used in food. The quantity of any substance employed to effect such modification Shall not exceed the amount reasonably required to accomplish the intended physical or technical effect, nor exceed any limitation prescribed. To insure safe use of the food starch-modified, the label of the food additive container shall bear the name of the additive "Food starch-modified" in addition to other information required by the Act. Food starch may be modified by treatment, prescribed as follows:

(a) Food starch may be acid-modified by treatment with hydrochloric acid or sulfuric acid or both.

(b) Food starch may be bleached by treatment with one or more of the following:

Use	Limitations
Active oxygen obtain from hydrogen peroxide and/or peracetic acid, not to exceed 0.45% of active oxygen.	
Ammonium persulfate, not to exceed 0.075% and sulfur dioxide, not to exceed 0.05%.	
Chlorine, as calcium hydrochlorite, not to exceed 0.0036 % of dry starch.	The finished food starch-modified is limited to use only as a component of batter for commercially processed foods.
Chorine, as sodium hypochlorite, not to exceed 0.0082 pound of chlorine per pound of dry starch.	
Potassium permanganate, not to exceed 0.2%.	Residual manganese (calculated as Mn), not to exceed 50 parts per million in food starch-modified.
Sodium chlorite, not to exceed 0.5%.	

(c) Food starch may be oxidized by treatment with chlorine, as sodium hypochlorite, not to exceed 0.055 pound of chlorine per pound of dry pound of dry.

(d) Food starch may be esterified by treatment with one of the following:

Use	Limitations
Acetic anhydride	Acetyl groups in food starch-modified not to exceed 2.5%.
Adipic anhydride, not to exceed 0.12 percent, and acetic anhydride.	Do.
Monosodium orthophosphate....	Residual phosphate in food starch-modified not to exceed 0.4% calculate as phosphorus.
1-octenyl succinic anhydride, not to exceed 3%.	
1-octenyl succinic anhydride, not to exceed 2%, and aluminum sulfate, not to exceed 2%.	
1-octenyl succinic anhydride, not to exceed 3%, followed by treatment with a beta-amylase enzyme that is either an approved food additive or is generally recognized as safe......	Limited to use as stabilizer or emulsifier in beverages and beverage base as defined in170.3 (n)(3) of this chapter.
Phosphorus oxychloride, not to exceed 0.1%.	
Phosphorous oxychloride, not to exceed 0.1%, followed by either acetic anhydride, not to exceed 8%, or vinyl acetate, not to exceed 7.5%.	Acetyl groups in food starch-modified not to exceed 2.5%.
Sodium trimetaphosphate	Residual phosphate in food starch-modified not to exceed 0.04%, calculated as ~P.
Sodium tripolyphosphate and sodium trimetaphosphate	Residual phosphate in food starch-modified not to exceed 0.4%, calculated as ~P.
Succinic anhydride, not to exceed 4%.	
Vinyl acetate............................	Acetyl groups in food starch-modified not to exceed 2.5%.

TABLE 8.1 (CONTINUED)

Regulations for Food Starch — Modified in the United States

(e) Food starch may be etherified by treatment with one of the following:

Use	Limitations
Acrolein, not to exceed 0.6%.	
Epichlorohydrin, not to exceed 0.3%.	
Epichlorohydrin, not to exceed 0.1%, and propylene oxide, not to exceed 10% added in combination or in any sequence.	Residual propylene chlorohydrin not more than 5 ppm in food starch-modified.
Epichlorohydrin, not to exceed 0.1%, followed by propylene oxide, not to exceed 25%.	Do.
Propylene oxide, not to exceed 25%.	Do.

(f) Food starch may be modified and etherified by treatment with one of the following:

Use	Limitations
Acrolein, not to exceed 0.6% and vinyl acetate, not to exceed 7.5%.	Acetyl groups in food starch-modified not to exceed 2.5%.
Epichlorohydrin, not to exceed 0.3%, and acetic anhydride.	Acetyl groups in food starch-modified not to exceed 2.5%.
Epichlorohydrin, not to exceed 0.3%, and succinic anhydride, not to exceed 4%.	
Phosphorus oxychloride, not to exceed 0.1%, and propylene oxide, not to exceed 10%.	Residual propylene chlorodrin not more than 5 ppm in food starch-modified.

(g) Food starch may be modified by treatment with one of the following:

Use	Limitations
Chlorine, as sodium hypochlorite, not to exceed 0.055 pound of chlorine per pound of dry starch; 0.45% of active oxygen obtained from hydrogen peroxide; and propylene oxide, not to exceed 25%.	Residual propylene chlorohydrin not more than 5 ppm in food starch-modified.
Sodium hydroxide, not to exceed 1%.	

(h) Food starch may be modified by a combination of the treatments prescribed by paragraphs (a), (b), and/or (i) of this section and any one of the treatments prescribed by paragraph ©, (d), (e), (f), or (g) of this section, subject to any limitations prescribed by the paragraphs named.

(i) Food starch may be modified by treatment with the following enzyme:

Enzyme	Limitations
Alpha-amylase (E.C. 3.2.1.1)	The enzyme must be generally recognized as safe or approved as a food additive for this purpose. The resulting nonsweet nutritive saccharide polymer has a dextrose equivalent of less than 20.
Bata-amylase (E.C. 3.2.1.2) Glucoamylase (E.C.3.2.1.3) Isoamylase (E.C.3.2.1.68) Pullulanase (E.C.3.2.1.41)	

Source: From Code of Federal Regulations, 2001.[3]

$$H-C-O-Cl \xrightarrow{\text{Slow}} C=O \ + \ HCl \tag{8.2}$$

2. In alkaline medium, both starch and hypochlorite will carry negative charges. The two negatively charged ions would repel each other; as a result, the reaction rate is decreased with the increase of pH (Equation 8.3 and Equation 8.4).

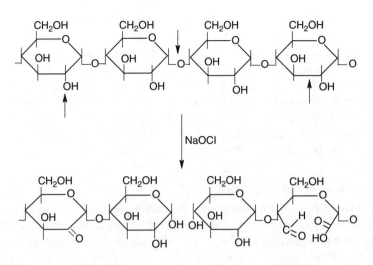

FIGURE 8.1
Hypochlorite oxidation of starch, showing carbonyl and carboxyl formation.

$$H-\overset{|}{\underset{|}{C}}-OH \ + \ NaOH \ \longrightarrow \ H-\overset{|}{\underset{|}{C}}-O^- Na^+ \ + \ H_2O \qquad (8.3)$$

$$2 \ H-\overset{|}{\underset{|}{C}}-O^- \ + \ OCl^- \ \longrightarrow \ 2 \ \overset{|}{\underset{|}{C}}=O \ + \ H_2O \ + \ Cl^- \qquad (8.4)$$

3. In neutral or mild (slightly acidic or basic) media, starch is neutral and hypochlorite is mainly undissociated (hypochlorous acid). The reaction of hypochlorous acid with starch can easily occur, and then produce the oxidized starch. Neutral starch will react with any hypochlorite anion to form the oxidized product (Equation 8.5 to Equation 8.7).

$$H-\overset{|}{\underset{|}{C}}-OH \ + \ HOCl \ \longrightarrow \ H-\overset{|}{\underset{|}{C}}-OCl \ + \ H_2O \qquad (8.5)$$

$$H-\overset{|}{\underset{|}{C}}-OCl \ \longrightarrow \ \overset{|}{\underset{|}{C}}=O \ + \ HCl \qquad (8.6)$$

$$H-\overset{|}{\underset{|}{C}}-OH \ + \ OCl^- \longrightarrow \ \overset{|}{\underset{|}{C}}=O \ + \ H_2O \ + \ Cl^- \qquad (8.7)$$

8.2.1.2 Preparation Procedures

The oxidation of starch for food use is performed by adding sodium hypochlorite solution with 5 to 10% available chloride to an agitated starch slurry under controlled temperatures (25 to 40°C) and pH. The maximum level of hypochlorite allowed by the Food and Drug Administration of the USA, is equivalent to 0.25 moles of active chlorine per mole of D-glucose unit. Dilute alkali is added continuously to neutralize hydrochloric acid formed during the reaction. The reaction rate of oxidation is most rapid at pH 7.[2]

Calcium hypochlorite powder can be used to bleach starch by simply dry blending with starch at a concentration of 0.036% active chlorine based on a starch dry weight. The mixture is allowed to stand for one week; the lightly oxidized starch is ready for use in food.[5]

8.2.1.3   Functional Properties

Oxidized starches retain Maltese crosses and exhibit the same x-ray diffraction pattern as that of the native starch, which indicates that oxidation occurs mainly in the amorphous phases of the granule.[2] Oxidized starches are normally whiter than unmodified starches because pigments on the granule surface are bleached. Oxidation increases the diameter of wheat starch granules by approximately 16%, while no change in granule size was found for waxy maize starch.[2]

Oxidation of starch mainly causes the scission of the glucosidic linkages and oxidation of hydroxyl groups to carbonyl and carboxyl groups. The scission of the glucosidic linkage results in depolymerization of amylose and amylopectin, hence decreases swelling power and paste viscosity. However, treatment with low levels of hypochlorite has been reported to increase paste viscosity.[4] Formation of the carbonyl and carboxyl groups discontinuously along the chains reduces gelatinization temperature, increases solubility, and decreases gelation. The carbonyl and carboxyl groups also reduce the thermal stability of the oxidized starch that causes browning. Since the bulkiness of the carboxyls and carbonyls sterically interfere with the tendency of amylose to associate and retrograde, oxidized starches produce pastes of greater clarity and stability than those of unmodified starch.

8.2.1.4   Applications

Oxidized starches are used as binding agents in foods, such as batters applied to meats before frying. The tenacious coatings formed on frying may result from the salt bridges between the protein molecules and anionic oxidized starch or the reaction of amino groups with the aldehyde groups on the oxidized starch. An acid-thinned maltodextrin may be oxidized to give a double modification, resulting in low viscosity and nongelling starch which can be used as fillings for chocolate confections.[2,4]

8.2.2   Cross-Linking

Starch contains two types of hydroxyls, primary (6-OH) and secondary (2-OH and 3-OH). These hydroxyls are able to react with multifunctional reagents resulting in cross-linked starches. Cross-linking is done to restrict swelling of the starch granule under cooking conditions or to prevent gelatinization of starch. Cross-linking of starch is effected by a low level of reagent. Starch molecules are long chain polymers which occur in close proximity within granules. Starch molecules can be interconnected by reactions with trace amounts of a multifunctional reagent.[6]

The reagents permitted by FDA for making cross-linking food grade starch are phosphoryl chloride, sodium trimetaphosphate, adipic acetic mixed anhydride, and mixtures of sodium trimetaphosphate and tripolyphosphates. Epichlorohydrin is no longer used by starch manufacturers in the U.S. because chlorohydrins are carcinogens.[6]

8.2.2.1 Preparation and Chemical Reactions

Cross-linking of starch with phosphorous oxychloride is a rapid reaction which produces a distarch phosphate. The reaction is especially efficient above pH 11 and in the presence of sodium sulphate (2% based on starch).[5,7] When phosphoryl chloride is added to a slurry of starch the first chloride ion of phosphoryl chloride reacts with water at 25°C immediately (with half life of 0.01 sec)[8] to form a phosphorous dichloride, which is likely the cross-linking agent. There is no time for phosphoryl chloride to diffuse into starch granules. The remaining two chlorides of phosphorous dichloride react with water almost simultaneously with a half life of ~4 minutes at 25°C. Practically, phosphoryl chloride can be added as rapidly as possible to a starch slurry in order to generate phosphorous dischloride *in situ*, which then diffuses into the granule to affect cross-linking[6] (Equation 8.8 and Equation 8.9).

$$\underset{\underset{Cl}{|}}{\overset{\overset{O}{\|}}{Cl-P-Cl}} + H_2O \xrightarrow{\text{pH 11.3, 25°C}} \underset{\underset{Cl}{|}}{\overset{\overset{O}{\|}}{Cl-P-O^-}} + Cl^- \qquad (8.8)$$

Phosphorus dichloride

$$\underset{\underset{Cl}{|}}{\overset{\overset{O}{\|}}{Cl-P-O^-}} + Starch \xrightarrow[\substack{20\text{-}60\text{min} \\ \\ Na_2SO_4}]{\text{pH 11.3, 25°C}} \underset{\underset{O_-}{|}}{\overset{\overset{O}{\|}}{Starch-O-P-O-Starch}} + 2\ Cl^- \qquad (8.9)$$

distarch phosphate

Sodium trimetaphosphate, a nonhazardous solid, is another cross-linking agent. The cross-linking starch with sodium trimetaphosphate proceeds slowly to give distarch phosphate (Equation 8.10). However, this reaction can be accelerated by increasing the pH and the concentration of sodium sulphate.[6]

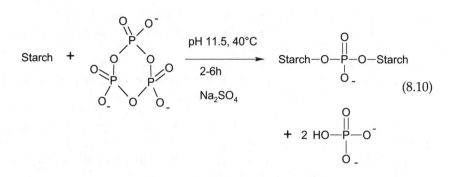

$$(8.10)$$

Cross-linking of starch with adipate is a rapid reaction at pH 8 (Equation 8.11). The mixed acetic/adipic acid anhydride is added slowly (exothermic reaction) to a starch slurry along with aqueous sodium hydroxide to maintain pH. Often, adipate cross-linking is done along with acetylation of hydroxyls to produce a doubly modified starch.[6]

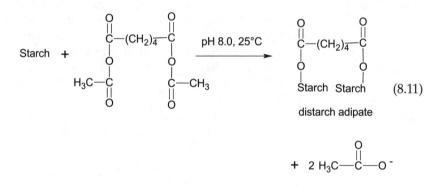

$$(8.11)$$

Since low levels of reagents are used in cross-linking, chemical measurement of the extent of cross-linking is difficult. In the case of cross-linking with phosphorylating reagents, ^{31}P-NMR could be used to measure the extent of cross-linking.[9] The physical methods related to the change in swelling of starch granules are also used to measure cross-linking. Three popular methods include measurement of fluidity, optical clarity, and pasting curves are available in the literature.[6]

8.2.2.2 Functional Properties

Starch with a low level of cross-linking shows a higher peak viscosity than that of native starch and reduced viscosity breakdown. The chemically bonded cross-links may maintain granule integrity to keep the swollen granules intact, hence, prevents loss of viscosity and provides resistance to mechanical shear. Increasing the level of cross-linking eventually will reduce granule swelling and decrease viscosity. At high cross-linking levels, the cross-links

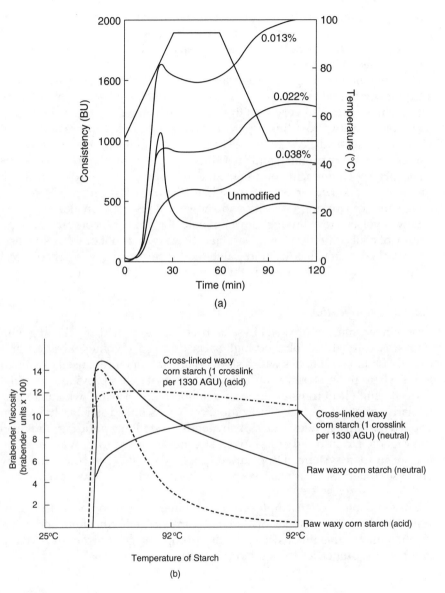

FIGURE 8.2
(a) Amylograms (7.5% starch solids) of cross-linked waxy wheat. (Adapted from Reddy and Seib, 2000.)[11] (b) Effect of mild cross-linking on viscosity of waxy corn starch and raw waxy corn starch. (Adapted from Wurzburg, 1986.)[12]

completely prevent the granule from swelling and the starch cannot be gelatinized in boiling water even under autoclave conditions.[10] Figure 8.2a shows the amylograms of cross-linked waxy wheat starches at 7.5% starch solids.[11] Starch cross-linked with 0.013% phosphoryl chloride gives a higher viscosity and the viscosity continues to increase on prolonged cooking compared to unmodified waxy wheat starch. Increasing levels of phosphoryl chloride from 0.013 to 0.038% caused a steady decline in both peak viscosity and cool paste viscosity.

High swelling starches, such as waxy corn starch and root or tuber starch, are quite fragile and tend to be fragmented by prolonged heating or agitation. These starches are also very sensitive to acid, which results in a rapid breakdown in viscosity. Cross-linking prevents starch granule rupture and loss of viscosity under acidic conditions. Figure 8.2b shows the Brabender viscosity curves for waxy corn starch and cross-linked waxy corn starch cooked at pH 5 and pH 3.[12] Under acid conditions, the viscosity of raw waxy corn starch drops sharply. However, waxy corn starch with one cross-link per 1330 AGU retains higher working viscosity and shows less viscosity breakdown.

Low level of cross-linking can eliminate the rubbery, cohesive, stringy texture of cooked native starch, particularly waxy and root or tuber starches. Cross-linked starch has a short salve-like paste texture with rheological properties particularly suited as thickener in food applications.[12]

8.2.2.3 Applications

Cross-linked starches are used in salad dressings to provide thickening with stable viscosity at low pH and high shear during the homogenization process. Cross-linked starches with a slow gelatinization rate are used in canned foods where retort sterilization is applied; such starches provide low initial viscosity, high heat transfer, and rapid temperature increase, which are particularly suitable for quick sterilization.[2,13,14] Cross-linked starches have been applied in soups, gravies, sauces, baby foods, fruit filling, pudding, and deep fried foods. Drum dried cross-linked starches are used to provide a pulpy texture in food systems. Drum dried cross-linked starches containing low amylose content, such as waxy corn, have been used to improve cake volume and crumb softness.[2]

Cross-linking is often employed in combination with other methods for starch modification, such as oxidation, hydrolysis, etherification, and esterification (monosubstituents), to provide appropriate gelatinization, viscosity, and textural properties for food applications.

8.2.3 Esterification

Starch ester is a group of modified starches in which some hydroxyl groups have been replaced by ester groups. The level of substituents of the hydroxyl groups along the starch chains is often expressed as average degree of substitution (DS). The average degree of substitution is the moles of substituent per mole of D-glucose repeat residue (anhydroglucose unit). The maximum possible DS is 3.0 when all three hydroxyls are substituted on each glucose unit along a starch chain. The DS can be calculated by the following equation:

$$DS = \frac{162\ W}{100\ M - (M-1)\ W}$$

where W = % by weigh of substituent and M = formula weight of substituent.

The reagents approved by the FDA for preparation of organic and inorganic monoesters of starch intended for food use are acetic anhydride, vinyl acetate, succinic anhydride, 1-octenyl succinic anhydride, and sodium tripolyphosphate (Table 8.1). In this section, three types of starch esters are introduced:

- Starch acetates prepared by reacting starch with acetic anhydride
- Starch succinate and starch alkenylsuccinate, which are produced by the reaction of starch with succinic anhydride and alkenyl substituted succinic anhydride, respectively
- Starch phosphate resulting from the reaction of starch with tripolyphosphate and/or trimetaphosphate

8.2.3.1 Starch Acetate

8.2.3.1.1 Chemical Reactions

Acetylation of starch in aqueous suspension by acetic anhydride at alkaline pH is used commercially to produce starch acetates of low DS. Under alkaline conditions, starch is indirectly reacted with carboxylic anhydride. An alkali starch complex forms first, which then interacts with the carboxylic anhydride to form a starch ester with the elimination of carboxylate ion and one molecule of water[15,17] (Equation 8.12). During the acetylation process there are two side reactions including the deacetylation of starch (Equation 8.13), and the formation of sodium acetate (by product) (Equation 8.14).

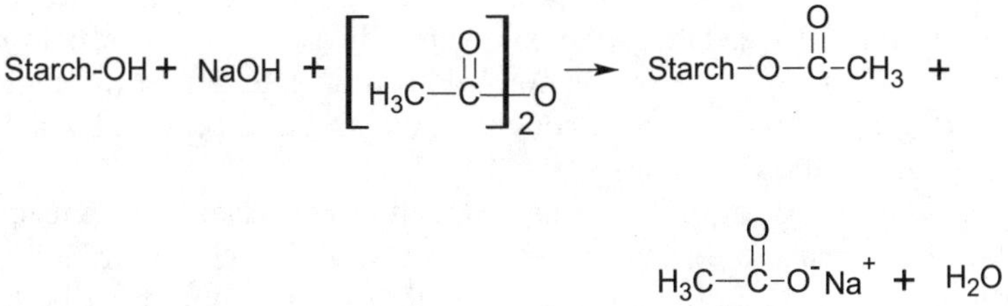

(8.12)

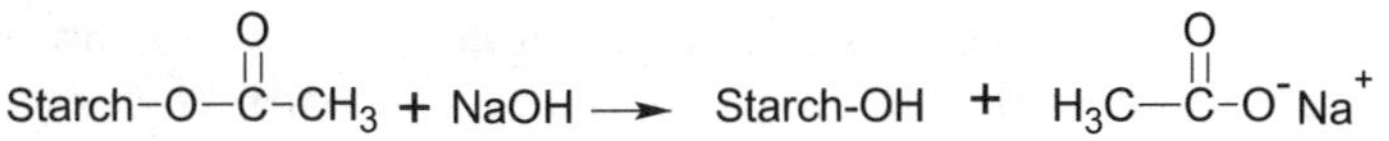

(8.13)

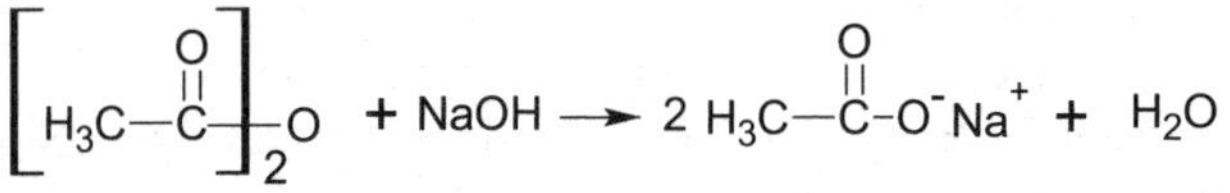

(8.14)

8.2.3.1.2 Preparation Procedures

A slurry of starch is prepared at pH 8 and 25 to 30°C, the acid anhydride is added slowly to the starch slurry while maintaining pH 8 and temperatures of 25 to 30°C. After addition of reagents, the mixture is stirred for 20 to 30 min, and then adjusted to pH 6 to 7. The modified starch is isolated by washing and drying. The amount of acetic anhydride legally permitted in food starch modification in the U.S. can be up to 8% of the weight of starch, providing the acetylated starch contains less than 2.5% acetyl groups, which corresponds to DS 0.1.[2,6]

Sodium acetate formed as a by product must be completely removed by the washing process because it causes acetyl odors in puddings made from these starch acetates.[18]

Doubly modified starches may be prepared by first cross-linking with up to 0.1% phosphoryl chloride then acetylating with up to 8% acetic anhydride. Alternatively, the doubly modified starch may be made with low levels of adipic/acetic mixed anhydride in combination with acetic anhydride.[6]

The level of organic esters on starch can be determined by saponification or proton-NMR methods.[19,20]

8.2.3.1.3 Functional Properties of Starch Acetates

Starch granules with DS up to 0.5 show no changes from native starch under the light microscope. X-ray diffraction of the modified starches indicated the reaction mainly occurred in the amorphous regions that are more accessible to reagent and catalyst. Granular acetylated corn starch containing 1.85% acetyl can be completely deacetylated at 25°C for four hours at pH 11. The deacety-lated starch granules are practically identical to the native starch granules.[2,15]

Gelatinization temperature of starch acetate is markedly lower than that of the native starch. The pasting properties of the native and acetylated corn starches are shown in Figure 8.3.[21] Normal corn acetate starch has a ~7°C lower gelatinization temperature and a slightly higher peak viscosity, reached at ~7°C lower than that of the unmodified starch. In the cooling cycle, the viscosity of normal corn acetate starch is lower, indicating improved cold stability.[2] The reduction in viscosity of normal corn acetate starch on cooling is probably due to acetate interference within the amylose portion by reducing the association of the molecules.

The reduction in gelatinization temperature on acetylation is particularly important in the case of high amylose corn starch. Starches containing 50% amylose or higher do not gelatinize in boiling water; they must be heated to 160°C under pressure to achieve gelatinization.[3,15] Acetylation of high amylose starch to a DS of 0.1 to 0.2 lowers gelatinization temperature; therefore, this starch can be dispersed in a boiling water bath. Figure 8.3 shows acetate starch containing 66% amylose (DS = 0.130) exhibiting a pasting curve compared to native starch.

Starch containing 0.5 to 2.5% acetyl usually improves the stability and clarity of sols by increasing the degree of swelling and dispersion of starch granule; it also reduces retrogradation.[2]

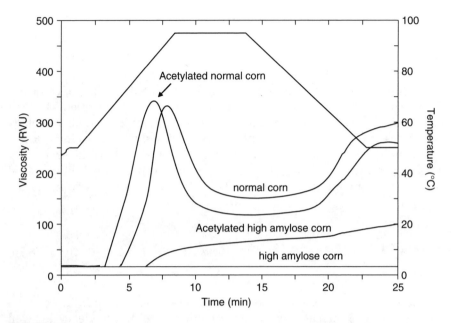

FIGURE 8.3
RVA pasting curves of normal, high amylose corn (66% amylose) and their acetylated starches. (Redrawn from Liu et al., 1997.)[21]

Acetate starches normally lack resistance to mechanical shear and acids. Cross-linked starches are resistant to viscosity breakdown and acidic conditions but exhibit poor clarity and cold aging stability. By combining acetylation with cross-linking, both viscosity stability and clarity can be achieved, as well as low temperature storage stability. Cross-linking also provides desired textural properties.[6,15]

Some food applications may require a high concentration of starch with suitable handling viscosity. Such starches can be produced through acid or oxidative treatments, first to reduce starch molecule size, and then **to** stabilize the starch by acetylation.

8.2.3.1.4 Applications

Cross-linked acetate starches are used as thickeners in baked, canned, frozen, and dry foods. They are also used in fruit and cream pie fillings, tarts, salad dressings, and gravies. Frozen fruit pies, pot pies, and gravies made from acetate starch can maintain their stability under low temperature storage. Acetylated cross-linked tapioca and waxy maize starches have excellent stability and thickening properties, which are suitable in making baby foods.[15] The root and root-type starch acetates were developed as thick-thin canning media for better heat penetration and flavor retention in canned food retorting.[22,23] Pregelatinized acetate starches are used in dry mixes, instant gravies, and pie fillings.[2]

8.2.3.2   Starch Succinate and Starch Alkenylsuccinate

8.2.3.2. 1   Starch Succinate

8.2.3.2.1.1   Preparation and Chemical Reaction — Starch succinate is a half-ester produced by the reaction of starch with succinic anhydride (Equation 8.15). The level of succinic anhydride legally permitted by the U.S. Food and Drug Administration in food starch modification is not to exceed 4%. Low DS starch succinate can be prepared by a similar procedure as starch acetate. Starch suspension is prepared at pH ~ 8. Succinic anhydride is added slowly to a starch suspension while maintaining pH. Finally, starch is isolated by washing and drying.[2,6] High DS starch succinate can be prepared in glacial acetic acid containing sodium acetate at 100°C or using pyridine as the medium.[24]

$$\text{Starch-OH} + \underset{\substack{\text{O=C} \qquad \text{C=O} \\ \diagdown \text{O} \diagup}}{\overset{\text{H}_2\text{C}\!-\!-\!-\!\text{CH}_2}{}} \xrightarrow[\text{pH 8}]{\text{OH}^-} \text{Starch-O}\!-\!\overset{\text{O}}{\overset{\|}{\text{C}}}\!-\!\text{CH}_2\!-\!\text{CH}_2\!-\!\overset{\text{O}}{\overset{\|}{\text{C}}}\!-\!\text{O}^-\,\text{Na}^+ \quad (8.15)$$

8.2.3.2.1.2   Functional Properties — Starch succinate contains a free carboxylate group that increases the water holding power and tendency to swell in cold water. The capability of cold water swelling increases with greater levels of succinic anhydride. Cold water swelling capacities for succinic anhydride treatment at 0, 2, and 4% are 34ml, 40ml, and 49ml, respectively, which is determined by measuring the volume of starch in water in a graduated cylinder.[25]

The presence of hydrophilic succinate groups in corn starch increased viscosity significantly. Cooked starch succinates show excellent viscosity, stability, and clarity, as well as freeze-thaw stability. The Brabender visco-amylogram (Table 8.2) showed the increase in viscosity of corn starch resulted from succinic anhydride treatment: the peak viscosity and the final viscosity increase, whereas, gelatinization temperature decreases with increasing DS from 0.05 to 0.20.[26]

The pH has a significant effect on the viscosity of starch succinate by affecting the carboxylic acid group. The substituent group is mostly in the acid form at lower pH, which is less hydrophilic than the salt form. Normally,

TABLE 8.2
Brabender Characteristics of Corn Starch Succinate at Neutral pH

	Gelatinization Temperature/ Pasting Temperature (°C)	Peak Viscosity (BU)	Viscosity after Cooling to 30°C (BU)
Native	78–85	120	190
DS = 0.05	78–84	130	210
DS = 0.10	78–82	130	220
DS = 0.15	80	140	240
DS = 0.20	76	150	290

Source: From Bhandari and Singhal, 2002.[26]

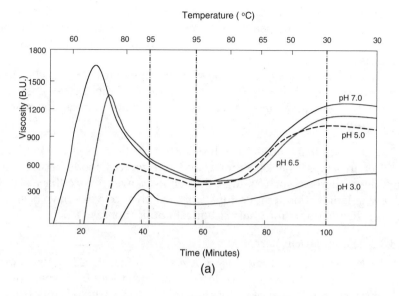

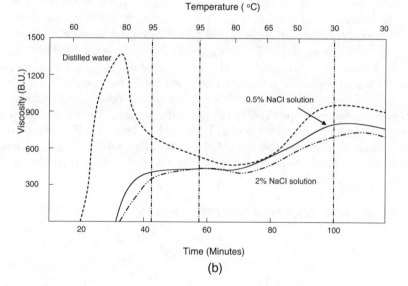

FIGURE 8.4
(a) Effect of pH on Brabender viscosity of corn starch treated with 3% succinic anhydride (5% anhydrous solids). (b) Effect of sodium chloride on Brabender Amylograph of corn starch with 3% succinic anhydride (5% anhydrous solids, pH = 6.5). (Adapted from Trubiano, 1986.)[25]

the highest viscosity is obtained at neutral pH. The peak and final viscosity reduced considerably when pH is dropped from 7 to 3 (Figure. 8.4a).[25]

Since starch succinate is an anionic polymer, the viscosity of starch succinate is greatly affected by the amount of salt present in the solution. Figure 8.4b shows the comparison of the viscosities of corn starch succinate (3% treatment of succinic anhydride) in distilled water, 0.5, and 2% sodium chloride solutions, respectively.[25] The peak viscosities were greatly reduced by the presence of salt. The salt also protects the starch from swelling which results in an increase in gelatinization temperature.

Cooked starch succinates are somewhat gummy in texture, which is undesirable for food applications. Succinate starches with light cross-linking can improve the cooking texture, as well as the resistance to breakdown at high temperature.[25]

8.2.3.2.1.3   Applications — Starch succinate has been used as a binder and thickener in soups, snacks, canned, and refrigerated foods because of its desirable properties, such as high thickening power, freeze-thaw stability, and low gelatinization temperature. Corn starch succinates have been used as tablet disintegrants for pharmaceutical applications.[25]

8.2.3.2.2   Starch Alkenylsuccinate

8.2.3.2.2.1   Preparation and Chemical Reaction — Starch alkenylsuccinates sodium salt, such as octenyl and decenyl, are produced by suspending 100 parts of starch in 150 parts of water containing 2 to 5 parts sodium carbonate. The suspension is agitated while slowly adding 5 parts of decenyl succinic acid anhydride or 0.1 to 0.3 part of octenyl succinic acid anhydride. Agitation is continued for 14 h for decenyl and 12 h for octenyl at room temperature, then the pH is adjusted to 7.0 using dilute hydrochloric acid. The starch is finally filtered, washed, and dried. For the aluminum salt, the starch acid ester is resuspended in 100 parts water with agitation, and 2% aluminum sulphate is dissolved in the starch suspension with continuous agitation for 4 h at room temperature. The starch derivative is then filtered, washed, and dried.[27,28] 1-Octenylsuccinic acid anhydride has been legally permitted for food starch modification by the U.S. Food and Drug Administration, with a maximum level of 3%. The reaction of starch with 1-octenyl succinate is as follows (Equation 8.16):

$$Starch\text{-}OH \ + \ CH_3(CH_2)_5\text{-}CH{=}CH\text{-}CH\text{-}CH_2 \qquad OH^- \longrightarrow$$

$$\underset{\displaystyle O}{O{=}C \qquad C{=}O}$$

$$(8.16)$$

$$CH_3(CH_2)_5\text{-}CH{=}CH\text{-}CH\text{-}CH_2\text{-}COO^-\ Na^+$$
$$O{=}C\text{-}O\text{-}Starch$$

8.2.3.2.2.2   Functional Properties — The sodium starch (1-octenyl)-succinate gives markedly higher peak and cold paste viscosities in comparison with native starch. Table 8.3 shows the effect of 1-octenyl succinate group on amylograph properties of large granular wheat starch.[20] The increase in paste viscosity throughout the heating, holding, and cooling cycles is attributed

TABLE 8.3

Effect of 1-Octenyl Succinate Group on Amylograph Properties of Large Granular Wheat Starch

		Amylograph Viscosity, B.U.				
Starch	Pasting Temperature (°C)	Peak	Heated to 95°C	Final at 95°C	Cooled to 50°C	Final at 50°C
Untreated	84.9	290	245	260	520	495
Sodium starch (1-Octenyl Succinate)	62.4	1090	1055	625	845	780

Source: From Maningat, 1986.[20]

to both hydrocarbon and carboxylic groups of the alkenyl succinate, which causes mutual repulsion among polymer chains, hence resulting in greater starch swelling.[20]

Starch alkenylsuccinates contain hydrophobic and hydrophilic groups. In an emulsion system, these starch molecules are attracted to the interface of the water by its hydrophilic group and to the oil droplets by its hydrophobic group and form very strong films at the oil-water interface, therefore reducing the tendency to coalescence and separate in the dispersed phase.[25] These starch derivatives with the desired viscosity and balanced hydrophobic and hydrophilic groups have been utilized in the food industry to stabilize emulsions and to encapsulate flavors.[29]

When granular starch alkenylsuccinates are treated with a polyvalent metal ion, such as aluminum sulphate, they resist wetting and flow freely when added to water. The aluminum salt of starch 1-octenylsuccinate can continue to flow even after absorbing 30% moisture from the atmosphere.[20]

8.2.3.2.2.3   Applications — Starch alkenylsuccinates are used in nonalcoholic beverages for the stabilization of flavors. Cold water soluble, low viscosity octenylsuccinate derivatives have been very successfully used for the stabilization of carbonated beverages in replacing gum arabic because of their superior emulsion stabilizing properties. High viscosity octenylsuccinate starches are used as stabilizers in high viscosity high oil systems, such as salad dressings. Encapsulation of water insoluble substances, both volatile and nonvolatile, is another application of starch alkenylsuccinate. Free flowing, water repellent, starch derivatives have been found applications as anticaking agents for dry food formulations and antisticking agents for dried fruits.[25]

8.2.3.3   Starch Phosphate

8.2.3.3.1   Preparation and Chemical Reaction

Two steps are involved in the preparation of starch phosphate. Starch is first sprayed or soaked with an aqueous solution of a mixture of sodium tripolyphosphate and sodium sulphate together with a low level of sodium trimetaphosphate, then dried at ~100°C. Sodium hydroxide or hydrochloric acid is used to control the starting pH. In the second step, the impregnated starch is roasted at 130°C, washed with water and dried. Starch reacts with tripolyphosphate at pH 6 to 10 to produce starch monophosphate. At pH above 10, cross-linking occurs and a distarch phosphate is produced. Starch reacts with a mixture of tripolyphosphate/trimetaphosphate at pH 9.5; starch cross-linking by the trimetaphosphate occurs rapidly at pH > 9.5.[6,30] A wheat starch phosphate prepared by reaction with sodium tripolyphosphate was found to contain mostly the 6-phosphate ester with lower and almost equal levels of 2- and 3-phosphate ester.[31] The levels of residual phosphate in food grade modified starches are not to exceed 0.04% for sodium trimetaphosphate and 0.4% for sodium tripolyphosphate and sodium trimetaphosphate, respectively. FDA in U.S. also allows monosodium orthophosphate (not to exceed 0.4% as phosphorous) and phosphorus oxychloride (not to exceed 0.1%) to be used for food grade starch modification. The reaction scheme of phosphate monoester of starch with sodium tripolyphosphate is as follows (Equation 8.17):

$$\text{Starch-OH} + \text{Na}^+{}^-\text{O}-\overset{\overset{\displaystyle O}{\|}}{\underset{\underset{\displaystyle Na^+}{|}}{P}}-\text{O}-\overset{\overset{\displaystyle O}{\|}}{\underset{\underset{\displaystyle Na^+}{|}}{P}}-\text{O}-\overset{\overset{\displaystyle O}{\|}}{\underset{\underset{\displaystyle Na^+}{|}}{P}}-\text{O}^-\,\text{Na}^+ \longrightarrow \text{Starch}-\text{O}-\overset{\overset{\displaystyle O}{\|}}{\underset{\underset{\displaystyle Na^+}{|}}{P}}-\text{O}^-\,\text{Na}^+ + \text{Na}_2\text{HP}_2\text{O}_7$$

$$(8.17)$$

A method for measuring mono- and diester substitution in low DS products has been developed based on their equivalence points in alkali titrations. Starch monophosphate has two equivalence points in the range of pH 4 to 9, whereas, starch diester phosphate has only one.[32]

8.2.3.3.2   Functional Properties

Starch phosphates are strongly anionic polymers, which yield higher viscosity, more clear and stable dispersions with long cohesive texture and resistance to retrogradation.[3] The viscosities are reduced by the presence of salts. The gelatinization temperature decreases with the increasing degree of substitution, and the monoester becomes cold water swelling when DS is increased to 0.07. Dispersion of starch phosphates has superior freeze-thaw stability than other modified starches.[33,34] Starch phosphates are good emulsification agents because of the ionic properties.[2]

8.2.3.3.3   Applications

Starch phosphates are used in foods as emulsion stabilizers for vegetable oil in water and thickening agents with good freeze-thaw stability. A combination of starch phosphate, guar gum, and propylene glycol has been used as an emulsion stabilizer for vinegar and vegetable oil in water.[35] Cold water swelling starch phosphate is dry mixed with sugar and a flavoring agent, which is added to cold milk to form puddings with a smooth, soft, even texture, and superior eating quality.[36] Some starch phosphates have been used in breads to improve the baking quality.[35]

8.2.4 Etherification

Unlike ester linkage such as starch acetate, which tends to deacetylate under alkaline condition, ether linkages are more stable even at high pH. Etherification gives starch excellent viscosity stability. Hydroxyalkyl starches, including hydroxyethyl and hydroxypropyl, are mainly produced for industrial applications. Hydroxyethyl starch has not yet been approved as a direct food additive, but can be used as an indirect food additive, such as a sizing agent for paper contacting food, whereas hydroxypropyl starch is primarily used for the food industry. Therefore, the focus of this section will be on hydroxypropyl starch which is made by reacting propylene oxide with starch under alkaline conditions.

8.2.4.1 Chemical Reaction

Propylene oxide is an unsymmetrical epoxide and a very reactive molecule because of its highly strained three-membered epoxide ring with bond angels of ~60°. Opening of the unsymmetrical epoxide in base occurs by attacking at the sterically less hindered end of epoxide to produce 2-hydroxypropyl starch ether.[37]

The reaction of propylene oxide with starch under alkaline conditions is a substitution, nucleophilic, bimolecular, or S_N2 type.[37] Nucleophiles react with oxiranes in S_N2 fashion to give ring opened products. The reaction equation is as follows (Equation 8.18 and Equation 8.19):

$$\text{Starch-OH} + \text{NaOH} \longrightarrow \text{StarchO}^- \text{Na}^+ + \text{H}_2\text{O} \qquad (8.18)$$

$$\text{StarchO}^- \text{Na}^+ + \text{H}_3\text{C}-\text{HC}\underset{\text{O}}{\overset{}{\diagdown\diagup}}\text{CH}_2 \overset{\text{HOH}}{\longrightarrow} \text{Starch}-\text{O}-\text{CH}_2-\underset{\underset{\text{OH}}{|}}{\text{CH}}-\text{CH}_3 + \text{NaOH}$$

propylene oxide   hydroxypropyl starch   (8.19)

The hydroxypropyl groups are predominantly located at 2-O position on the glucose units. The NMR data indicate that the hydroxypropyl groups are distributed with a ratio of 7:2:1 on the 2-O, 3-O, and 6-O positions.[38] The possible reason for the higher reactivity of HO-2 over HO-6 is the higher relative acidity of the HO-2 group due to its proximity to the anomeric center.[37]

8.2.4.2 Preparation Procedures

Depending on the degree of modification, hydroxypropylation of starch is achieved in one of two ways. At DS ≤ 0.25, starch is slurried in aqueous sodium hydroxide (pH ~ 11.5). Sodium sulphate, 5 to 15% (based on dry starch weight) is added to protect the starch from swelling. Propylene oxide is added up to 10% (based on starch dry weight). The mixture is warmed

(45 to 50°C) and stirred ~ 24 hours. After the etherification reaction, cross-linking with phosphoryl chloride is usually done to give hydroxypropyl distarch phosphate. Without cross-linking, the hydroxypropyl derivatives of starches swell excessively and give stringy pastes.[6]

The preparation of hydroxypropyl starch with a high DS is carried out in a semi-dry state. Starch is first sprayed with aqueous sodium hydroxide until it contains ~ 20% moisture, then the alkali starch is stirred in a closed reactor with gaseous propylene oxide up to a maximum level of 25% (based on dry starch weight) (Table 8.1). Upon completion of the reaction in 24 to 36 hours, the highly hydroxypropylated starch is isolated by special washing procedures to avoid gelatinization. The end product is a cold water swelling starch.[6,37]

The level of hydroxypropylation is conveniently determined by proton-NMR. Commercial food grade starches have hydroxypropyl levels from 3.3 to 11.5% (DS ≈ 0.1 to 0.3).[38]

8.2.4.3   Functional Properties

Introduction of hydroxypropyl groups has affected the enzyme digestibility of starches. Substitution of potato, corn, waxy maize, and high-amylose maize starches with hydroxypropyl groups decreased the extent of enzymatic hydrolysis of raw and gelatinized starches with increasing DS.[39] The hydroxypropyl groups on starch chains may hinder enzymatic attack and also make neighboring bonds resistant to degradation.[40]

Substitution of hydroxypropyl groups on starch chains disrupts the internal bond structure resulting in the reduction of the amount of energy needed to solubilize the starch in water. As a result, the pasting temperature of starch decreases with an increase in the level of hydroxypropyl substitution. When the DS reaches a certain level, the starch becomes cold water swelling.[37]

Hydroxypropyl groups interfere with the retrogradation of starch molecules. The substituent groups inhibit gelation and syneresis, apparently by sterically interfering with formation of junction zones and double helices in starch.[6] Thus, the solution of hydrxypropyl starches showed improved clarity, viscosity stability, cold storage stability, and a reduced tendency to retrogradation.

Hydroxypropyl starches usually are cross-linked to improve viscosity stability at high temperature, low pHs, and mechanical shear, and to obtain desired texture.

8.2.4.4   Applications

Hydroxypropyl starches are being widely used in food products where they provide viscosity stability and freeze-thaw stability. These starches are normally combined with cross-linking to provide the desired viscosity, texture, and stability for processing and storage. Hydroxypropyl starches are used as thickeners in fruit pie fillings, puddings, gravies, sauces, and salad dressings. Hydroxypropyl tapioca starch has been successfully used in frozen pudding.[41] Hydroxypropyl high-amylose starch produces an excellent edible film coating on edible foods.[42]

8.2.5 Cationization

Cationic starches are important industrial derivatives in which the starch is given a positive ionic charge by introducing ammonium, amino, imino, sulfonium, or phosphonium groups. Cationic starches are used in large scale by the paper industry as wet-end additives, surface size, and coating binders. Cationic starches containing teriary amino or quaternary ammonium groups are the most important commercial derivatives.[2]

Cationic starches made from specific reagents permitted by FDA can be used to make paper or paperboard in contact with foods. These reagents include 2-diethylaminoethyl chloride, 2,3-(epoxypropyl)trimethylammonuim chloride, and (4-chlorobutene-2)-trimethylammonium chloride.[43]

8.2.5.1 *Preparation Procedures*

8.2.5.1.1 Tertiary Aminoalkyl Starch Ethers

Cationic starches can be made with various dialkylaminoalkyl chloride reagents. One example is to mix starch with 2-diethylaminoethyl chloride in water at pH 10.5 to 12.0 and 25 to 50°C.[44,45] Under alkaline reaction conditions, the 2-diethylaminoethyl chloride forms a cyclic ethyleneimmonium intermediate (Equation 8.20), which is a highly reactive molecule and undergoes nucleophilic attack by the alkali-starch complex, resulting in a starch tertiary amino ether as the free base (Equation 8.21). Neutralization of the reaction with acid converts the free amine to the cationic tertiary ammonium salt[43] (Equation 8.22). During the reaction, sodium sulphate may be added to prevent starch swelling. Cationic starches are usually analyzed by the Kjeldahl method. Nitrogen content can be monitored to measure the extent of cationization.[2]

$$Cl-CH_2CH_2N(CH_2CH_3)_2 \xrightarrow{OH^-} \left[\begin{array}{c} (C_2H_5)_2N-CH_2 \\ \diagdown \diagup \\ CH_2 \end{array} \right]^+ Cl^- \qquad (8.20)$$

$$Starch-O^- + \left[\begin{array}{c} (C_2H_5)_2N-CH_2 \\ \diagdown \diagup \\ CH_2 \end{array} \right]^+ Cl^- \xrightarrow{OH^-} Starch-O-CH_2CH_2N(CH_2CH_3)_2 \qquad (8.21)$$

$$Starch-O-CH_2CH_2N(CH_2CH_3)_2 \xrightarrow{HCl} \left[\begin{array}{c} H \\ Starch-O-CH_2CH_2\overset{|}{N}(CH_2CH_3)_2 \end{array} \right]^+ Cl^- \qquad (8.22)$$

8.2.5.1.2   Quaternary Ammonium Starch Ethers

2,3-Epoxypropyltrimethylammonium chloride is normally used for adding quaternary ammonium groups to the starch molecule.[46,48] The reagent can be stored in water in the chlorohydrin form (3-chloro-2-hydroxy-propyltri-methylammonium chloride) and rapidly converted to the epoxide form by adding sodium hydroxide (Equation 8.23 and Equation 8.24). An aqueous reaction of chlorohydrin form with corn starch was investigated; the reaction efficiency was 84% under the conditions of a 2.8 mole ratio of sodium hydroxide to reagent, 50°C for 4 h, and a 35% starch concentration.[2]

$$\left[\underset{Cl-CH_2\overset{|}{C}H-CH_2N(CH_3)_3}{\overset{OH}{}} \right]^{+} Cl^- + NaOH \longrightarrow \left[H_2C\underset{O}{\diagup\diagdown}CH\,CH_2N(CH_3)_3 \right]^{+} Cl^- + NaCl \tag{8.23}$$

$$Starch-OH + \left[H_2C\underset{O}{\diagup\diagdown}CH\,CH_2N(CH_3)_3 \right]^{+} Cl^- \longrightarrow \left[Starch-O-CH_2\underset{OH}{\overset{|}{C}H}-CH_2-N(CH_3)_3 \right]^{+} Cl^- \tag{8.24}$$

Tertiary amines, particularly those containing at least two methyl groups, can react with epichlorohydrin to form quaternary ammonium reagents. However, residual epichlorohydrin or 1,3-dichloropropanol (by-products) must be removed by vacuum distillation or solvent extraction, because they can cross-link the starch under alkaline conditions and reduce the dispers-ibility and performance of cationic starch.[2]

Nonvolatile quaternary ammonium reagents are used in dry heat reactions with starch. The reaction efficiencies for epoxide forms of the reagents are 50 to 60% at 120 to 150°C for one hour without alkaline catalyst.[49,50] When alkaline catalyst is added, 75 to 100% reaction efficiencies are obtained at 60 to 80°C for 1 to 6 h.[51,52]

8.2.5.1.3   Aminoethylated Starches

Starch reacts with ethyleneimine to produce the 2-aminoethyl ether (Equation 8.25) in dry or solvent state. In dry reaction, the dry or semi-dry starch can mix with gaseous ethyleneimine at 75 to 120°C without catalysts,[53,55] but the temperature and pressure of reaction have to be carefully controlled to eliminate the formation of polyethyleneimine (a by-product). In solvent reaction, aminoethylated starch can be obtained by treatment with an aziridine-sulfur dioxide complex in inert solvents, such as carbon tetrachloride and benzene, or by reaction with ethyleneimine in water in the presence of alkylene oxide or organic halides and an alkaline catalyst.[2] Tertiary amino starch derivatives showed higher reaction efficiency with ethyleneimine and improved performance as wet-end additives in the manufacture of paper.[56]

$$\text{Starch-OH} \; + \; \underset{\underset{H}{\overset{|}{N}}}{\overset{H_2C\text{------}CH_2}{\diagdown\!\diagup}} \; \longrightarrow \; \text{Starch-O-CH}_2\,\text{CH}_2\,\text{NH}_2 \tag{8.25}$$

8.2.5.1.4 Iminoalkyl Starches

Iminoalkyl starch can be produced by the reaction of starch with a cyanamide or dialkyl cyanamides at pH 10 to 12 (Equation 8.26). The imino nitrogen atom forms the ionic salt after acidification with acids, which makes the cationic starch stable to hydrolysis. Granular products tend to be cross-linked and cooked pastes thicken with time, while addition of phosphate salts or aluminum sulphate can stabilize the pastes.[57] Adjusting the reacted starch to pH 1.0, followed by gelatinization and drum drying have been used to produce storage-stable cationic cyanamide starch.[58]

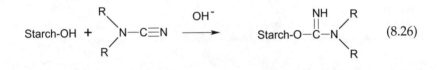

$$\text{Starch-OH} \; + \; \underset{R}{\overset{R}{\diagdown}}N\!-\!C\!\equiv\!N \; \xrightarrow{\;\;OH^-\;\;} \; \text{Starch-O}-\overset{\overset{\displaystyle NH}{\|}}{C}-N\underset{R}{\overset{R}{\diagup}} \tag{8.26}$$

8.2.5.1.5 Cationic Dialdehyde Starch

Cationic dialdehyde starches have been achieved by the reaction of dialdehyde starch with substituted hydrazine or hydrazide compounds which contain tertiary amino or quaternary ammonium groups,[59] and also have been produced by the oxidation of tertiary amino or quaternary ammonium starches by periodic acid.[60] The reaction of dialdehyde starch with betaine hydrazide hydrochloride is as follows (Equation 8.27).[2]

$$\text{Starch}-\text{CHO} + \left[H_2NNH\overset{\overset{\displaystyle O}{\|}}{C}CH_2N(CH_3)_3 \right]^{+} Cl^- \longrightarrow \left[\text{Starch}-\text{CH}=NNH\overset{\overset{\displaystyle O}{\|}}{C}CH_2N(CH_3)_3 \right]^{+} Cl^- \tag{8.27}$$

8.2.5.2 Functional Properties

Cationic starches show a decrease in peak intensity in x-ray diffraction patterns with an increase in DS. Gelatinization temperature of cationic starches decrease with increasing DS and starches become cold water swelling at DS 0.07. The effect of cationization on the pasting properties of potato and tapioca starches is shown in Figure 8.5. Cationization of potato starch (DS = 0.041) has markedly reduced the pasting temperature from 67.5°C to 46.5°C, the same as for cationized tapioca starch (DS = 0.039).[61] Cationic starches exhibit higher solubility and dispersibility with improved clarity and stability.[43]

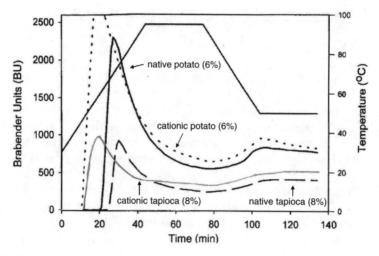

FIGURE 8.5
Effect of cationization on the pasting properties of potato and tapioca starches in 6% and 8% suspensions respectively, at pH 5.5. (Adapted from Han and Sosulski, 1998.)[61]

8.2.5.3   Applications

Cationic starches are mainly used in the manufacture of paper. These starches are used as wet-end additives during formation of the paper to improve retention and drainage rate of the pulp and strength of the finished sheets. They are also used as coating binders in paper making, wrap sizing agents in textile manufacture, binders in laundry detergents, and flocculants for suspension of inorganic or organic matter containing a negative charge. Tertiary aminoalkyl high amylose starches are used as an ingredient in hair holding spray. A derivative of cationic starch is made for pharmaceutical applications as bactericides.[44]

8.3   Physical Modification

Physical modification of starch can be applied alone or with chemical reactions to change the granular structure and convert native starch into cold water soluble starch or into small crystallite starch. Cold water soluble starch is prepared by pregelatinization of native starch slurry, followed by drum drying. Because of pregelatinization and drying, the granular integrity is lost and paste viscosity of starch is reduced. Therefore, the modified starch is cold water soluble.

Small granular starches (diameter < 5 μm) have been used as a good fat substitute. Small particle starch can be made by a combination of acid hydrolysis and mechanical attrition of the native starch.[62]

Heat-moisture and annealing treatments induce the rapid migration or rearrangement of the amylose molecules in the granules to form intermolecular bonds between the amylose molecules and/or between the amylose molecules and the amylopectin molecules.

Extrusion modification of starch is a process that uses the molten phase of high solid concentration to transform starch while maintaining a macromolecular structure.

In this section, the preparation, properties, and application of pregelatinized starch, heat-moisture treated starch, annealed starch, and extruded starch will be introduced.

8.3.1 Pregelatinization

8.3.1.1 Preparation Procedures

Pregelatinized starch can be produced by spray cooking, drum drying, solvent-based processing, and extrusion. During a spray cooking process,[63] a starch slurry enters through one special nozzle and is atomized (fine spray) in a chamber. At the same time, hot steam is injected into the chamber through a second nozzle to cook out the starch. This method is particularly useful in producing a uniformly gelatinized starch with minimum shear and heat damage. Pregelatinized starch can be made by the drum drying method, in which a cooked starch sheet is produced from a starch slurry on a hot drum; the starch is ground after drying to a desired particle size.[6] Cold water swelling starch can be produced by a solvent-based method. For example, 20% starch in aqueous alcohol (20 to 30% water) is heated to 160 to 175°C for 2 to 5 min. Unlike drum dried starch or extruded starch, the end product by solvent-based method maintains granular integrity, but loses its birefringence.[64]

8.3.1.2 Properties

The most important property of pregelatinized starch is that it instantly hydrates and swells in water at room temperature. However, finely ground products of pregelatinized starches are difficult to disperse in water homogeneously since they hydrate rapidly at contact with water and form lumps or fish eyes. The hydration rate can be slowed down by premixing the pregelatinized starches with other ingredients. At room temperature, the pastes of the drum cooked/dried starch reduce consistency and have a dull grainy appearance. The gel also has reduced strength. These negative attributes appear to result from the leached amylose molecules establishing a partially intractable network during drying.[6]

8.3.1.3 Applications

Pregelatinized starch is used as a thickening agent for pie fillings, puddings, sauces, and baby foods. Starches that generate a pulpy texture are used to modify the texture of soups, gravies, and sauces. Such starches are made by cross-linking a normal starch followed by drum cooking/drying and grinding to a specific particle size.[65,66] The particles swell upon rehydration, but do not disperse under heat and shear. The swollen particles resemble the cell wall material dispersed in food ingredients prepared from vegetables and fruits.[6]

8.3.2   Heat-Moisture Treatment

8.3.2.1   Definition and Preparation

Heat-moisture treatment of starch is a physical treatment in which starches are treated at varying moisture levels (<35%) for a certain period of time at a temperature above the glass transition temperature but below the gelatinization temperature. However, the temperature is often chosen without considering the gelatinization temperature.[67]

8.3.2.2   Properties

Heat-moisture treatment of maize, wheat, yam, lentil, potato, and mucuna bean starches has no effect on the shape and size of granules. However, the wide-angle x-ray scattering patterns are altered from B-type to A- (or C-) type for potato and yam starches and C-type to A-type for arrowroot and cassava starches after heat-moisture treatment. The A-type pattern of cereal starches is not changed by this treatment, but the x-ray diffraction pattern may sharpen or intensify.[68,69] This indicates that some double helices are moving into a more perfect position in the crystalline phase. The perfection of crystallites in heat-moisture treated waxy maize or dull waxy maize starches results in the 1 to 2°C increase in melting of crystallites (Table 8.4). In some cases, for cereal starches, amylose-lipid complexes are formed by heat-moisture treatment as indicated by a decrease in the apparent amylose content.[67,68]

Heat-moisture treatment normally increases the gelatinization temperature, broadens the gelatinization temperature range (Table 8.4),[68,69] and decreases swelling power. Enthalpy of gelatinization (ΔH_{gel}) for B-type starch, such as potato or yam, decreases. It indicates that some double helices are unraveled after heat moisture treatment. The enthalpy of gelatinization for A-type starch (wheat, normal, and waxy maize starches) did not change upon heat-moisture treatment, even though the onset temperature of gelatinization increased 2 to 11°C (Table 8.4), indicating that an insignificant number of double helices unraveled (unchanged ΔH_{gel}), whereas a significant number of double helices moved into more perfect crystalline position (increased gelatinization temperature).[6] Pasting curves (Figure 8.6) showed that heat-moisture treatment reduced the 95°C paste consistency except for wheat starch. This reduction could be caused by chain-cleavage, amylose-lipid complexing, or amylose-amylose association during heat-moisture treatment.[6]

Heat-moisture treatment also decreases susceptibility of starch to acid hydrolysis for maize and pea starches. Impact of heat-moisture treatment on

TABLE 8.4

Gelatinization Endotherms of Native and Heat-Moisture Treated Starches

Source	Treatment	T_o (°C)	T_p (°C)	T_c (°C)	$T_c - T_o$ (°C)	H (cal/g)
Wheat	Native	56	61	67	11	2.3
	HMT	65	70	78	13	2.3
Oat	Native	60	64	70	10	2.5
	HMT	64	75	80	16	2.5
Lentil	Native	55	61	68	13	1.8
	HMT	64	71	78	14	1.8
Potato	Native	54	59	64	10	3.8
	HMT	65	71	80	15	2.7
Yam	Native	72	77	83	11	5.0
	HMT	77	84	90	13	3.6
Normal maize	Native	59	66	73	14	3.4
	HMT	62	71	82	20	3.4
Waxy maize	Native	65	73	82	17	3.8
	HMT	67	74	84	17	3.8
Dull waxy maize	Native	62	73	83	21	3.7
	HMT	63	75	84	21	3.7

Note: HMT: heat-moisture treatment at 100°C, 16h and 30% moisture. T_o, T_p, and T_c indicate the temperature of the onset, midpoint, and end of gelatinization. H: enthalpy of gelatinization.

Source: From Hoover and Vasanthan, 1994; Hoover and Manuel, 1996.[18,19]

the enzyme hydrolysis of starch varies depending on botanical origin and treatment conditions; both decreased and increased rates of enzyme hydrolysis were observed after heat-moisture treatment.[67]

8.3.2.3 Applications

Heat-moisture treated potato starch can be used to replace maize starch in times of shortage of corn starch[70] and to improve the baking quality.[71] Heat-moisture treated cassava starch exhibits excellent freeze-thaw stability and could be used in pie filling with good organoleptic properties.[69] Resistant starch type 3 (RS₃) is the retrograded amylose formed after heat-moisture treatment, which occurs in cooked and cooled potatoes, for example, and in ready-to-eat breakfast cereals. Commercially developed RS₃ starches such as NOVELOSE by National Starch and Chemical Company are derived from heat-moisture treated high amylose corn starch and are ready for the market as a function food ingredient.

8.3.3 Annealing

8.3.3.1 Definition and Preparation

Annealing of starch is a physical treatment whereby the starch is incubated in excess water (>60% w/w) or intermediate water content (40 to 55% w/w)

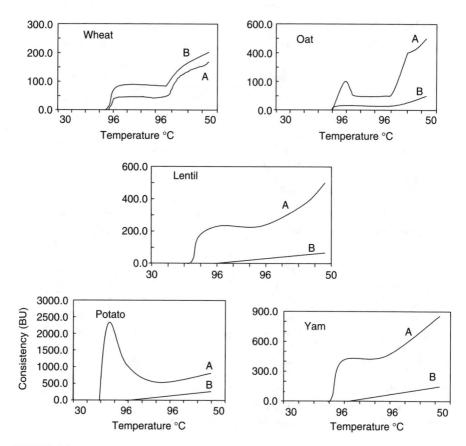

FIGURE 8.6

Pasting curves of (A) native and (B) heat-moisture treated starches (16h, 100°C, 30% moisture). (Adapted from Hoover and Vasanthan, 1994.)[68]

at a temperature between the glass transition temperature and the gelatinization temperature for a certain period of time.[67,73] Annealing increases starch gelatinization temperature and sharpens the gelatinization range. However, there are few commercial processes to be used to generate starches with higher gelatinization temperatures. Often annealing is applied unintentionally, such as the steeping step used in the maize wet-milling process.

8.3.3.2 *Properties*

Annealing modifies the physicochemical properties of starch without destroying the granule structure. For annealed wheat and potato starches, no changes were found in the wide-angle and small-angle x-ray scattering patterns, and no significant changes were found in crystalline type and degree of crystallinity. Annealing may induce formation of the amylose-lipid complex, but it is unlikely to affect existing amylose-lipid complexes because their dissociation temperature is 95 to 125°C, which is far beyond the annealing temperature.[67]

Annealing elevates starch gelatinization temperature, decreases gelatinization temperature range, and reduces swelling power. Annealed starch granules contain more glassy amorphous regions and greater alignment of amylopectin double helices, resulting in the restriction of starch granule hydration during gelatinization and elevation of gelatinization temperature.[64] The effects of annealing on viscosity are complex. Annealed potato and maize starches exhibit a decrease in viscosity peak with an increased onset temperature of swelling, while annealed rice and pea starches exhibit an increased viscosity.[69] The effects of annealing on pasting characteristics for different starch sources are overviewed in Table 8.5.

TABLE 8.5

Changes in Starch Pasting Characteristics after Annealing

Technique	Starch	Onset Temp[a]	Visc Max[b]	Visc Cooling[c]	Reference[67]
Ostwald viscometer[d]	Potato	+	−	0	Wiegel (1933)
Corn Industries viscometer[e]	Potato	+	−	+	Hjermstad (1971)
Viscoamylograph[e]	Potato	+	−	=	Kuge and Kitamura (1985)
		+	−	+	Stute (1992)
		+	−	+	Hoover and Vasanthan (1994b)
		+	−	+	Jacobs et al. (1995, 1996)
	Wheat	=	+	+	Hoover and Vasanthan (1994b)
		−	+	+	Jacobs et al. (1995, 1996)
	Lentil	+	−	−	Hoover and Vasanthan (1994b)
	Oat	+	+	−	Hoover and Vasanthan (1994b)
	Rye	+	=/−	=/−	Schierbaum and Kettlitz (1994)
	Pea	=	+	+	Jacobs et al. (1995, 1996)
	Rice	−	+	+	Jacobs et al. (1995, 1996)
Rapid viscoanalyzer[f]	Maize	=	−	−	Deffenbaugh and Walker (1989a)
	Wheat	+	−	=	Deffenbaugh and Walker (1989a)
		+/−	+	+	Jacobs et al. (1995, 1996)
	Potato	+	−	+	Jacobs et al. (1995, 1996) Eerlingen (1997)
	Pea	+/=	+/−	+	Jacobs et al. (1995, 1996)
	Rice	=	+	+	Jacobs et al. (1995, 1996)

Note: + is increase; − is decrease; = is unchanged; 0 is not determined.

[a] Onset temperature of viscosity development.
[b] Maximum viscosity during heating.
[c] Viscosity on cooling.
[d] 1% starch suspension.
[e] 5–6.6% starch suspension.
[f] 6.6–10% starch suspension.

Source: Adapted from Jacobs and Delcour, 1998.[67]

Annealing affects the degree of susceptibility of starch to acid and enzymatic hydrolysis and the susceptibility varies with different starch sources. For example, annealing was found to increase enzymatic susceptibility of barley, oat, and wheat starches, whereas the inverse was true for potato starch.[67,73,74]

8.3.4   Extrusion

8.3.4.1   Extrusion Process

Extrusion combines several unit operations, including mixing, kneading, shearing, heating, cooling, shaping, and forming. Material is compressed to form a semi-solid mass under a variety of controlled conditions and forced to pass through a restricted opening at a pre-determined rate. There are two main types of cooking extruders: single-screw extruder and twin-screw extruder. The latter is further classified into co-rotating and counter-rotating, depending on the direction of screw rotation. The temperature in the extruder can be as high as 200°C, while the residence time is short (~10 to 60 seconds).[75] Extrusion cooking is considered to be a high temperature short time (HTST) process. An extruder can be used as a bioreactor or chemical reactor for starch modifications, such as thermomechanical gelatinization, liquefaction, esterification, and etherification.

Extrusion process can be used to produce pregelatinized starch, in which granular starch with different moisture contents is compressed into a dense, compact mass, and disrupted by the high pressure, heat, and shear during the process. The extruded starch is further dried and ground to a desired particle size for food applications.

8.3.4.2   Functional Properties

8.3.4.2.1   Transformation of Extruded Starch

Starch polymers are degraded into smaller molecules during extrusion processing, and amylopectin is more influenced than amylose. Neither short oligosaccharide nor glucose is formed, however, the formation of amylose-lipid complexes occurrs during the extrusion process.[76]

Starch granule and its crystalline structure are destroyed either partially or completely, depending on the ratio of amylose/amylopectin and the extrusion variables, such as moisture, temperature, and shear. Cereal starches such as corn starch produce a V-type x-ray diffraction pattern when extruded at a lower temperature (135°C) and higher moisture (22%), whereas a new E-type or extruded structure is formed at a higher temperature (185°C) and lower moisture (13%). The E-type spectrum can be irreversibly transformed into the V-type at moisture content of 30%.[77]

8.3.4.2.2   Water Solubility and Water Absorption

Water solubility and absorption are two main functional properties of extruded starches when they are dispersed in water. Two indices have been used to determine the capacity of starch solubility and water absorption.

The water absorption index (WAI) is the weight of gel formed per gram of dry sample, which can be determined by the method of Anderson et al.[78] The water solubility index (WSI) is the percentage of the dry matter in the supernatant from the water absorption determination. Extruded starch shows an increase in water solubility, but a decrease in water absorption (less able to form a gel). Increasing severity of thermal treatment in the extruder will progressively increase WSI, indicating that more starch polymers have been degraded into smaller molecules.

8.3.4.2.3 Pasting Properties

The pasting characteristics of extruded starches are shown in Figure 8.7. Extruded starches lack a gelatinization peak during the heating process in excess water, whereas native starch exhibits a rapidly rising viscosity peak with the onset of gelatinization. Extruded starches exhibit higher initial uncooked paste viscosity and the consistency decreases sharply in the range of 90 to 96°C. Increasing severity of extrusion treatment (low moisture and high temperature) results in the decrease in initial cold viscosity, and increasing extrusion temperature causes a reduction in the pasting consistency.[75] Normally, the increase in specific energy mechanical input results in a reduction in viscosity and an increase in solubility.[75]

Extruded starch absorbs water rapidly to form a paste at room temperature. Gels of extruded starch have lower retrogradation values than nonextruded gelatinized starch.[75]

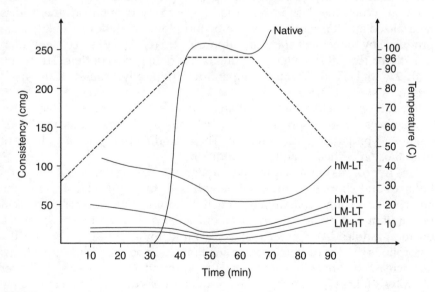

FIGURE 8.7
Brabender viscoamylograms of native and extruded wheat starches at different extrusion conditions. hM-LT: 125°C, 34.7% moisture content (db); hM-hT: 180°C, 34.7% moisture content (db); LM-LT: 130°C, 24.4% moisture content (db); LM-hT: 180°C, 23.9% moisture content (db). (Adapted from Colonna and Mercier, 1989.)[75]

### 8.3.4.3	Applications

Pregelatinized starches have been widely used in instant foods and have more recently been prepared mainly by extrusion cooking, because of various advantages of this process over traditional methods. Extrusion also has been used in the food industry to manufacture numerous products, such as ready-to-eat cereals, snacks, confectionery products, texturized vegetable protein, and macaroni, as well as pet foods.[79] Extruders are used as a bioreactor to promote enzymatic conversion of starch and to initiate starch liquefaction and saccharification in a single pass.[75] Extruders have been used as a chemical reactor for the chemical modification of starch, such as carboxymethylated and cationic potato starch,[80] starch phosphates,[81] anionic starch (ester of various dicarboxylic acids),[82] oxidized starches,[83] and starch-fatty acid esters.[84]

## 8.4.	Starch Hydrolyzates and Their Applications

### 8.4.1	Acid Hydrolysis

Acid hydrolysis involves suspending starch in an aqueous solution of hydrochloric acid (HCl) or sulfuric acid (H_2SO_4) and maintaining at a temperature between ambient and just below the pasting temperature to prevent gelatinization. When the desired reduction in viscosity is obtained, the acid is neutralized and the starch recovered by filtration. Amylodextrins Nägeli are produced by immersing starch in sulphuric acid (typically at a concentration of 1.7 M or 15%) at room temperature for a certain period of time. Treatment with 2.2 M HCl at elevated temperatures (typically around 30 to 40°C) produces starch Lintner.

   In addition to acid treatment in an aqueous solution, the acid hydrolysis of starch can also be prepared in anhydrous alcohols such as methanol, ethanol, 2-propanol, and 1-butanol. These modified starches are different from those modified in aqueous conditions.[85]

   Acid hydrolysis of starch proceeds randomly, cleaving both α-1,4 and α-1,6 linkages and shortening the chain length with time. α-1,4 linkage and the amorphous regions containing α-1,6 linkage are more accessible to acid penetration and hydrolysis. Acid modification includes a two-stage attack on the granules. At the early stage, a rapid initial attack occurs on the amorphous regions of starch containing branching points with α-1,6 linkage; an increase in linear fraction in starch is found in this stage. A slower hydrolysis takes place on the more crystalline areas during the second stage. Factors influencing acid modification include acid type and concentration, temperature and time, type of alcohol, and type of starch.[85]

8.4.2 Enzyme Hydrolysis

There are many enzymes used in starch hydrolysis to alter starch structure and to achieve desired functionality. Enzymes hydrolyze (1→4) or (1→6) linkages between α-D-glucopyranosyl residues. The most common enzymes for starch modification include α-amylase, β-amylase, glucoamylase, pullulanase, and isoamylase. These enzymes have been isolated from fungi, yeasts, bacteria, and plant kingdoms.

Starch hydrolysis involves liquefaction and saccharification of starch. Liquefaction is a process to convert a concentrated suspension of starch granules into a solution of low viscosity by α-amylase. Saccharification converts starch or intermediate starch hydrolysis products to D-glucose by enzymes such as glucoamylase (amyloglucosidase). The reducing power of the product is a measure of the degree of starch breakdown. Dextrose equivalent (DE) is used to define the reducing power. It is measured in a specific way and calculated as dextrose, expressed as a percentage of the dry substance. By definition the DE of dextrose is 100.

8.4.2.1 *α-Amylase (1, 4-α-D-Glucan Glucanohydrolase, EC 3.2.1.1)*

This class of α-amylase can hydrolyse the (1→4)- α-glucosidic bonds of starch in an endo-action. Endo-action is when hydrolysis occurs in a random fashion at any (1→4)-linkage within the starch chain to rapidly reduce the molecular size of starch and the viscosity of the starch solution. It results in a mixture of linear and branched oligosaccharides, and eventually maltotriose, maltose, glucose, and a range of branched α-limit dextrins. α-Amylase can be extracted from animal, plant, and microbial kingdoms. The majority of α-amylases have molecular weights in the range of 47,000 to 52,000. Maximum activities of the α-amylases are usually achieved in acidic conditions between pH 4.8 and 6.5, but the enzyme activities vary with enzyme sources. The pH optimum for plant and microbial α-amylase is generally lower than that of the animal α-amylase. The thermostability of enzyme also varies with enzyme sources. Bacterial amylases are the most thermostable. Fungal amylases are the most thermolabile. Plant amylases are intermediate. Although the α-amylase from microbial sources are the basis for many commercial applications, the plant enzymes have particular importance in the brewing, distilling, and baking industries.

8.4.2.2 *β-Amylase (1, 4-α-D-Glucan Maltohydrolase, EC 3.2.1.2)*

β-Amylase exists in many types of plants such as barley, wheat, sweet potatoes, and soybeans. β-Amylase acts in an exo-fashion. It hydrolyzes from the nonreducing ends of the outer chains of starch or related polysaccharides. Amylose is converted almost completely to maltose by β-amylase. The hydrolysis of amylopectin by β-amylase results in conversion about 50 to

60% to maltose. β-Amylase is unable to hydrolyze (1→6)-linkages in the amylopectin. Thus, it produces high molecular weight limit dextrins containing all the original branch points. β-Amylase from plant sources is used in the production of high maltose syrups of 80% maltose content. However, β-amylase from microbial sources is widely used in industry because of its easy production by fermentation.

8.4.2.3   Glucoamylase (Amyloglucosidase) (1, 4-α-D-Glucan Glucohydrolase, EC 3.2.1.3)

Glucoamylase occurs almost exclusively in fungi such as *Aspergillus niger*. It is an exo-acting enzyme that releases D-glucose from the nonreducing ends of the starch chain. Glucoamylase catalyzes the hydrolysis of both the α-D-(1→4) and α-D-(1→6)-linkages to convert starch to D-glucose. However, the rate of hydrolysis of α-D-(1→4)-linkages differs from that of α-D-(1→6)-linkages. Glucoamylase is the enzyme employed in starch hydrolysis to produce high conversion syrups and high dextran syrups.

8.4.2.4   Pullulanase (Pullulan 6-Glucanohydrolase, EC 3.2.1.41)

Pullulanase is an endo-acting enzyme that hydrolyzes α-D-(1→6)-linkages of pullulan. Pullulan is a linear polysaccharide composed of maltotriose units linked via α-D-(1→6) bonds. This enzyme can also catalyze the hydrolysis of α-D-(1→6) linkage in starch, amylopectin, and limit dextrins to give products containing no branch points. However, the hydrolysis of the (1→6)-linkages in the substrate depends on enzyme sources. For example, enzymes from *Bacillus macerans* and *Bacillus cereus* var. *mycoides* have very little action on amylopectin but are able to degrade its β-limit dextrin, while the enzyme from *Aerobacter aerogenes* almost totally debranches amylopectin and its β-limit dextrin but has little or no action on rabbit liver glycogen.

8.4.2.5   Isoamylase (Glycogen 6-D-Glucanohydrolase, EC 3.2.1.68)

Isoamylase is a starch-debranching enzyme. It hydrolyzes the α-1,6-glucosidic linkages in amylopectin and glycogen, but rarely in pullulan. The enzyme exists in yeast and fungi such as *Pseudomonas* and *Cytophaga*. The optimum conditions for isoamylases are in the range of pH 5.0 to 6.4 and in the range of temperature 40 to 50°C.

8.4.2.6   Cyclomaltodextrin D-Glucotransferase (1,4-α-D-Glucan 4-α-D-[1,4-α-D-Glucano]-Transferase [Cyclizing] (CGTase) EC 2.4.1.19)

This enzyme is capable of hydrolyzing starch, amylose, and other polysaccharides to a series of nonreducing cyclomalto-oligosaccharides that contain six, seven, or eight D-glucopyranosyl residues. The seven residue cyclomaltoheptaose is the most common cyclodextrin. The enzyme is generally found in bacteria and was also discovered in archaea. The enzyme (CGTase) isolated from *Bacillus macerans* has the biggest market share of the commercially available CGTases.[86]

8.4.3 Functional Properties and Applications of Starch Hydrolyzates

8.4.3.1 Starch Hydrolyzates

Starch hydrolyzate is the product obtained by the acid and/or enzymatic hydrolysis of starch, producing a mixture of low molecular weight polysaccharides, oligosaccharides, and simple sugars. Although starch hydrolyzates can be obtained by acid treatment, high conversion hydrolyzates are obtained almost exclusively by the use of enzymes. The use of enzyme modification has a number of advantages including fewer by-products, more specific hydrolysis products and high yield. Another advantage is that enzyme modification ensures better control of the process and end products with particular properties.

Enzyme modification of starch may be carried out using one or more of the above enzymes under appropriate conditions of temperature and pH. The optimum conditions for a particular enzyme depend on the enzyme source and may change over the year. Enzyme hydrolysis is conducted at relatively low temperatures and pH 4 to 5 where reducing sugars have maximum stability.

8.4.3.2 Properties and Applications of Enzyme Hydrolyzates

Depending on the enzymes and hydrolysis conditions, the products of starch hydrolyzate include maltodextrins, syrups, and cyclodextrins.

Maltodextrins have a DE of 20 or less. Syrups have a DE of 20 or more. Cyclodextrins are cyclic compounds. The cyclic structure of these hydrolyzates does not have a reducing end group. They are more stable than the common dextrins to degradation by acid, alkali, or enzymes.

Maltodextrins have many important functional properties including bulking, crystallization inhibition, promotion of dispersibility, freezing control, and binding. They have been widely employed in the food industry as fat replacers and bulking agents, encapsulation of flavor, colorants, in bakery and confectionery, beverages, dairy, desserts, meats and gravies. Maltodextrins are also used in coffee whiteners. Maltodextrins of DE 5 to 10 are used as encapsulants and bulking agent for flavorings.[87]

Syrups are widely used in the food industry. Syrups with DE 40 to 50 are used to produce marshmallows and caramels. Syrup with a DE of 62 and fructose syrup are used in the baking industry to make bread, rolls, and buns. It is easy to use syrups in development of these products because of their low viscosity. In these products, the reducing sugars contribute to the brown crust coloration. Syrup of DE 42 is used in table syrups, jams, and jellies, and in processed meats. This syrup stabilizes the fat-water emulsion

TABLE 8.6
Properties and Industrial Applications of Corn Syrups

Type of Syrup	Properties	Industrial Applications
Acid converted		
43 DE	High viscosity, moderately sweet	Sugar confectionery, ice cream, frozen dessert, jams, jellies, chewing-gum
55 DE	Increased sweetness	Baking, snack-cakes, fillings, cookies
Acid-enzyme converted		
63 DE	Increased moisture holding, higher sweetness, higher fermentability and low viscosity	Brewing, canned and frozen fruits or vegetables, confectionery, bar-candy, jams, jellies
70 DE	Increased sweetness, reduced content of higher sugars, reduced viscosity	Baking, brewing, yeast-raised goods
95 DE	Commercial liquid "dextrose"	Baking, brewing
High maltose		
43 DE	Reduced crystallization, improved color stability and taste	Confectionery, jams, jellies, ice cream, hard candy
52 DE	Increased maltose content	Confectionery, ice cream

Source: From Kennedy et al., 1987.[88]

and inhibits oxidative degradation. Syrups are also widely used in dairy foods to give texture to ice cream and improve shelf life. The properties and industrial applications of corn syrup are listed in Table 8.6.[88]

Starch hydrolyzates from enzyme hydrolysis are also used in the pharmaceutical industry. For example, glucose syrup of low DE can be used to carry calcium and other nutrients into the lower digestive tract. In addition, starch hydrolyzate such as glucose is an essential carbohydrate in industrial fermentation processes to produce ethanol, acetone, *n*-butanol, organic acids, and other industrial chemicals.

The application of cyclodextrins has also been developed. Cyclodextrins have different molecular sizes. The presence of the cavity within the cyclic structure offers unique properties for cyclodextrins. The cavity permits the formation of crystalline inclusion complexes with a wide variety of organic and inorganic molecules. Cyclodextrins have been used to stabilize aroma compounds during extrusion and thermal processing of foods.

8.4.3.3   *Properties and Applications of Acid Hydrolyzates*

The x-ray diffraction pattern of acid modified starches remains unchanged after different treatments. Figure 8.8 shows similar x-ray diffraction patterns for regular maize starch treated with 2.2 M HCl at different times. However,

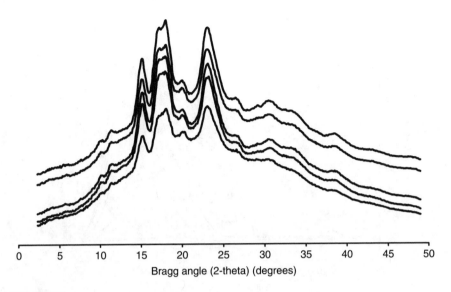

FIGURE 8.8
The time effect of acid treatment (from top to bottom 0, 1, 3, 8 and 12 weeks) of regular corn starch on x-ray diffraction pattern.

the percentage for crystallinity of the starch increased with the increase of acid modification time.[89] It is found that A-type starches, such as cereal starch, are usually more susceptible to acid treatment than B-type ones such as potato starch.

The average molecular weight of acid modified starch decreases as acid hydrolysis time increases. Figure 8.9 shows the high performance size exclusion chromatography (HPSEC) profiles of regular maize starch after hydrochloric acid treatment for different time periods. The average molecular weight (M_w) of starch was 4.1×10^5, 6.2×10^4, 2.2×10^4, 1.5×10^4, and 1.2×10^4 after treated with 2.2 N HCl for 0, 1, 3, 6, and 12 weeks, respectively. As the molecular size decreases, the intrinsic viscosity decreases, whereas gelatinization temperature and hot water solubility increase. The hot pastes of acid modified starches from corn and wheat are relatively clear fluids with lower hot paste viscosity. However, opaque rigid gels are formed on cooling due to retrogradation.

Acid modification in various alcohols produced starches with different average degree of polymerization (DP) values — the highest value obtained in methanol and the lowest value in 1-butanol.[85] The size distribution of acid modified starch chains is narrower and more homogeneous than those of native starch. When potato starch is modified by acid in 2-propanol and 1-butanol, the amylose is completely absent and a minimum amount of alcoholysis occurred.[85]

Acid modified starch (hydrolyzate) has been widely applied in the food industry. Due to the need of low hot viscosity and gelling on cooling, acid hydrolyzate are used extensively in the preparation of confectionery gums, pastilles, and jellies. Acid hydrolyzate is fully cooked with water, sugar, and

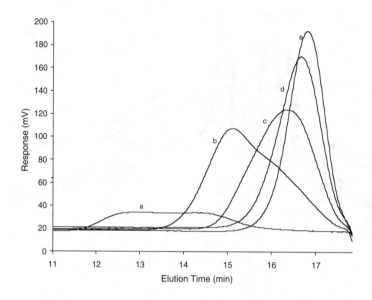

FIGURE 8.9
HPSEC profiles of regular maize starch after hydrochloric acid treatment for different times (a: native corn starch; b: 1 week; c: 3 weeks; d: 6 weeks; e: 12 weeks).

glucose syrup to gelatinize starch before adding flavor and color agents. The mix of gelatinized starch is immediately poured into the moulds of the product. When the mix cools, the starch gels take on the shape of the molds. The gelled confection is then allowed to dry to obtain the final product.

Acid hydrolyzates are also suitable for film formation. Because they are much lower in viscosity than native starch, they can be cooked and pasted at much a higher concentration than native starch. Acid hydrolyzates absorb less water. Thus, the starch film will dry faster and provide adhesion.

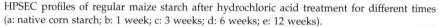

8.5   Biotechnological Modification

One alternative to physical and/or chemical modification is to identify sources of native starch with desired functionality through extensive breeding programs and the characterization of mutant varieties. Biotechnological modification of crops may allow the production of desired starch products without chemical and/or physical modification. The structure of starch is controlled by the genetic make up of the plant, but the exact effects of many genes are not very clear due to the complexity of the system. Many studies have been carried out to identify those genes responsible for the production of starch. However, there are major concerns within the scientific community and the public at large about the safety of genetic modification of starches.

Biotechnological developments in starch synthesis have a great impact on modifying the basic structural composition of starch. The pathway of starch synthesis is currently known to involve the inter-conversion of sugars, sugar-phosphates, and nucleotide-sugars. Four enzymes involved in starch biosynthesis have been identified for maize: ADP-glucose pyrophosphorylase, starch synthase, starch branching enzyme, and starch debranching enzymes. Some isoforms of these enzymes were also identified. However, the exact role of each isoform in the biosynthesis pathway is not clear. Nevertheless, amylopectin is synthesized by starch synthase and branched by a branching enzyme. Amylose is synthesized by granule bound starch synthase (GBSS). These enzymes have many isoforms that play various roles in starch synthesis. Thus, biotechnological modification could result in altering starch structure in terms of the ratio of amylose to amylopectin; amylopectin branching patterns and branch chain lengths; phosphate, lipid, and protein content; cross-linking between chains. The structural changes may be achieved by an empirical approach in which individual enzymes in the pathway are under and over expressed. For those interested in this area, please refer to attached references for details.[90–93]

This section mainly introduces physicochemical properties of several maize mutant starches that have been employed in industrial products. The typical mutant starches described in this section include starches from single, double, and triple mutants. The fine structure and functional properties of these starches are discussed. For information on other mutant starches, please see related literature.[94,95]

8.5.1 Mutant Maize Starches

Several mutants and various multiple mutant combinations are found to affect the quantity and quality of starches and proteins in maize, pea, sorghum, barley, and rice.[94] The changes include the proportion of amylose and amylopectin, granular morphology, physicochemical properties, chain length and its distribution of the fine structure of amylopectin in the mutant starches. Since chemical and gene transfer modification of starches are limited due to strict food regulations, mutation selection provides an opportunity to produce novel specialty starches to meet the needs of new product development in the food industry.

In maize, the major mutants altering starch properties include *waxy* (*wx*), *amylose-extender* (*ae*), *sugary* (*su*), *sugary-2* (*su2*) and *dull* (*du*). The greater diversity of mutants can be created by making double or triple mutants, and by making crosses between mutants (intermutants). The *waxy* (*wx*) produces starch granules in the endosperm with nearly 100% amylopectin. Starch and dry weight production in *wx* kernel are equal to that in *normal* kernels and increase at similar rates. The *amylose-extender* (*ae*) genes cause an increase in apparent amylose content in starch of maize as well as of pea cotyledons[95] and barley pollen. The mutants also alter the morphology of starch granules compared to *normal*. Starch and dry weight production in *ae wx* kernel are reduced and sugar is increased in the high amylose genotypes compared to normal kernels or seeds. The *sugary (su)* mutant results in increased sugar levels and decreased both amylose and amylopectin in starch. The *sugary-2 (su2)* causes a reduction in the molecular association between starch molecules of the granules. The *dull (du)* mutant results in increased apparent amylose content in the granules compared to the *normal*. In the double mutant *ae wx*, *wx* prevents the production of amylose, and the *ae* reduces the degree of branching. As a result, it has the accumulation of loosely branched polysaccharides (*aewx* starch).

8.5.2 Chemical Composition and Semi-crystalline Structure of Mutant Starches

By combining the mutations, a broad range of granular structure can be obtained in maize starch with different amylose content and crystalline structure as shown in Table 8.7.[97] Apparent amylose content in the mutant starches varied from 0 to 100%. The *wx* and *wxdu* starches contain almost 100% amylopectin. The *wxae* starch contains almost 100% amylose. Different apparent amylose contents result in new views for the structure of traditional amylopectin. The *normal*, *wx* and *duwx (wxdu)* starches are pure A-type semi-crystalline structure, a typical cereal starch structure. The *ae* maize starch showed a pure B-type x-ray diffraction pattern. However, some mutant starches showed x-ray diffraction pattern with combination of A-, B- and V-type (Table 8.7), indicating different mutation caused different packing in the starch granules with different crystallinity. The crystallinity of mutant starches varied from 19 to 48%.[97] The different structure of mutant starches

TABLE 8.7

Apparent Amylose Content, Crystallinity, A-, B-, and V-Type Fraction, and Final Hydrolysis Extent of Different Mutant Maize Starches

Mutant Starch	Apparent Amylose (%)	Crystallinity (%)	A-type Fraction	B-type Fraction	V-type Fraction	Final Hydrolysis Extent (%)
normal	24	31	1.00	0.00	0.00	97
wx	ND	48	1.00	0.00	0.00	98
ae	63	30	0.00	1.00	0.00	15
du	45	28	0.70	0.10	0.20	74
su2	50	19	0.15	0.15	0.70	99
aedu	56	28	0.25	0.45	0.30	60
dusu2	58	24	0.35	0.20	0.45	78
wxdu	ND	44	1.00	0.00	0.00	98
wxae	100	44	0.30	0.70	0.00	32

Note: ND is not detected

Source: From Gerard et al., 2001.[95]

also resulted in changes of amylolysis. For example, when mutant starch is hydrolyzed by α-amylase, the extent of hydrolysis is between 15 and 99%.[97]

8.5.3 Relationship between Molecular Structure and Functional Properties of Mutant Starches

In the presence of an excess amount of water (e.g., ~70%, w/w), mutant starches exhibited different thermal behavior upon heating as shown in Figure 8.10 and Table 8.8. Although the sources of mutant starches and experimental conditions are different in Figure 8.10 and Table 8.8, *aewx* starch demonstrated the highest gelatinization temperatures and enthalpy. The *aewx* starch also showed a broader DSC endotherm than the *duwx* and *wx* starch under the same retrogradation condition.[98,99] Retrogradation of *aewx* starch occurs more rapidly than other mutant starches. In contrast, retrogradation rate and extent of *su2wx* starch are very slow and small compared to other mutant starches.[100] It is believed that different external conditions such

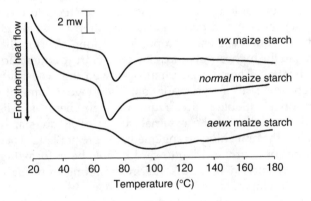

FIGURE 8.10
Typical DSC thermograms for gelatinization of 30% starch heated at 10°C/min from 5 to 180°C.

TABLE 8.8

Gelatinization Properties of Mutant Maize Starches

Mutant Starches	Onset Temperature (°C)	Peak Temperature (°C)	Enthalpy (ΔH) (J/g)
wx[a]	54.2	75.1	15.3
du wx[a]	51.3	75.7	15.4
ae wx[a]	61.8	95.4	18.4
su2 wx[b]	51.9	58.2	12.5
ae du wx[a]	52.6	81.1	15.1

Source: From (a) (25% starch content): Shi and Seib, 1994.[98] (b) (30% starch content): Liu and Thompson, 1998.[100]

TABLE 8.9

Areas (% of Total) of Groups of Unit Chains
in Maize Starches from Four Mutant Starches

Starch	Degree of Polymerization				
	6–11	12–16	17–20	21–30	31–60
wx	19.3	33.2	19.4	23.5	4.8
du wx	19.0	35.4	17.1	22.7	5.8
ae wx	10.4	26.0	20.1	30.1	13.1
ae du wx	14.7	31.0	16.2	27.4	10.7

Source: From Shi and Seib, 1994.[98]

as water content, initial heating conditions, and storage conditions influence the gelatinization and retrogradation properties of starch. However, the molecular structure of mutant starch is the major internal factor influencing the functionality of starch.

As shown in Table 8.9, *aewx* starch had the highest proportion of chains longer than DP 16 and the lowest proportion of chains shorter than DP 11. The higher amount of long chains in the *aewx* starch could result in double helices varying in size and perfection. In addition to overall chain length, *aewx* starch contained a greater proportion of long interior chains.[98] Thus, *aewx* starch had the highest gelatinization temperature and broad endotherm peak upon heating. The *aewx* starch also showed the fastest retrogradation and the highest extent of retrogradation. The longer chains of *aewx* starch could form longer double helices requiring high energy for dissociation.[99] Although the mobility of the external chains of *ae wx* starch is far less constrained by the branching arrangement, the greater length of these chains might account for a strong drive toward double helix formation, sufficient to rapidly overcome extensive disordering resulting from high dilution or heating to high temperatures.[101]

In contrast, *su2wx* starch had the highest proportion of short chains (DP 12 or below) among the mutant starches.[101] The large proportion of short chains could result in the lowest gelatinization temperature and enthalpy (Table 8.8) and the slowest retrogradation for *su2wx* starch. After the loss of the ordered granular structure on gelatinization, chains that are too short to form stable double helices in *su2wx* starch would serve to dilute the longer chains and sterically interfere with formation of double helices from longer chains, thus retarding the retrogradation process. The slow retrogradation rate could meet the quality required for frozen foods.

Similar gelatinization properties of *duwx* and *wx* starches (Table 8.8) could be due to their similar chain length and its distribution (Table 8.9). However, different retrogradation behavior was observed between *duwx* and *wx* starches.[101] This may relate to the periodicity difference in branching within the cluster for these two mutant starches.[101]

In summary, mutation affects the branch length and degree of branching of amylopectin resulting in altered physicochemical properties of starch.

Mutation alters not only the chemical composition, but more importantly, the chain length and its distribution in amylopectin, thus changing the formation of double helices and wide range of crystalline structure. These different mutant starches offer various functionality for food and nonfood applications.

References

1. Jobling, S.A., Schwall, G.P., Westcott, R.J., Sidebottom, C.M., Debet, M., Gidley, M.J., Jeffcoat, R., and Safford, R. A minor form of starch branching enzyme in potato (*Solanum tuberosum* L.) tubers has a major effect on starch structure: cloning and characterisation of multiple forms of SBE A. *Plant Journal*, 18, 163–171. 1999.
2. Rutenberg, M.W. and Solarek, D. Starch derivatives: Production and uses. In: R. Whistler , J.N. BeMiller and E.F. Paschall, eds. *Starch: Chemistry and Technology*. New York: Academic Press, Inc., 314–388. 1984.
3. Code of Federal Regulations, Title 21, Paragraph 172.892. Food Starch-Modified, U.S. Government Printing Office, Washington, D.C, 2001.
4. Wurzburg, O.B. Converted starches. In: O.B. Wurzburg. ed. *Modified Starches: Properties and Uses*. Boca Raton, FL: CRC Press, 17–29. 1986.
5. Furin, J.C. Batter Starch. U.S. Patent No. 3,767,826. 1973.
6. Seib, P.A. *Starch Chemistry and Technology, Syllabus*. Kansas State University, Manhattan, KS, 1996.
7. Wu, Y. and Seib, P.A. Acetylated and hydroxypropylated distarch phosphates from waxy barley: paste properties and freeze-thaw stability. *Cereal Chemistry*, 67, 202–208.1990.
8. Hudson, R.F. and Moss, G. The mechanism of hydrolysis of phosphorochloridates and related compounds. Part IV. Phosphoryl chloride. *Journal of Chemistry Society*, 703, 3599–3604. 1962.
9. Kasemsuwan, T. and Jane, J. Location of amylose in normal starch granules. II. Location of phosphodiester cross-linking revealed by phosphorus-31 nuclear magnetic resonance. *Cereal Chemistry*, 71, 282–287. 1994.
10. Srivastava, H.C. and Patel, M.M. Viscosity stabilization of tapioca starch. *Starch*, 25, 17–21. 1973.
11. Reddy, I and Seib, P.A. Modified waxy wheat starch compared to modified waxy corn starch. *Journal of Cereal Science*, 31, 25–39. 2000.
12. Wurzburg, O.B. Cross-linked starches. In: O.B. Wurzburg. ed. *Modified Starches: Properties and Uses*. Boca Raton, FL: CRC Press, 41–53. 1986.
13. Evans, R.B., Kruger, L.H. and Szymanski, C.D. Inhibited Starch Products. U.S. Patent 3,463,668. 1969.
14. Rutenberg, M.W., Tessler, M.M. and Kruger, L. Inhibited Starch Products Containing Labile and Non-Labile Cross-Links. U.S. Patent 3,899,602. 1975.
15. Jarowenko, W. Acetylated starch and miscellaneous organic esters. In: O.B. Wurzburg, ed. *Modified Starches: Properties and Uses*. Boca Raton, FL: CRC Press, 55–78. 1986.
16. Aszalos, A. and Prey, V. A new acetylation reaction for sugars. *Die Stärke*, 14, 50–52. 1962.

17. Grundschober, F. and Prey, V. Acetylation in aqueous solution. *Die Stärke*, 15, 225–227. 1963.
18. Katcher, J.H. and Ackilli, J.A. Process for Preparing an Odor-Free Acetylated Starch. U.S. Patent 4,238,604. 1980.
19. de Graaf, R.A., Lammers, G., Janssen, L.P.B.M., and Beenackers, A.A.C. Quantitative analysis of chemically modified starches by ^{1}H-NMR spectroscopy. *Starch*, 47, 469–475. 1995.
20. Maningat, C.C. Chemical Modification of Wheat Starch. Doctoral Dissertation. Kansas State University, Manhattan, KS. 1986.
21. Liu, H., Ramsden, L. and Corke, H. Physical properties and enzymatic digestibility of acetylated *ae, wx,* and normal maize starch. *Carbohydrate Polymers*, 34, 283–289. 1997.
22. Eastman, J.E. Single Step Filling Method for Retortable Canned Food Products. U.S. Patent 3,959,514. 1976.
23. Eastman, J.E. Retort Process Starch Derivatives. U.S. Patent 4,038,482. 1977.
24. McIntire, F.C. Cation Exchange Medium. U.S. Patent 2,505,561. 1950.
25. Trubiano, P.C. Succinate and substituted succinate derivatives of starch. In: O.B. Wurzburg, ed. *Modified Starches: Properties and Uses.* Boca Raton, FL: CRC Press, 131–148. 1986.
26. Bhandari, P.N. and Singhal, R.S. Effect of succinylation on the corn and amaranth starch pastes. *Carbohydrate Polymers*, 48, 233–240. 2002.
27. Caldwell, C.G. Free-Flowing Starch Esters. U.S. Patent 2,613,206. 1952.
28. Caldwell, C.G. and Wurzburg, O.B. Polysaccharide Derivatives of Substituted Dicarboxylic Acids. U.S. Patent 2,661,349. 1953.
29. King, W., Trubiano, P. and Perry, P. Modified starch encapsulating agents offer superior emulsification, film forming, and low surface oil. *Food Products Development*, Dec., 54–57. 1976.
30. Lim, S. and Seib, P.A. Preparation and pasting properties of wheat and corn starch phosphates. *Cereal Chemistry*, 70, 137–144. 1993.
31. Lim, S. and Seib, P.A. Location of phosphate esters in a wheat starch by ^{31}P-nuclear magnetic resonance spectroscopy. *Cereal Chemistry*, 70, 145–152. 1993.
32. Koch, H.D. and Koppers, J. Analytical investigations on phosphate cross-linked starches. *Die Stärke*, 34, 16–21. 1982.
33. Kerr, R.W. and Cleveland, F.C. Thickening Agent and Method of Making the Same. U.S. Patent 3,021,222. 1962.
34. Albrecht, J.J., Nelson, A.I., and Steinberg, M.P. Chrematistics of corn starch and starch derivatives as affected by freezing, storage, and thawing. I and II. *Food Technology*, 14, 57–63, 64–68. 1960.
35. Solarek, D.B. Phosphorylated starches and miscellaneous inorganic esters. In: O.B. Wurzburg, ed. *Modified Starches: Properties and Uses.* Boca Raton, FL: CRC Press, 97–112. 1986.
36. Neukom, H. Pudding Mix. U.S. Patent 2,865,762. 1958.
37. Tuschhoff, J.V. Hydroxypropylated starches. In: O.B. Wurzburg, ed. *Modified Starches: Properties and Uses.* Boca Raton, FL: CRC Press, 89–96. 1986.
38. Xu, A. and Seib, P.A. Determination of the level and position of substitution in hydroxypropylated starch by high-resolution ^{1}H-NMR spectroscopy of alpha-limit dextrins. *Journal of Cereal Science*, 25, 17–26. 1997.
39. Azemi, B.M.N.M. and Wootton, M. In vitro digestibility of hydroxypropyl maize, waxy maize and high amylose maize starches. *Starch*, 36, 273–275. 1984.

40. Björck, L., Gunnarsson, A. and Østergård, K. A study of native and chemically modified potato starch. Part II. Digestibility in the rat intestinal tract. *Starch*, 41, 128–134. 1989.

41. D'Ercole, A.D. Process of Preparing Frozen Pudding Composition. U.S. Patent 3,669,687. 1972.

42. Mitan, F.J. and Jokay, L. Process of Coating Food. U.S. Patent 3,427,951. 1969.

43. Solarek, D.B. Cationic starches. In: O.B. Wurzburg, ed. *Modified Starches: Properties and Uses*. Boca Raton, FL: CRC Press, 113–148. 1986.

44. Caldwell, C.G. and Wurzburg, O.B. Ungelatinized Tertiary Amino Alkyl Ethers of Amylaceous Materials. U.S. Patent 2,813,093. 1957.

45. Hullinger, C.H. and Uyi, N.H. Process for Preparing Amino Ethers of Starch. U.S. Patent 2,970,140. 1961.

46. Shildneck, P.A. and Hathaway, R.J. Preparation of Quaternary Ammonium Starch Ethers. U.S. Patent 3,346,563. 1967.

47. Doughty, J.B. and Klem, R.E. Process for Making Quaternary Amines of Epichlorohydrin. U.S. Patent 4,066,673. 1978.

48. Tasser, E.L. Process for Making Cationic Starch. U.S. Patent 4,464,528. 1984.

49. Caesar, G.V. Process for Forming Polysaccharide Cationic Ethers and Products. U.S. Patent 3,422,087. 1969.

50. Billy, J.M. and Seguin, J.A. Dry Heat Process for Preparation of Cationic Starch Ethers. U.S. Patent 3,448,101. 1969.

51. Rankin, J.C. and Phillips, B.S. Low pH Preparation of Cationic Starches and Flours. U.S. Patent 4,127,563. 1978.

52. Cuveller, G. and Wattiez, D. Dry Heat Reactions of Quaternized Epoxides and Quaternized Chlorohydrins. Hydroxylated Textiles. U.S. Patent 3,685,953. 1972.

53. Rankin, J.C. and Russell, C.R. Acidified Ethylenimine Modified Cereal Flours. U.S. Patent 3,522,238. 1970.

54. McClendon, J.C. and Berry, E.L. Aminoethylation of Flour and Starch with Ethylenimine. U.S. Patent 3,725,387. 1973.

55. McClendon, J.C. Cationic Flours and Starches. U.S. Patent 3,846,405. 1974.

56. Jarowenko, W. The Graft Polymerization of Ethylenimine onto Tertiary Amino Starch. U.S. Patent 3,331,833 (1067) and 3,354,034 (1967).

57. Elizer, L.H., Glasscock, G.C. and Seitz, J.M. Stabilized Cationic Starch Composition. U.S. Patent 3,316,646. 1964.

58. Ralph, J. and Nagy, D.E. Storage-Stable Cyanamide-Starch and Method for the Manufacture Thereof. U.S. Patent 3,438,970. 1969.

59. Mehltretter, C.L. Cationic Oxidized Starch Products. U.S. Patent 3,251,826. 1966.

60. Hofreiter, B.T., Hamerstrand, G.E., and Mehltretter, C.L. Cationic Polymeric Dialdehydes and Their Use in Making Wet Strength Paper. U.S. Patent 3,087,852. 1963.

61. Han, H.L. and Sosulski, F.W. Cationization of potato and tapioca starches using an aqueous alcoholic-alkaline process. *Starch*, 50, 487–492. 1998.

62. Jane, J., Shen L., Wang, L., and Maningat, C.C. Preparation and properties of small-particle corn starch. *Cereal Chemistry*, 69, 280–283. 1992.

63. Pitchon, E., O'Rourke, J.D., and Joseph, T.H. Process for Cooking or Gelatinizing Materials. U.S. Patent 4,280,851. 1981.

64. Thomas, D.J. and Atwell, W.A. Starch modifications. In: D.J. Thomas and W.A. Atwell, eds. *Starch Chemistry and Technology*. St. Paul, MN: American Association of Cereal Chemists, 43–48. 1999.

65. Trubiano, P.C. and Marotta, N.C. Pulpy Textured Food Systems Containing Inhibited Starches. U.S. Patent 3,579,341. 1971.

66. Marotta, N.C. and Trubiano, P.C. Pulpy Textured Food Systems Containing Inhibited Starches. U.S. Patent 3,443,964. 1969.

67. Jacobs, H. and Delcour, J.A. Hydrothermal modifications of granular starch, with retention of the granular structure: A review. *Journal of Agricultural and Food Chemistry*, 468, 2895–1905. 1998.

68. Hoover, R. and Vasanthan, T. Effect of heat-moisture treatment on the structure and physicochemical properties of cereal, legume, and tuber starches. *Carbohydrate Research*, 252, 33–53. 1994.

69. Hoover, R. and Manuel, H. The effect of heat-moisture treatment on the structure and physicochemical properties of normal maize, waxy maize, dull waxy maize and amylomaize V starches. *Journal of Cereal Science*, 23, 153–162. 1996.

70. Stute, R. Hydrothermal modification of starches: the difference between annealing and heat moisture treatment. *Starch*, 6, 205–214. 1992.

71. Lorenz, K., and Kulp, K. Heat-moisture treatment of starches. II. Functional properties and baking potential. *Cereal Chemistry*, 58, 49–52. 1981.

72. Abraham, T.E. Stabilization of paste viscosity of cassava by heat moisture treatment. *Starch*, 45, 131–135. 1993.

73. Tester, R.F. and Debon, S.J.J. Annealing of starch — a review. *International Journal of Biological Macromolecules*, 27, 1–12. 2000.

74. Tester, R.F., Debon, S.J.J. and Karkalas, J.J. Annealing of wheat starch. *Journal of Cereal Science*, 28, 259–272. 1998.

75. Colonna, J.T. and Mercier, C. Extrusion cooking of starch and starchy products. In: C. Mercier and J.M. Harper, eds. *Extrusion Cooking*. St. Paul, MN: American Association of Cereal Chemists, Inc., 247–320. 1989.

76. Mercier, C., Charbonniere, R., Gallant, D., and Delagueriviere, J.F. Formation of amylose-lipid complexes by twin-screw extrusion cooking of manioc starch. *Cereal Chemistry*, 57, 4–9. 1980.

77. Mercier, C., Charbonniere, R., Gallant, D., and Guilbot, A. Structural modifications of various starches by extrusion cooking with a twin-screw French extruder, In: J.M.V. Blanshard and J.R. Mitchell, eds. *Polysaccharides in Food*. London: Butterworths, 153–170. 1979.

78. Anderson, R.A., Conway, H.F., Pfeifer, V.F., and Griffin, L.E.J. Gelatinization of corn grits by roll- and extrusion cooking. *Cereal Science Today*, 14, 4–7, 11–12. 1969.

79. Harper, J.M. Food extruders and their applications. In: C. Mercier and J.M. Harper, eds. *Extrusion Cooking*. St. Paul, MN: American Association of Cereal Chemists, Inc., 1–15. 1989.

80. Gimmler, N., Lawn, F., and Meuser, F. Influence of extrusion cooking conditions on the efficiency of the cationization and carboxymethylation of potato starch granules. *Starch*, 46, 268–276. 1994.

81. Chang, Y.H. and Lii, C.Y. Preparation of starch phosphates by extrusion. *Journal of Food Science*, 57, 203–205. 1992.

82. Tomasik, P., Wang, Y.J., and Jane, J.L. Facile route to anionic starches. Succinylation, maleination and phthalation of corn starch on extrusion. *Starch*, 47, 96–99. 1995.

83. Wing, R.E. and Willett, J.L. Water soluble oxidized starches by reactive extrusion. *Industrial Crops and Products*, 7, 45–52. 1997.

84. Miladinov, V.D. and Hanna, M.A. Starch esterification by reactive extrusion. *Industrial Crops and Products*, 11, 51–57. 2000.

85. Ma, W. and Robyt, J.F. Preparation and characterization of soluble starches having different molecular sizes and composition, by acid hydrolysis in different alcohols. *Carbohydrate Research*, 166, 283–297. 1987.

86. Biwer, A., Antranikian, G., and Heinzle, E. Enzymatic production of cyclodextrins. *Applied Microbiol Biotechnology*, 59, 609–617. 2002.

87. Blanchard, P.H. and Kate, F.R. Starch Hydrolysates. In: A.M. Stephen, ed. *Food Polysaccharides and Their Applications*. New York: Marcel Dekker, Inc., 99–122. 1995.

88. Kennedy, J.F., Cabral, J.M.S., Sá-Correia, I., and White, C.A. Starch biomass: a chemical feedstock for enzyme and fermentation processes. In: T. Galliard, ed. *Starch: Properties and Potential*. Chichester: John Wiley & Sons, 4. 1987.

89. Muhr, A.H., Blanshard, J.M.V., and Bates, D.R. The effect of lintnerisation of wheat and potato starch granules. *Carbohydrate Polymers*, 4, 399–425. 1984.

90. Buléon, A., Colonna, P., Planchot, V., and Ball, S. Starch granules: structure and biosynthesis. *International Journal of Biological Macromolecules*, 23, 85–112. 1998.

91. Davis, J.P., Supatcharee, N., Khandelwal, R.L., and Chibbar, R.N. Synthesis of novel starches in plants: opportunity and challenges. *Starch*, 55, 107–120. 2003.

92. Jobling, S. Improving starch for food and industrial applications. *Current Opinion in Plant Biology*, 7, 1–9. 2004.

93. Smith, A.M. Making starch. *Current Opinion in Plant Biology*, 2, 223–229. 1999.

94. Shannon, J. and Garwood, D.L. Genetics and physiology of starch development. In: R.L. Whistler, ed. *Starch: Chemistry and Technology*. New York: Academic Press, Inc., 25. 1984.

95. Gerard, C., Colonna, P., Bouchet, B., Gallant, D.J., and Planchot, V. A multistages biosynthetic pathway in starch granules revealed by the ultrastructure of maize mutant starches. *Journal of Cereal Science*, 34, 61–71. 2001.

96. Lloyd, J.R., Hedley, C.L., Bull, V.J., and Ring, S.G. Determination of the effect of *r* and *rb* mutations on the structure of amylose and amylopectin in pea (*Pisum sativum* L.). *Carbohydrate Polymers*, 29, 45–49. 1996.

97. Gerard, C., Colonna, P., Buleon, A., and Planchot, V. Amylolysis of maize mutant starches. *Journal of the Science of Food and Agriculture*, 81, 1281–1287. 2001.

98. Shi, Y.-C. and Seib, P.A. Fine structure of maize starches from four *wx*-containing genotypes of the W64A inbred line in relation to gelatinization and retrogradation. *Carbohydrate Polymers*, 26,141–147. 1994.

99. Yuan, R.C., Thompson, D.B., and Boyer, C.D. Fine structure of amylopectin in relation to gelatinization and retrogradation behaviour in maize starches from three *wx*-containing genotypes in two inbred lines. *Cereal Chemistry*, 70, 81–89. 1993.

100. Liu, Q. and Thompson, D.B. Retrogradation of *du wx* and *su2 wx* maize starches after different gelatinization heat treatments. *Cereal Chemistry*, 75, 868–874. 1998.

101. Liu, Q. and Thompson, D.B. Effects of moisture content with different initial heating temperature on retrogradation from different maize starches. *Carbohydrate Research*, 314, 221–235. 1998.

Index

H

X

Y

Z